Aquatic Ecosystem and its Management

Aquatic Ecosystem and its Management

— Editors —

Dr. K. Vijaykumar
Professor
Department of Zoology,
Gulbarga University,
Gulbarga – 585 106, Karnataka

Dr. B. Vasanthkumar
S.S. Lecturer and Head
Department of Zoology and Fish and Industrial Fishery
Government First Grade College of Arts and Science
(Karnataka University affiliation)
Karwar-Uttar Kannada

2010
DAYA PUBLISHING HOUSE
Delhi - 110 035

© 2010, K. VIJAYKUMAR (b. 1963–   )

B. VASANTHKUMAR (b. 1968–   )
ISBN 9788170359562

Published by   :   **Daya Publishing House**
                            A Division of
                            Astral International Pvt. Ltd.
                            – ISO 9001:2008 Certified Company –
                            4760-61/23, Ansari Road, Darya Ganj,
                            New Delhi-110 002
                            Ph. 011-43549197, 23278134
                            E-mail: info@astralint.com
                            Website: www.astralint.com

Laser Typesetting   :   **Classic Computer Services**
                            Delhi - 110 035

Printed at   :   **Chawla Offset Printers**
                            Delhi - 110 052

PRINTED IN INDIA

Dedicated to
MY PARENTS

Preface

The achievement of sustainable development entails the integration of scientific approaches and environmental management, legislation, and policy making realities. Aquatic ecosystem management is a key element in this endeavour for sustainability, since water quality is an important prerequisite for human health and development. The health and integrity of aquatic ecosystems should therefore not only be regarded as important in its own right, but also be seen as an indicator, or barometer, of success in achieving sustainability.

Addressing this issue of integration within environmental assessment as well as between scientific assessment and policy making strongly depends on spatial scale. Indeed, it seems likely that the accuracy of our comprehension of processes decreases with increasing spatial scale. Therefore, the "scale issue" raises, among others, two types of questions regarding (i) the appropriateness of methods to the stakes, and (ii) the relevance of either the quality criteria or the remediation objectives.

Iterations between these scales followed by integration are necessary, and they need to be achieved consistently in order to produce relevant managerial approaches and to promote efficient remediation and préservation of aquatic ecosystems.

Aquatic ecosystems are very valuable to mankind since they perform many ecosystem functions like ground water recharging, nutrient recycling, water purification, etc.; provide habitat for biodiversity (flora, fauna and microbial organisms); augment and maintain stream flow, water supply, fisheries, and provide recreation and tourism opportunities and are also part of religious, cosmological and folklore beliefs.

These fragile aquatic ecosystems and their biodiversity (known to support specialist species) are being lost at a phenomenal rate due to various threats like population explosion; unsustainable use of resources; unplanned, ecologically unfriendly development activities (urbanisation, industrialisation, mining); habitat alteration for transportation, agriculture, aquaculture, and dam construction for water supply and hydroelectricity generation; pollution and nutrient inputs from point and non-point sources (silt deposition carrying inorganic fertilizer, pesticides and herbicides applied to crops etc.); infilling for solid and liquid waste disposal; dredging and channelisation; over-fishing and dynamite fishing and introduction of exotic species. These threats are compounded by institutional and legal factors such as: lack of legislation or its enforcement; lack of scientific knowledge and the dearth of information in the field; destructive, unscientific policies; land ownership; official attitudes towards these ecosystems; lack of adequate trained staff for proper maintenance; lack of appropriate planning between various government departments in undertaking developmental activities; fund scarcity or misuse of available funds, etc. These threats have great impact on physical, chemical and biological processes in the ecosystem leading to catastrophic declines in the coverage of aquatic ecosystems around the world.

A growing concern is that the increasing availability of restoration technology and improved best-practice must not be used by developers to justify the further damage of the natural ecosystem. *There is definitely no replacement for in situ conservation of natural and seminatural habitats.* Hence, potential aquatic ecosystems need to be identified and safeguarded by the planning process. Restoration should be seen as a vital component of integrated watershed management planning.

The need of the hour is to develop an integrated and accelerated effort towards conservation, restoration and proper management of the aquatic ecosystems to stop further degradation. If not, it may result in sharply increased environmental cost, the extinction of species, permanent ecological damage and loss of ecosystem or ecosystem services essential for the very survival of mankind.

Professor K. Vijaykumar   *Dr. B. Vasanthkumar*

Contents

Chapter 1

Freshwater Fish Diversity in India: Issues of Conservation and Management

Tapas Paria, Chinmoy Chakraborty & Chayan Roy

Central Inland Fisheries Research Institute,
Barrackpore, Kolkata – 700 120, West Bengal

ABSTRACT

India is a full of diversified faunal in teristrial as well as aquatic ecosystems. The present discussion depends on the aquatic resources maily freshwater fish faunal diversity in different aquatic ecosystem. Indian freshwater aquatic ecosystem can be divided in to three categories, warm water, cold water and brackish water it comprised of rivers, reservoirs, lakes, lagoogs, floodplain wetlands, ponds and estuaries. The Indian subcontinent occupying a position at the confluence of three biogeographic realms *viz.*, the Paleartic, Afro-tropical and Indo-Malayan exhibits a great variety of ecological habitats, harbouring rich icthyofaunal diversity. In total 678 fish fauna present in Indian territory. Most of the freshwater ecosystems are being changed very rapidly by man made activities, *viz.*, deforestation, resulting in increase in floods, deposition of fine sediments, increase in suspended matter, introduction of exotic species; exploitative fishing and pollution. Conservation of endangered fishes in their natural habitat is the best option available for conservation. Biological diversity is the central tenet of nature, one of its key defining features. The survival of human societies and cultures is dependent on biological diversity. It provides the essential ecosystem services including hydrological and geochemical cycles and climatic regulation that form the basis for human survival. It also meets the myriad survival and livelihood needs of fisherfolk, farmers, forest-dwellers, pastoralists, craftspersons, and others. This wonderful diversity and each of its components are worthy of highest respect and conservation in their own right. Most importantly biodiversity is the basis for the continuous evolution of species and ecosystems.

Keywords: Biodiversity, Fish fauna, Conservation, Management.

Introduction

India is the seventh largest country in the world and Asia's second largest nation with an area of 3,287,263 square km. India is fortunate to posses vast and varied fish germplasm resources distributed widely in vivid aquatic ecosystem. The Indian mainland stretches from 8°4' to 37°6' N latitude and from 68°7' to 97°25' E longitude. India is fortunate to posses the vast and varied fish germplasm resources consisting about 11 per cent of the total fish biodiversity of 20,000 species of the world.It has a land frontier of some 15,200 kms and a coastline of 7,516 km (Government of India, 1985). India's northern frontiers are with Xizang (Tibet) in the Peoples Republic of China, Nepal and Bhutan. In the north-west, India borders on Pakistan; in the north-east, China and Burma; and in the east, Burma. The southern peninsula extends into the tropical waters of the Indian Ocean with the Bay of Bengal lying to the south-east and the Arabian Sea to the south-west. The total land coverage of the region is 2.55 lakh sq. km. which is about 7.76 per cent of the country, out of which 70 per cent is hilly terrain. For administrative purposes India is divided into 28 states and 7 union territories and the country is home to around 107 million people.

Fish culture in the Indian subcontinent is perhaps as old as Kautilya's Arthasasthra written between 321 and 300 B.C., in which are indications that fish culture flourished at that time. End of 19th Century saw warm water fish culture in ponds involving collection and transport of carp spawn from river, and pond management confined originally to Bengal, Bihar and Orissa, subsequently spreading to other states. The importance of fish culture as a source of food production was realized resulting in fish culture activities throughout the country. The beginning of the 20ᵗʰ century marked the introduction of several exotic species in Indian water. (Jhingran, 1983).

Physically the massive country is divided into four relatively well defined regions–the Himalayan mountains, the Gangetic river plains, the southern (Deccan) plateau, and the islands of Lakshadweep, Andaman and Nicobar. The Himalayas in the far north include some of the highest peaks in the world. The highest mountain in the Indian Himalayas is Khangchenjunga (8586 m) which is located in Sikkim on the border with Nepal. To the south of the main Himalayan massif lie the Lesser Himalaya, rising to 3,600–4,600 m, and represented by the Pir Panjal in Kashmir and Dhaula dhar in Himachal Pradesh. Further south, flanking the Indo-Gangetic Plain, are the Siwaliks which rise to 900-1,500 m.

The aqua resources of the country include 2.02 million square km. area of Exclusive Economic Zone (EEZ) of the surrounding seas, more than 29,000 km. length of rivers, about 1, 13,000 km. of canals, about 1.75 million ha of existing water spread in the form of reservoirs, about 1 million ha in the form of tanks and ponds and about 0.6 million ha of stagnant, derelict swampy water spread areas. About 2, 118 finfish species have been recorded from different ecosystem like coldwater, warmwater, brackishwater and marine environment of India (Table 1.1). Biodiversity was coined as a contraction of biological diversity in 1985, but new term arguably has taken on a meaning and import all its own. Ten years further on it would be hard to count how many times biodiversity is used everyday by scientists, policy makers and others.

The Convention on Biological Diversity (CBD) has made it obligatory on part of India to document our resources. For equitable sharing of benefits derived from biodiversity, a database is essential.Fishes are cold-blooded vertebrates typically with backbones, gills and fins. They are the oldest group of vertebrates, having evolved some 450 million years ago, dating from deposits of the early Ordovician age. Fishes are also the most numerous of vertebrates constituting more than half of the vertebrate fauna of the world. During the long course of their evolution fishes have kept pace with the development of a variety of aquatic environments both on the surface and in subterranean waters, from the depths of the seas to the highest mountain ranges from the tropics to polar regions, penetrating into hot springs and soda lakes. They range in size from a few mm to over 5 meters and are found in all shapes, from typical torpedo shape, to round and flat and bizarre forms. Nelson (1996) estimated 24,618 valid species and mentions that the eventual number of extinct species may be close to 28,500. The fish species of the world are included under 4258 genera, 482 families and 55 orders. About 9966 species or 40 per cent of all species, normally live in freshwater lakes and rivers that cover only 1 per cent of the earth's surface and account for a little less than 0.01 per cent of its water.

Table 1.1: Species Diversity in Indian Ecosystem

Ecosystem	Fish Species (No.)
Cold water	154[1]
Warm water	433[2]
Brackish water	171[3]
Marine	1360
Total	2118

[1]   Of these, 34 fishes are common to cold and warm water.

[2]   Of these, 67 fishes are common to warm and brackish water.

[3]   Of these, 16 fishes are found only in brackish water, 73 are common to warm, brackish and marine water and another 82 are common to brackish and marine water.

The Indian subcontinent occupying a position at the confluence of three biogeographic realms *viz.*, the Paleartic, Afro-tropical and Indo-Malayan exhibits a great variety of ecological habitats, harbouring rich icthyofaunal diversity. The Indian species represent 11.72 per cent of the known fish species of the world, 23.96 per cent of the genera (make it 24 per cent), 57 per cent of families and 80 per cent of the orders (Barman, 1998); Jayaram (1999) listed 852 freshwater species of fishes under 272 genera, 71 families, and 16 orders, including both primary and secondary freshwater fishes (can you tell us here or as foot note what is primary and what is secondary) from India, Bangladesh, Myanmar, Nepal, Pakistan, and Sri Lanka. A recent checklist of Menon (1994) lists 446 primary freshwater species under 33 families and 11 orders from the Indian region alone.

Information on fish genetic resources is essential for undertaking programmes on conservation of fish germplasm and long term sustainable commercial utilization of fish resources. Each passing day makes it clear that biological diversity, the natural heritage of our world, is endangered and fast dwindling. Diversity means variations among some sort of entities, while biodiversity refers to the variety of biological entities inhabiting the given region, along with ecological role they play and genetic diversity they contain (Figure 1.1).

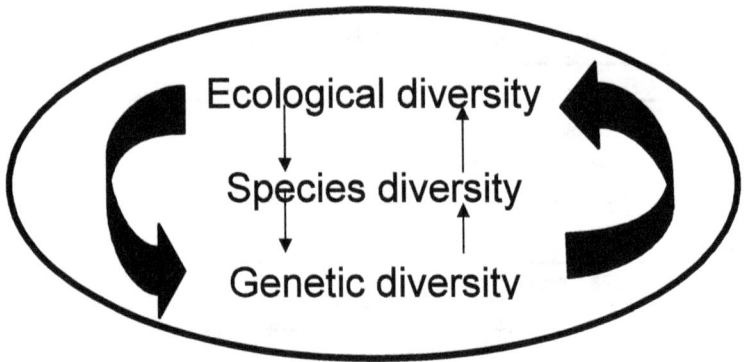

Figure 1.1: Biodiversity Components and their Interaction

A more sophisticated definition of biodiversity could therefore be–an ensamble and hierarchical interaction of the genetic, taxonomic and ecological scales of organization at different levels of interaction.

According to the IUCN Red list categories, 1994, any one of five criteria within the categories has to be satisfied for a taxon to be categorised as "threatened". The criteria that are used in categorisation of threat are

    ☆ Population reduction,

    ☆ Restricted distribution,

    ☆ Population estimates,

    ☆ Restricted population, and

    ☆ Probability of extinction.

The degree of threat depending on each or any of these five criteria determines the threat category or status. The participants also formulated post-assessment research and management recommendations for every fish taxon based on their status and information contributed in the working groups. Participants identified subject areas that need prioritisation as indicated in the recommendation section. Survey and monitoring for understanding distribution and trends of fish populations were the most frequently recommended research and management actions.

The Special Issue Working Groups convened after completion of the assessments were

- ☆ Endemism,
- ☆ Legal issues including Wildlife Protection Act and its implementation with respect to freshwater fishes,
- ☆ Taxonomy,
- ☆ Education and awareness,
- ☆ Introduced fishes,
- ☆ Sanctuaries for fishes,
- ☆ Suggestions to IUCN categories as applicable to fishes and
- ☆ Research working group
- ☆ Working group reports are included at the end of the main report section.

The northern plains of India stretch from Assam in the east to the Punjab in the west (a distance of 2,400 km), extending south to terminate in the saline swamplands of the Rann of Kachchh (Kutch), in the state of Gujarat. Some of the largest rivers in India including the Ganga (Ganges), Ghaghara, Brahmaputra, and the Yamuna flow across this region. The delta area of these rivers is located at the head of the Bay of Bengal, partly in the Indian state of west Bengal but mostly in Bangladesh. The plains are remarkably homogenous topographically: for hundreds of kilometres the only perceptible relief is formed by floodplain bluffs, minor natural levees and hollows known as 'spill patterns', and the belts of ravines formed by gully erosion along some of the larger rivers. In this zone, variation in relief does not exceed 300 m (FAO/UNEP, 1981) but the uniform flatness conceals a great deal of pedological variety.

Freshwater Resources

The inland fishery waters can be categorized into aquacultural waters, rivers, reservoirs, floodplain wetlands and estuaries. The details of their expanse in different states, production and production potential, yield, etc. (Sinha and Katiha, 2002) are summarized below.

Freshwater Aquacultural Water Bodies or Ponds

By virtue of its geographical situation in monsoon belt, India is endowed with good rainfall and consequently extensive aquacultural water bodies. The inland aquacultural water resources in the form of ponds and tanks have been distributed almost over all the states of India (Table 2). These bodies have over 2.85 million ha area, with maximum in the state of Tamil Nadu (0.69 million ha) followed by Andhra (0.52 million ha) and Karnataka (0.41 million ha). These states account for about 57 per cent of aquacultural waters of the country. Despite immense efforts for horizontal expansion of this industry, only 0.8 million ha area could be brought under scientific fish culture. In early seventies, with World Bank assistance, Fish Farmers Development Agency (FFDA) has been set up, with the objectives of promotion of pond fish culture and adoption of modern aquacultural techniques to achieve high fish production. This agency has adopted over half of the area covered under fish culture. The maximum percentage of area covered by FFDA was in the states of Punjab and Haryana, where fish farmers were taking more than one crop, more evident from higher cropped area than the actual. The productivity was also highest for Punjab at 4,085 kg ha^{-1} year^{-1} followed by Haryana at 3,501 kg ha^{-1} yr^{-1}. The national average productivity from FFDA supported ponds has increased from 50 kg ha^{-1} yr^{-1} in 1974-75 to about 2,135 kg ha^{-1} yr^{-1}in 1994-95. Yet lots of efforts are needed to harness unexploited water bodies. During Ninth plan the stress is on strengthening the technical wings of FFDA to assist fish farmers to adopt various improved package and practices of aquaculture (Anon., 1996a).

In this regard suitable strategies for enhancement of area coverage and productivity have been suggested. In summary, to achieve the targeted fish production, an increment of 45.2 per cent in area coverage and 50.9 per cent in productivity is needed. Therefore, the strategy for aquacultural development has considered components of horizontal and vertical expansion in concurrence with the potential and problems in different states. The state-wise distribution of area under ponds and tanks according to their production potential and projected productivity is summarized in Table 1.2.

Table 1.2: Aquacultural Water Bodies in India

State	Total Area (million ha)		Covered Area (000 ha)	% of State Area Covered	Produ- ction (t)	Yield (kg per ha per year)
Andhra Pradesh	0.517	(18.13)	13.72	2.65	26074	1900
Arunachal Pradesh	0.001	(0.04)	0.16	16.40	180	1098
Assam	0.023	(0.81)	3.44	14.97	6368	1850
Bihar	0.095	(3.33)	22.31	23.49	47527	2130
Goa	0.003	(0.11)				
Gujarat	0.071	(2.49)	30.93	43.57	34027	1100
Haryana	0.01	(0.35)	18.57	185.70	65005	3501
Himachal Pradesh	0.001	(0.04)	0.26	26.30	658	2502
Jammu and Kashmir	0.017	(0.60)	1.56	9.15	2022	1300
Karnataka	0.414	(14.52)	21.70	5.24	31898	1470
Kerala	0.03	(1.05)	4.00	13.34	7202	1800
Madhya Pradesh	0.119	(4.17)	54.96	46.19	86292	1570
Maharastra	0.05	(1.75)	11.31	22.63	6109	540
Manipur	0.005	(0.18)	1.79	35.82	2507	1400
Meghalaya	0.002	(0.07)	0.03	1.25	18	720
Mizoram	0.002	(0.07)	0.15	7.30	219	1500
Nagaland	0.05	(1.75)	1.16	2.33	1163	1000
Orissa	0.114	(4.00)	39.84	34.95	75698	1900
Punjab	0.007	(0.25)	12.15	173.57	49628	4085
Rajasthan	0.18	(6.31)	4.17	2.32	7211	1730
Sikkim			0.06		196	3500
Tamil Nadu	0.691	(24.23)	12.15	1.76	16521	1360
Tripura	0.012	(0.42)	3.33	27.78	6666	2000
Uttar Pradesh	0.162	(5.68)	69.21	42.72	138410	2000
West Bengal	0.276	(9.68)	98.78	35.79	296349	3000
Pondicherry			0.07		75	1119
Other	0.003	(0.11)				
Total	2.852	(100.00)	425.82	14.93	908023	2135

Source: Anon., 1996a, b and Sinha and Katiha, 2002.

The covered area, production and yield are under farms adopted by FFDA. Figures in parentheses represent per cent of total area.

Rivers

The river systems of India may be classified into two major groups, namely, Himalayan or extra-peninsular rivers and peninsular rivers (Table 1.3). The general profile of these groups is mentioned below.

Table 1.3: The Profile of Various River Systems in India

River System	Name of Main Rivers	Approx. Length (km)	States
Himalayan or Extra- Peninsular rivers			
Ganga	Ganga	2525	Uttar Pradesh, Bihar, West Bengal
	Ramganga	569	Uttar Pradesh
	Gomti	940	Uttar Pradesh
	Ghaghra	1080	Uttar Pradesh, Bihar
	Gandak	300	Bihar
	Kosi	492	Bihar
	Subarnarekha	395	Bihar, Orissa, West Bengal
	Yamuna	1376	Punjab, Haryana, Delhi, Uttar Pradesh
	Chambal	1080	Madhya Pradesh, Uttar Pradesh, Rajasthan
	Tons	264	Uttar Pradesh
	Son	784	Uttar Pradesh
	Ken	360	Madhya Pradesh
Braha-mputra	Brahamputra, Dibang, Siang, Lohit, Manas, Buri Dihang, Dhansiri, Koppili	4000	Arunachal Pradesh, Assam, Nagaland, Sikkim, Manipur
Indus	Jhelum	400	Jammu & Kashmir
	Chenab	330	Jammu & Kashmir, Himachal Pradesh
	Beas	460	Himachal Pradesh, Punjab
	Sutlej		Himachal Pradesh, Punjab
	Ravi		Jammu & Kashmir, Himachal Pradesh , Punjab

Contd...

Table 1.3–Contd...

River System	Name of Main Rivers	Approx. Length (km)	States
		Peninsular rivers	
East Coast	Mahanadi	851	Orissa, Madhya Pradesh
	Brahmani	799	Orissa, Bihar
	Godavari	1465	Maharastra, Andhra Pradesh
	Krishna	1401	Maharastra, Andhra Pradesh, Karnataka
	Cauvery	800	Karnataka, Tamil Nadu
	Pennar	597	Karnataka, Andhra Pradesh
	Bhima	861	Karnataka
West Coast	Narmada	1322	Maharastra, Gujarat, Madhya Pradesh
	Tapti	720	Gujarat, Maharastra
	Mahi	583	Gujarat
	Sabarmati	371	Gujarat, Rajasthan

Source: Jhingran, 1991; Rao, 1979; Sinha and Katiha, 2002.

The Himalayan or Extra-Peninsular Rivers

Originating from the Himalayas to transverse great alluvial Indo-Gangatic plains, these snow and rainfed rivers are characterized by complicated flood regimes and seasonal variations in volume of flow. Descending on the plains, they become sluggish and inundate vast land area. These rivers may be categorized into three systems, the Ganga, the Brahamputra and the Indus. The Ganga river system has combined length of 12,500 kms and a catchments area of 97.6 million ha. The Ganga, Ghagra, Gomti, Ramganga, Kosi, Gandak, Yamuna, Chambal, Sone and Tons are the major rivers of this system. These rivers are spread over most of the north Indian states (except the hilly states) to extend upto West Bengal through Bihar. In the upland waters of system commercial fisheries is virtually absent, due to inaccessible terrain and other exploitation problems. The stretch of river Ganga from Haridwar to Lalgola is recognised as one of the richest source of capture fisheries in India, comprising highly priced major carps, hilsa and catfishes. Mid September to June are peak months for fishing. During lean period of monsoon months the fishing activities generally confined to riverbanks. The combined length of the Brahamputra river system is 4,023 km. with catchment area of 51 million ha. Originating from Tibet the river flows through northern slopes of Himalayas to enter India at northeast corner of Arunachal Pradesh. It has 918 km stretch in India, including 730 km in Assam alone. Its northern tributaries Subansiri, Kameng and Manas are large with steep, shallow-braided channels, whereas those on the southern bank, Buri Dihing, Dhansiri and Kopilli are deeper with meandering channels and low gradient. The Brahamputra valley is marked for its abandoned river beds (beels) supporting rich fishery. Catfishes and major and minor carps dominate the commercial catches of upper middle and lower stretches, while the commercial catch in lower-middle stretch primarily composed of catfish and miscellaneous catch.

The Peninsular Rivers

The torrential and rain fed, peninsular rivers have well defined stable course. These include two river systems, the East Coast and the West Coast. The East Coast river system has vast expanse of water in the states of Orissa, Madhya Pradesh, Maharastra, Andhra Pradesh, Karnataka and Tamil Nadu. This river system mainly has four constituent rivers, the Mahanadi, the Godavari. The Krishna and the Cauvery have a combined length of 6,437 km and catchment area of 121 million ha. This system drains entire Peninsular India and east of Western Ghats in the west and south parts of central India. Besides its own fish fauna of several carps, catfishes, murrels, and prawn, the system is repeatedly enriched by transplantation of Gangatic carps.

The combined length of rivers of West Coast river system and catchment area are 3,380 km and 69.16 million ha, respectively. The Narmada and the Tapti are the longest rivers of system along with 600 small rivers. Its rivers are distributed in the states of Gujarat, Maharastra and Madhya Pradesh. The fish fauna of the system consists of carps, catfishes, mahseers, prawns, etc. The riverine resources had major share in inland capture fish production. But, during past few decades riverine ecosystem witnessed marked alterations, due to mammoth human interventions in the form of water abstraction, dam construction, sedimentation, and irrational fishing. These have discerningly disturbing affect on natural riverine fish production, which showed continuous declining trends. The studies of Central Inland Fisheries Research Institute, Barrackpore revealed that the average yield of major carps from Ganga river system has declined from 26.62 kg ha^{-1} year^{-1} during1958-61 to 2.55 kg ha^{-1} year^{-1} during 1989-95. The fisheries of anadromous hilsa have declined by 96 per cent above Farakka after construction of Farakka barrage in 1974. These examples of Ganga river system may be extended to depict the status of fish production in all the rivers of India. The restoration of riverine fisheries would entail an integrated approach encompassing the requirements of fisheries alongwith other uses of land and water.

Floodplain Wetlands

India has extensive floodplains in the form of oxbow lakes (mauns, beels, chaurs and jheels) especially in the states of Assam, Bihar and West Bengal (Table 1.4). These are shallow, nutrient rich water bodies formed due to change in course of the river. Some of these retain connection with the main river, atleast in monsoons, while others have lost it permanently (Sinha, 1997). Due to their high production potential these are adopted for aquaculture based capture fisheries. The areas having river connections can be exploited optimally by keeping the deeper central zones exclusively for capture fisheries and renovating the marginal pockets for culture fisheries.

Table 1.4: The Floodplain Wetlands of India

State	River Basin	Local Name	Area (ha)
Arunachal Pradesh	Kameng, Subansiri, Siang, Dibang, Lohit, Dihang, Tira	Beel	2500
Assam	Brahamputra and Barak	Beel	100000
Bihar	Gandak and Kosi	Mauns and Chaurs	40000
Manipur	Iral. Imphal. Thoubal	Pat	16500
Meghalaya	Someshwari and Jinjiram	Beel	213
Tripura	Gomti	Beel	500
West Bengal	Ganga and Ichhamati	Beel	42500
Total			202213

Source: Sinha, 1997 and Sinha and Katiha, 2002.

Estuarine Inland Fisheries

The estuarine capture fishery forms an important component of inland fisheries (Sinha, 1997). The open estuarine system includes Hoogly-Matlah and Mahanadi estuarine systems (Table 1.5). Godavari estuary is the main estuary of peninsular India, with Adyar Mankanam and Mandovi as other estuaries and Chilka, Pulicat and Vembanad as important brackish water lagoons. These estuaries and lagoons are recognised as excellent sources of naturally occurring fish and prawn seed. The fisheries of the estuaries are considered as above the subsistence level. The average yield varies between 45-75 kg ha^{-1}.

Table 1.5: Major Estuaries of India

Estuarine System	Estimated Area (ha)	Production (t)
Hoogly-Matlah	234000	20000 to 26000
Godavary estuary	18000	c.5000
Mahanadi estuary	3000	c.550
Narmada estuary	30000	c.4000
Peninsular esturine system	–	c. 2000
Chilka lagoon	103600	c.4000
Pulicate lake	3900	760–1370
Vembanad lake and Kerala Backwaters	50000	14000-17000
Wetlands of West Bengal		
(a) Freshwater Bheries	9600	10–14
(b) Saline Bheries	33000	c. 25500
Mangroves	356500	–

Source: Sinha, 1997.

Reservoirs

During post independence period, large number of river valley projects created a chain of impoundments, which are highly amenable for fishery activities. These man made water bodies created by obstructing the surface flow, by erecting a dam of any description, on a river, stream or any water course are called reservoirs (Sugunan, 1995). These are generally classified into small (<1000 ha), medium (1000-5000 ha) and large (>5000 ha). The area under these water bodies is on a continuous increase by adopting more and more reservoirs for fisheries. At present in India total area under reservoirs is 3.15 million ha (Table 1.6), out of which small reservoirs occupy 1.49 million ha followed by large 1.14 million ha and medium 0.52 million ha. Among the states, maximum percentage area under reservoirs is in Madhya Pradesh (14.6) followed by Andhra Pradesh (14.54), Karnataka (13.87) and Tamil Nadu (11.38). Among different sized reservoirs, maximum annual production is from small reservoirs (49.9 kg ha^{-1}) followed by medium (12.3 kg ha^{-1}) and large 11.43 (kg ha^{-1}) with overall average of 20.13 kg ha^{-1}. Despite the amenability for fish production and a production potential in the range of 50-300 kg ha^{-1}, the present yield from reservoirs in India is very low. The large and medium reservoirs are generally managed as stocking cum capture fisheries resources. The management policies based on norms of stock manipulation through selective stocking and harvesting operations have been suggested to rectify the imbalances in species spectrum and to increase fish yield. For small reservoirs, culture based management is considered the best.

Factors Affecting Freshwater Fisheries Resources

Most of the freshwater ecosystems are being changed very rapidly by man made activities, *viz.*, deforestation, resulting in increase in floods, deposition of fine sediments, increase in suspended matter, introduction of exotic species; exploitative fishing and pollution. Some west flowing rivers need thorough surveys.

(*i*) Natural Habitat Destruction

Huge silt load from deforested, mined catchment areas of our major river systems have been destructing the breeding and feeding grounds of some prized fishes. Dams and weirs constructed at different stretches have affected the famous game fish–mahseer. The fishes have also been affected in some of tributaries like Sone, Damodar and Rupnarain after the commissioning of anicuts, dams are barrages (Joshi, 1988). Hilsa fishery in middle stretch of Ganga river has also been adversely affected after construction of Farakka barrage.

(*ii*) Over Exploitation of Spawns

Every year millions of spawns and fingerlings are collected every year during the post flood period for the purposes of stocking

Table 1.6: State and Size-wise Distribution of Reservoirs in India

State	Size											
	Small			Medium			Large			Total		
	Area (ha)	% of Total	Yield (kg/ha)	Area (ha)	% of Total	Yield (kg/ha)	Area (ha)	% of Total	Yield (kg/ha)	Area (ha)	% of Total	Yield (kg/ha)
Tamil Nadu	315941	21.27	48.50	19577	3.71	13.74	23222	2.04	12.66	358740	11.38	22.63
Karnataka	228657	15.39	–	29078	5.51	–	179556	15.75	–	437291	13.87	–
Madhya Pradesh	172575	11.62	47.26	169502	32.13	12.02	118307	10.38	14.53	460384	14.60	13.68
Andhra Pradesh	201927	13.58	188.00	66429	12.59	22.00	190151	16.68	16.80	458507	14.54	36.48
Maharastra	119515	8.05	21.09	39481	7.48	11.83	115054	10.09	9.28	273750	8.68	10.21
Gujarat	84124	5.66	–	57748	10.95	–	144358	12.66	–	286230	9.08	–
Bihar	12461	0.84	3.91	12523	2.37	1.90	71711	6.29	0.11	96695	3.07	0.05
Orissa	66047	4.45	25.85	12748	2.42	12.76	119403	10.47	7.62	198198	6.29	9.72
Kerala	7975	0.54	53.50	15500	2.94	4.80	6160	0.54	–	29635	0.94	23.37
Uttar Pradesh	218651	14.72	14.60	44993	8.53	7.17	71196	6.24	1.07	334840	10.62	4.68
Rajasthan	54231	3.65	46.43	49827	9.45	24.47	49386	4.33	5.30	153444	4.87	24.89

Contd...

Table 1.6–Contd...

State	Size											
	Small			Medium			Large			Total		
	Area (ha)	% of Total	Yield (kg/ha)	Area (ha)	% of Total	Yield (kg/ha)	Area (ha)	% of Total	Yield (kg/ha)	Area (ha)	% of Total	Yield (kg/ha)
Himachal Pradesh	200	0.01	–	–	–	–	41364	3.63	35.55	41564	1.32	35.55
North-east	2239	0.15	–	5835	1.11	–	–	–	–	8074	0.26	–
Haryana	282	0.02	–	–	–	–	–	–	–	282	0.01	–
West Bengal	732	0.05	–	4600	0.87	–	10400	0.91	–	15732	0.50	–
Total	1485557	47.11*	49.90	527541	16.73*	12.30	1140268	36.16*	11.43	3153366	100.00	20.13

*: Per cent of total area under reservoir in India.

Source: Sugunan, 1995 and Sinha and Katiha, 2002.

and selling it as fish seed. Bulk of spawn is destroyed during the process thus causing loss to good quality wild fish germplasm specially commercially important fishes like Catla, Rohu, Mrigal, Calbasu, Mystus sp. and other misc. fishes. Some of the important spawn collection centers along river Ganga are:

1. Koelwar, 2. Kurji, 3. Bans ghat, 4. Fathuha, 5. Bakhtiarpur, 6. Mokama, 7. Munger, 8. Sultanganj, 9. Bhagalpur, 10. Kahalgaon, 11. Kursela, 12. Karhagola, 13. Manihari, 14. Sahibganj (Bijli ghat), 15. Sakrigali, 16. Maharajpur, 17. Rajmahal, 18. Manikchak ghat, 19. Farakka. Around Patna region about 375 lakh spawns are collected per day from more than 29 spawn collection sites. Other spawn collection centres are located around Allahabad and Varanasi.

(*iii*) Over Exploitation of Fishes

Over exploitation of fishes from river systems, lakes and reservoirs are the major factors responsible for dwindling fish resources. Due to dependence of fishermen and other communities for their lively-hood on the capture fisheries, large-scale fishing of brood fishes including juveniles takes place round the year including breeding season. The Indian Fisheries Act is very old which needs to be revised. Population of *Chitala chitala*, *Pangasius pangasius*, *Eutropichthys vacha*, *Semiplotus semiplotus* is declining fast in the rivers of India. It has been observed that the use of illegal methods of fishing like dynamites, chemicals and poisons is causing heavy mortality of fishes In this course ban on fishing during the closed season and restriction on mesh-size of the nets is overlooked.

(*iv*) Abstraction of Water

Due to abstraction of water for various purposes reduce the flow appreciably resulting in an inadequate flow. A number of dams have come up in the basin to provide water for irrigation, hydropower development and flood control. In the Ganga basin about 33.5 billion m^3 of water is presently stored in reservoirs, the largest one is the Rihand reservoir (Govind Ballab Pant Sagar), holding nearly 9 billion m^3 of water.

In all there are 25 dams having 30m height and above, built of the Ganga river, in different States for the purpose of irrigation, hydel power, flood control, navigation, water supply and recreation.

(*v*) Wanton Destruction

Huge killing of brood fishes by the use of dynamites and poisons, especially in the spawning season has affected a number of fishes of especially in upland streams.

(*vi*) Introduction of Exotic Species

The role of exotics for enhancing agricultural production is now well established. For the last 100 years, more than 350 fish species were introduced in India for the purposes of sport, culture, ornamental and for biological control. Fisheries department introduced these fishes, traders of fish seed and aquarium purposes. Occurrence of more exotic fish species not cleared for introduction in Indian culture systems has been a concern of NBFGR. Based on the reports of various state governments, fisheries departments, other than those used in composite exotic fish culture, following fish species are understood to be presently cultured in many parts of India.

☆ *Aristichthys nobilis*

☆ *Clarias gariepinus*

☆ *Oreochromis mossambicus*

☆ *Oreochromis niloticus*

☆ *Gambusia affinis*

☆ *Puntius javanicus*

☆ *Lebistes reiculata*

☆ *Carassius auratus*

☆ *Ctenopharyngodon idella*

☆ *Hypopthalmichthys molitrix*

☆ *Cyprinus carpio*

The effect of introduced exotic fishes on indigenous fishes has not been thoroughly assessed at national level. Except for few fishes, other fishes are not very useful for enhancing production. Escape of exotic fishes in the natural waters has been reported from Govind Sagar reservoir in Punjab, where silver carp has displaced Catla population and has established itself in the reservoir. Like wise escape of Thai Magur in Eastern Uttar Pradesh and establishment of Common Carp in river Yamuna at Allahabad has been reported.

Such escape of exotic fishes in natural waters may create havoc by establishing itself and displacing the local fish fauna competing for common food and niche. Accidental escape of exotic species in the natural waters has resulted in the drastic decline of indigenous fishes competing for similar niche. At present culture of two exotic fish species *viz.* Bighead (*Aristichthys nobilis*) and Thai magur (*Clarias gariepinus*) is spreading in various parts of the India is a matter of concern.

The introduction of exotics like *Oreochromis mossambica, Cyprinus carpio* have no doubt contributed much to the country's fish production, but they have also had a negative impact on the indigenous fishes in the form of competition, predation, environmental modification and so on. The unprecedented spread ad colonization of the hardy *O.mossambica* has been posing severe threat to our indigenous fishes competing with the latter for space and food and quickly replacing them. The exotic common carp *C. carpio*, a popular aquaculture and angling species cause harm to other species by stirring up bottom sediments and creating turbid conditions in impoundments. The introduction of the common carp has already eliminated the indigenous Schizothoracine fishes from the Kashmir valley, Osteobrama belangari fishes from the Loktak lake in Manipur, and the large indigenous *H. periyarensis* from the Periyar lake, Kerala (Menon, 1989-92). Likewise the introduction of the larvicidal fishes *Gambusia affinis, Poecilia reticulata* have almost replaced the indigenous *Aplocheilus parvus* and *Oryzias melastigma*. Also the introduction of the Gangetic carps into the peninsular rivers has affected the indigenous carps of the peninsula. These introductions generally cause zoogeographical pollution loss of genetic identity of local populations, a high level of hybridization, and extinction or reduction of local communities of endemic species. Parasites and diseases have also been introduced with alien species.

(*vii*) Anthropogenic Pressure

The Ganga basin witnessed, rapid industrialization and urbanization during the last of few decades. The icthyo-fauna of the river system is under ecological stress from these anthropogenic developmental activities including the uncontrolled discharge of sewage and industrial effluents. Important factors effecting fish germplasm of the Ganga river system are as follows (Menon, 1994).

(*viii*) Aquatic Pollution

The river system is under severe ecological stress, due to continuous onslaughts of effluents ranging from industrial and municipal wastewaters to agricultural runoff. The industrial wastewater, although less in volume than the urban sewage, are more deleterious to ichthyo-fauna.

Indiscriminate use of poison to collect fish from pools and refugial pockets where fish take shelter when rivers dry up, and dynamiting to collect fish in large numbers, will result in complete elimination of the fish species, since both juveniles and breeding fish and other non-target species all fall prey to such destructive methods. Few sources of aquatic pollution are described below.

Municipal Wastewater

Along the river course of Ganga River system, more than 100 towns are situated, which discharge untreated sewage into the river, Sewage effluents received by the Ganga are 16 million litres per day (MLD) from each of the cities like Farukhabad, Mirzapur and Bhagalpur. 100 MLD from each of Allahabad and Varanasi, 154 MLD from Patna, 275 MLD from Kanpur and 850 MLD from Calcutta (Jhingran, 1989).

Industrial Wastes

The river system become heavily polluted due to untreated effluents from tanneries, refineries textile wool and jute mills, sugar mills, distilleries, pulp and paper factories, rubber factories, coal washeries, thermal plants etc. mushrooming along the river course.

Thermal Pollution

Thermal plants in the country generate hot water (42-52 °C) and fly ash, which also aggravate deterioration of water quality.

Agricultural Run Off

With the increased use of the organic fertilizers (9 million t/yr) and pesticides (0.08 million t/yr) the runoff from the land has been adding a variety of organic and inorganic pollutants. A substantial load of nutrients (NPK' and toxic pesticides is carried into the river Ganga through the runoff from the fertile plains of U.P., Bihar and West Bengal).

Disposal of Dead Bodies

Dead bodies of human being and animals are another major source of organic pollution all along the course of River of Ganga. Lakhs of Hindus are cremated every year. In Varanasi itself which is a major centre of cremation if near by districts about 50,000 pyres are to want every year and the residual burnt and help burnt body along with ash is dumped in the river. Besides this lakhs of human dead bodies are immersed in the river (Bilgrami 1991).

Construction of Canals, Dams

At Haridwar, the upper Ganga canal abstracts huge amounts of water from its main course and just 250 km. downstream at Narora another canal draws the same amount of water. Thus a series of canals and irrigation channels along its course, reduce its flow by nearly 1,000 million m^3/yr, which alters the water quality by reducing the load bearing capacity of the river.

Siltation on the River Bed

The runoff from agricultural fields, denuded forests and spent mine areas results in siltation of the riverine beds. The rapid siltation of the Ganga river system poses threat to the riverine ecology in the Gangetic West Bengal.

Land Use Impact

Of the total Ganga basin area about 62.45 per cent *i.e.*, 509,994 sq. km. comes under cultivation; 23.2 per cent *i.e.* 189,646 sq. km is the non-arable land. Over 5 per cent of the total area is used as human habitation area. Statistical data of 1976-77 reveals that in the region 80 per cent of the total population depend on agriculture and they utilized 1.15 million tones of chemical fertilizers; 2,600 tons of pesticides of different types. Present data will be still higher. On account of high intensive cultivation in the rural area, high population density and factories in the urban areas and consequently huge amount of organic and inorganic pollutants are generated, a major portion of which finds its way into the aquatic system. BOD analysis of the Ganga basin shows that the in the rural sector it is 24 grams per person per day while it is 64 grams per person per day in the urban areas.

Impacts of Anthropogenic Stresses

The industrial and urban activities have been found to influence the primary productivity and ultimately the fish availability. The primary productivity of the river stretches decreases with increase in the magnitude of pollution. In the tidal stretch of the river Ganga at Kolkata, the net organic production was found to be minimum in the industrial belt (15-33 mg C/m^3/hr) followed by 25-47 mg C/m^3/hr in the upper non-industrial belt and 43-100 mg C/m^3/hr in the lower estuarine belt (Anon, 1987). The sustained pollution results destruction of feeding and spawning grounds, destruction of fish eggs, to the gills and finally mass mortality of the fish.

Over-exploitation

Collection of fish in numbers more than what can be recuited will ultimately destroy the whole population. Avoiding catching of breeding fish and juveniles with mesh regulations and restrictions on catches of smaller sized fish will protect the progeny. Overfishing has resulted in the decline of several cat fishes like *Bagarius bagarius, Silonia childreni, Pangasius pangasius, Aorichthys aor* and *A.seenghala* in the lower reaches of the river Godavari.

Closed season *i.e.,* restriction on fishing during certain periods is followed in Bihar, Madras, Jammu and Kashmir, Madhya Pradesh, Mysore, etc., In all large reservoirs fishing is closed from June–to–July, to end of September so that fishes are not destructed during their breeding migrations.

Present Status of Freshwater Fish Biodiversity

Fishery Survey of India has listed total 678 inland fish species in India. The dominant fish groups in the freshwaters are the cyprinids followed by Siluroids and a few representatives of the Perciforms and Synbranchiform fishes. The names of these species are given in Table 1.7 and some common fishes showed in Figure 1.2.

A recent study by Menon (1994) provides a comprehensive account of our *"Endangered", "Vulnerable"* and *"Rare"* species occurring in Peninsular India. In this, he has included 45 species which needs immediate conservation measures. The causes, which drive these species into these categories especially the large peninsular barbs, and the Mahseers, are:

1. Construction of dams hampering the spawning migratory movements,
2. Change in the physical qualities of the water bodies owing to siltation resulting from deforestation and agricultural practices,
3. Introduction of exotic carps, and
4. Transplantation of Gangetic carps into peninsular rivers.

Table 1.7: Inland Fish Fauna in India

Sl.No.	Species	Family
1.	*Aborichthys elongates*	Balitoridae
2.	*Aborichthys garoensis*	Balitoridae
3.	*Aborichthys kempi*	Balitoridae
4.	*Aborichthys tikaderi*	Balitoridae
5.	*Acanthocobitis botia*	Balitoridae
6.	*Acanthocobitis rubidipinnis*	Balitoridae
7.	*Acanthocobitis zonalternans*	Balitoridae
8.	*Acantopsis choirorhynchos*	Cobitidae
9.	*Acentrogobius bontii*	Gobiidae
10.	*Acentrogobius griseus*	Gobiidae
11.	*Acentrogobius madraspatensis*	Gobiidae
12.	*Aetomylaeus nichofii*	Myliobatidae
13.	*Ailia coila*	Schilbeidae
14.	*Ailia punctata*	Schilbeidae
15.	*Ambassis buton*	Ambassidae
16.	*Ambassis commersonii*	Ambassidae
17.	*Ambassis dussumieri*	Ambassidae
18.	*Ambassis gymnocephalus*	Ambassidae
19.	*Ambassis interrupta*	Ambassidae
20.	*Ambassis miops*	Ambassidae
21.	*Ambassis nalua*	Ambassidae
22.	*Ambassis urotaenia*	Ambassidae
23.	*Amblyceps laticeps*	Amblycipitidae
24.	*Amblyceps mangois*	Amblycipitidae

Contd...

Table 1.7–Contd...

Sl.No.	Species	Family
25.	*Amblypharyngodon atkinsonii*	Cyprinidae
26.	*Amblypharyngodon chakaiensis*	Cyprinidae
27.	*Amblypharyngodon melettinus*	Cyprinidae
28.	*Amblypharyngodon microlepis*	Cyprinidae
29.	*Amblypharyngodon mola*	Cyprinidae
30.	*Anabas cobojius*	Anabantidae
31.	*Anabas testudineus*	Anabantidae
32.	*Anguilla bengalensis*	Anguillidae
33.	*Anguilla bicolor bicolor*	Anguillidae
34.	*Aplocheilus blockii*	Aplocheilidae
35.	*Aplocheilus kirchmayeri*	Aplocheilidae
36.	*Aplocheilus lineatus*	Aplocheilidae
37.	*Aplocheilus panchax*	Aplocheilidae
38.	*Aplocheilus parvus*	Aplocheilidae
39.	*Apocryptes bato*	Gobiidae
40.	*Apocryptodon madurensis*	Gobiidae
41.	*Aristichthys nobilis*	Cyprinidae
42.	*Arius gagorides*	Ariidae
43.	*Arius jatius*	Ariidae
44.	*Arius satparanus*	Ariidae
45.	*Arius subrostratus*	Ariidae
46.	*Arius sumatranus*	Ariidae
47.	*Aspidoparia jaya*	Cyprinidae
48.	*Aspidoparia morar*	Cyprinidae
49.	*Awaous grammepomus*	Gobiidae
50.	*Awaous guamensis*	Gobiidae
51.	*Awaous melanocephalus*	Gobiidae
52.	*Awaous ocellaris*	Gobiidae
53.	*Badis assamensis*	Nandidae
54.	*Badis badis*	Nandidae
55.	*Badis blosyrus*	Nandidae

Contd...

Table 1.7–Contd...

Sl.No.	Species	Family
56.	*Badis kanabos*	Nandidae
57.	*Bagarius bagarius*	Sisoridae
58.	*Bagarius yarrelli*	Sisoridae
59.	*Balitora brucei*	Balitoridae
60.	*Balitora mysorensis*	Balitoridae
61.	*Bangana almorae*	Cyprinidae
62.	*Bangana diplostoma*	Cyprinidae
63.	*Barbodes bovanicus*	Cyprinidae
64.	*Barbodes carnaticus*	Cyprinidae
65.	*Barbodes hexagonolepis*	Cyprinidae
66.	*Barbodes sarana*	Cyprinidae
67.	*Barbodes wynaadensis*	Cyprinidae
68.	*Barbonymus gonionotus*	Cyprinidae
69.	*Barilius bakeri*	Cyprinidae
70.	*Barilius barila*	Cyprinidae
71.	*Barilius barna*	Cyprinidae
72.	*Barilius bendelisis*	Cyprinidae
73.	*Barilius canarensis*	Cyprinidae
74.	*Barilius cocsa*	Cyprinidae
75.	*Barilius dimorphicus*	Cyprinidae
76.	*Barilius dogarsinghi*	Cyprinidae
77.	*Barilius evezardi*	Cyprinidae
78.	*Barilius gatensis*	Cyprinidae
79.	*Barilius lairokensis*	Cyprinidae
80.	*Barilius modestus*	Cyprinidae
81.	*Barilius radiolatus*	Cyprinidae
82.	*Barilius shacra*	Cyprinidae
83.	*Barilius tileo*	Cyprinidae
84.	*Barilius vagra*	Cyprinidae
85.	*Batasio batasio*	Bagridae
86.	*Batasio tengana*	Bagridae

Contd...

Table 1.7–Contd...

Sl.No.	Species	Family
87.	*Batasio travancoria*	Bagridae
88.	*Bathygobius ostreicola*	Gobiidae
89.	*Batrachocephalus mino*	Ariidae
90.	*Bengala elanga*	Cyprinidae
91.	*Bhavania australis*	Balitoridae
92.	*Boleophthalmus boddarti*	Gobiidae
93.	*Botia almorhae*	Cobitidae
94.	*Botia berdmorei*	Cobitidae
95.	*Botia birdi*	Cobitidae
96.	*Botia dario*	Cobitidae
97.	*Botia dayi*	Cobitidae
98.	*Botia histrionica*	Cobitidae
99.	*Botia lohachata*	Cobitidae
100.	*Botia macrolineata*	Cobitidae
101.	*Botia rostrata*	Cobitidae
102.	*Botia striata*	Cobitidae
103.	*Brachirus orientalis*	Soleidae
104.	*Brachyamblyopus brachysoma*	Gobiidae
105.	*Brachygobius nunus*	Gobiidae
106.	*Butis amboinensis*	Eleotridae
107.	*Butis gymnopomus*	Eleotridae
108.	*Caragobius urolepis*	Gobiidae
109.	*Carassius auratus auratus*	Cyprinidae
110.	*Carassius carassius*	Cyprinidae
111.	*Carcharhinus leucas*	Carcharhinidae
112.	*Carcharhinus melanopterus*	Carcharhinidae
113.	*Carinotetraodon imitator*	Tetraodontidae
114.	*Carinotetraodon travancoricus*	Tetraodontidae
115.	*Catla catla*	Cyprinidae
116.	*Chaca chaca*	Chacidae
117.	*Chagunius chagunio*	Cyprinidae
118.	*Chagunius nicholsi*	Cyprinidae

Contd...

Table 1.7–Contd...

Sl.No.	Species	Family
119.	*Chanda nama*	Ambassidae
120.	*Channa amphibius*	Channidae
121.	*Channa barca*	Channidae
122.	*Channa bleheri*	Channidae
123.	*Channa diplogramma*	Channidae
124.	*Channa gachua*	Channidae
125.	*Channa marulius*	Channidae
126.	*Channa micropeltes*	Channidae
127.	*Channa orientalis*	Channidae
128.	*Channa punctata*	Channidae
129.	*Channa stewartii*	Channidae
130.	*Channa striata*	Channidae
131.	*Channa theophrasti*	Channidae
132.	*Chanos chanos*	Chanidae
133.	*Chela cachius*	Cyprinidae
134.	*Chela dadiburjori*	Cyprinidae
135.	*Chela fasciata*	Cyprinidae
136.	*Chela laubuca*	Cyprinidae
137.	*Chitala chitala*	Notopteridae
138.	*Cirrhinus cirrhosus*	Cyprinidae
139.	*Cirrhinus fulungee*	Cyprinidae
140.	*Cirrhinus macrops*	Cyprinidae
141.	*Clarias batrachus*	Clariidae
142.	*Clarias dayi*	Clariidae
143.	*Clarias dussumieri*	Clariidae
144.	*Clarias gariepinus*	Clariidae
145.	*Clupisoma bastari*	Schilbeidae
146.	*Clupisoma garua*	Schilbeidae
147.	*Clupisoma montana*	Schilbeidae
148.	*Coilia dussumieri*	Engraulidae
149.	*Coilia reynaldi*	Engraulidae

Contd...

Table 1.7–Contd...

Sl.No.	Species	Family
150.	*Coius quadrifasciatus*	Coiidae
151.	*Colisa fasciatus*	Belontiidae
152.	*Colisa lalia*	Belontiidae
153.	*Congresox talabonoides*	Muraenesocidae
154.	*Conta conta*	Erethistidae
155.	*Corica laciniata*	Clupeidae
156.	*Corica soborna*	Clupeidae
157.	*Crossocheilus burmanicus*	Cyprinidae
158.	*Crossocheilus diplochilus*	Cyprinidae
159.	*Crossocheilus latius*	Cyprinidae
160.	*Crossocheilus periyarensis*	Cyprinidae
161.	*Ctenopharyngodon idellus*	Cyprinidae
162.	*Ctenops nobilis*	Belontiidae
163.	*Cyprinion semiplotum*	Cyprinidae
164.	*Cyprinus carpio carpio*	Cyprinidae
165.	*Danio aequipinnatus*	Cyprinidae
166.	*Danio dangila*	Cyprinidae
167.	*Danio frankei*	Cyprinidae
168.	*Danio malabaricus*	Cyprinidae
169.	*Danio regina*	Cyprinidae
170.	*Danio rerio*	Cyprinidae
171.	*Dario dario*	Nandidae
172.	*Dayella malabarica*	Clupeidae
173.	*Dermogenys pusilla*	Hemiramphidae
174.	*Devario acuticephala*	Cyprinidae
175.	*Devario assamensis*	Cyprinidae
176.	*Devario devario*	Cyprinidae
177.	*Devario fraseri*	Cyprinidae
178.	*Devario horai*	Cyprinidae
179.	*Devario manipurensis*	Cyprinidae
180.	*Devario naganensis*	Cyprinidae

Contd...

Table 1.7–Contd...

Sl.No.	Species	Family
181.	*Devario neilgherriensis*	Cyprinidae
182.	*Diptychus maculatus*	Cyprinidae
183.	*Drombus globiceps*	Gobiidae
184.	*Ehirava fluviatilis*	Clupeidae
185.	*Elops machnata*	Elopidae
186.	*Enobarbichthys maculatus*	Cobitidae
187.	*Erethistes pusillus*	Erethistidae
188.	*Erethistoides montana montana*	Erethistidae
189.	*Erethistoides montana pipri*	Erethistidae
190.	*Esomus barbatus*	Cyprinidae
191.	*Esomus danricus*	Cyprinidae
192.	*Esomus lineatus*	Cyprinidae
193.	*Esomus thermoicos*	Cyprinidae
194.	*Etroplus canarensis*	Cichlidae
195.	*Etroplus maculatus*	Cichlidae
196.	*Etroplus suratensis*	Cichlidae
197.	*Eutropiichthys goongwaree*	Schilbeidae
198.	*Eutropiichthys murius*	Schilbeidae
199.	*Eutropiichthys vacha*	Schilbeidae
200.	*Exostoma labiatum*	Sisoridae
201.	*Gagata cenia*	Sisoridae
202.	*Gagata gagata*	Sisoridae
203.	*Gagata itchkeea*	Sisoridae
204.	*Gagata sexualis*	Sisoridae
205.	*Gagata youssoufi*	Sisoridae
206.	*Gambusia affinis*	Poeciliidae
207.	*Gambusia holbrooki*	Poeciliidae
208.	*Gangra viridescens*	Sisoridae
209.	*Garo khajuriai*	Sisoridae
210.	*Garra annandalei*	Cyprinidae
211.	*Garra bicornuta*	Cyprinidae

Contd...

Table 1.7–Contd...

Sl.No.	Species	Family
212.	*Garra elongata*	Cyprinidae
213.	*Garra gotyla gotyla*	Cyprinidae
214.	*Garra gotyla stenorhynchus*	Cyprinidae
215.	*Garra hughi*	Cyprinidae
216.	*Garra kalakadensis*	Cyprinidae
217.	*Garra kempi*	Cyprinidae
218.	*Garra lamta*	Cyprinidae
219.	*Garra lissorhynchus*	Cyprinidae
220.	*Garra litanensis*	Cyprinidae
221.	*Garra manipurensis*	Cyprinidae
222.	*Garra mcclellandi*	Cyprinidae
223.	*Garra menoni*	Cyprinidae
224.	*Garra mullya*	Cyprinidae
225.	*Garra naganensis*	Cyprinidae
226.	*Garra nasuta*	Cyprinidae
227.	*Garra notata*	Cyprinidae
228.	*Garra periyarensis*	Cyprinidae
229.	*Garra rupecula*	Cyprinidae
230.	*Garra surendranathi*	Cyprinidae
231.	*Gilchristella aestuarius*	Clupeidae
232.	*Glossogobius giuris*	Gobiidae
233.	*Glyphis gangeticus*	Carcharhinidae
234.	*Glyptosternon maculatum*	Sisoridae
235.	*Glyptosternon reticulatum*	Sisoridae
236.	*Glyptothorax alaknandi*	Sisoridae
237.	*Glyptothorax anamalaiensis*	Sisoridae
238.	*Glyptothorax annandalei*	Sisoridae
239.	*Glyptothorax brevipinnis*	Sisoridae
240.	*Glyptothorax cavia*	Sisoridae
241.	*Glyptothorax coheni*	Sisoridae
242.	*Glyptothorax conirostre conirostre*	Sisoridae

Contd...

Table 1.7–Contd...

Sl.No.	Species	Family
243.	*Glyptothorax conirostre poonaensis*	Sisoridae
244.	*Glyptothorax dakpathari*	Sisoridae
245.	*Glyptothorax dorsalis*	Sisoridae
246.	*Glyptothorax garhwali*	Sisoridae
247.	*Glyptothorax gracilis*	Sisoridae
248.	*Glyptothorax housei*	Sisoridae
249.	*Glyptothorax indicus*	Sisoridae
250.	*Glyptothorax kapuri*	Sisoridae
251.	*Glyptothorax kashmirensis*	Sisoridae
252.	*Glyptothorax lonah*	Sisoridae
253.	*Glyptothorax madraspatanum*	Sisoridae
254.	*Glyptothorax nelsoni*	Sisoridae
255.	*Glyptothorax pectinopterus*	Sisoridae
256.	*Glyptothorax ribeiroi*	Sisoridae
257.	*Glyptothorax saisii*	Sisoridae
258.	*Glyptothorax sinense*	Sisoridae
259.	*Glyptothorax stoliczkae*	Sisoridae
260.	*Glyptothorax striatus*	Sisoridae
261.	*Glyptothorax telchitta*	Sisoridae
262.	*Glyptothorax trewavasae*	Sisoridae
263.	*Glyptothorax trilineatus*	Sisoridae
264.	*Gobiopterus chuno*	Gobiidae
265.	*Gonialosa manmina*	Clupeidae
266.	*Gudusia chapra*	Clupeidae
267.	*Gymnocypris biswasi*	Cyprinidae
268.	*Hara hara*	Sisoridae
269.	*Hara horai*	Sisoridae
270.	*Hara jerdoni*	Sisoridae
271.	*Hara serratus*	Sisoridae
272.	*Hemibagrus maydelli*	Bagridae
273.	*Hemibagrus menoda*	Bagridae

Contd...

Table 1.7–Contd...

Sl.No.	Species	Family
274.	*Hemibagrus microphthalmus*	Bagridae
275.	*Hemibagrus punctatus*	Bagridae
276.	*Hemigobius hoevenii*	Gobiidae
277.	*Heteropneustes fossilis*	Heteropneustidae
278.	*Heteropneustes microps*	Heteropneustidae
279.	*Himantura bleekeri*	Dasyatidae
280.	*Himantura chaophraya*	Dasyatidae
281.	*Himantura fluviatilis*	Dasyatidae
282.	*Himantura gerrardi*	Dasyatidae
283.	*Himantura imbricata*	Dasyatidae
284.	*Himantura uarnak*	Dasyatidae
285.	*Hippichthys penicillus*	Syngnathidae
286.	*Hippocampus fuscus*	Syngnathidae
287.	*Homaloptera manipurensis*	Balitoridae
288.	*Homaloptera menoni*	Balitoridae
289.	*Homaloptera montana*	Balitoridae
290.	*Homaloptera pillaii*	Balitoridae
291.	*Horabagrus brachysoma*	Bagridae
292.	*Horabagrus nigricollaris*	Bagridae
293.	*Horadandia atukorali*	Cyprinidae
294.	*Horaglanis krishnai*	Clariidae
295.	*Horaichthys setnai*	Adrianichthyidae
296.	*Horalabiosa joshuai*	Cyprinidae
297.	*Horalabiosa palaniensis*	Cyprinidae
298.	*Hypophthalmichthys molitrix*	Cyprinidae
299.	*Hyporhamphus limbatus*	Hemiramphidae
300.	*Hyporhamphus xanthopterus*	Hemiramphidae
301.	*Hypselobarbus curmuca*	Cyprinidae
302.	*Hypselobarbus dobsoni*	Cyprinidae
303.	*Hypselobarbus dubius*	Cyprinidae
304.	*Hypselobarbus jerdoni*	Cyprinidae

Contd...

Table 1.7–Contd...

Sl.No.	Species	Family
305.	*Hypselobarbus kolus*	Cyprinidae
306.	*Hypselobarbus kurali*	Cyprinidae
307.	*Hypselobarbus lithopidos*	Cyprinidae
308.	*Hypselobarbus micropogon*	Cyprinidae
309.	*Hypselobarbus periyarensis*	Cyprinidae
310.	*Hypselobarbus pulchellus*	Cyprinidae
311.	*Hypselobarbus thomassi*	Cyprinidae
312.	*Ichthyocampus carce*	Syngnathidae
313.	*Ilisha megaloptera*	Clupeidae
314.	*Indoreonectes evezardi*	Balitoridae
315.	*Istigobius diadema*	Gobiidae
316.	*Johnius gangeticus*	Sciaenidae
317.	*Ketengus typus*	Ariidae
318.	*Kuhlia rupestris*	Kuhliidae
319.	*Kurtus indicus*	Kurtidae
320.	*Labeo angra*	Cyprinidae
321.	*Labeo ariza*	Cyprinidae
322.	*Labeo bata*	Cyprinidae
323.	*Labeo boga*	Cyprinidae
324.	*Labeo boggut*	Cyprinidae
325.	*Labeo calbasu*	Cyprinidae
326.	*Labeo dussumieri*	Cyprinidae
327.	*Labeo dyocheilus*	Cyprinidae
328.	*Labeo fimbriatus*	Cyprinidae
329.	*Labeo gonius*	Cyprinidae
330.	*Labeo kawrus*	Cyprinidae
331.	*Labeo kontius*	Cyprinidae
332.	*Labeo nandina*	Cyprinidae
333.	*Labeo nigrescens*	Cyprinidae
334.	*Labeo pangusia*	Cyprinidae
335.	*Labeo porcellus*	Cyprinidae

Contd...

Table 1.7–Contd...

Sl.No.	Species	Family
336.	*Labeo potail*	Cyprinidae
337.	*Labeo rajasthanicus*	Cyprinidae
338.	*Labeo rohita*	Cyprinidae
339.	*Labeo udaipurensis*	Cyprinidae
340.	*Laguvia shawi*	Erethistidae
341.	*Leiognathus dussumieri*	Leiognathidae
342.	*Lepidocephalichthys berdmorei*	Cobitidae
343.	*Lepidocephalichthys manipurensis*	Cobitidae
344.	*Lepidocephalus annandalei*	Cobitidae
345.	*Lepidocephalus coromandelensis*	Cobitidae
346.	*Lepidocephalus guntea*	Cobitidae
347.	*Lepidocephalus irrorata*	Cobitidae
348.	*Lepidocephalus menoni*	Cobitidae
349.	*Lepidocephalus thermalis*	Cobitidae
350.	*Lepidopygopsis typus*	Cyprinidae
351.	*Liza macrolepis*	Mugilidae
352.	*Liza tade*	Mugilidae
353.	*Longischistura bhimachari*	Balitoridae
354.	*Longischistura striata*	Balitoridae
355.	*Macrognathus aculeatus*	Mastacembelidae
356.	*Macrognathus aral*	Mastacembelidae
357.	*Macrognathus guentheri*	Mastacembelidae
358.	*Macrognathus malabaricus*	Mastacembelidae
359.	*Macrognathus pancalus*	Mastacembelidae
360.	*Mastacembelus armatus*	Mastacembelidae
361.	*Megalops cyprinoides*	Megalopidae
362.	*Mesonoemacheilus guentheri*	Balitoridae
363.	*Mesonoemacheilus herrei*	Balitoridae
364.	*Mesonoemacheilus pulchellus*	Balitoridae
365.	*Mesonoemacheilus triangularis*	Balitoridae
366.	*Microphis brachyurus brachyurus*	Syngnathidae

Contd...

Table 1.7–Contd...

Sl.No.	Species	Family
367.	*Microphis cuncalus*	Syngnathidae
368.	*Microphis deocata*	Syngnathidae
369.	*Microphis insularis*	Syngnathidae
370.	*Monopterus albus*	Synbranchidae
371.	*Monopterus cuchia*	Synbranchidae
372.	*Monopterus eapeni*	Synbranchidae
373.	*Monopterus fossorius*	Synbranchidae
374.	*Monopterus hodgarti*	Synbranchidae
375.	*Monopterus indicus*	Synbranchidae
376.	*Monopterus roseni*	Synbranchidae
377.	*Moringua raitaborua*	Moringuidae
378.	*Mugil cephalus*	Mugilidae
379.	*Mystus armatus*	Bagridae
380.	*Mystus bleekeri*	Bagridae
381.	*Mystus cavasius*	Bagridae
382.	*Mystus gulio*	Bagridae
383.	*Mystus keletius*	Bagridae
384.	*Mystus malabaricus*	Bagridae
385.	*Mystus montanus*	Bagridae
386.	*Mystus oculatus*	Bagridae
387.	*Mystus tengara*	Bagridae
388.	*Mystus vittatus*	Bagridae
389.	*Nandus nandus*	Nandidae
390.	*Nangra carcharhinoides*	Sisoridae
391.	*Nangra nangra*	Sisoridae
392.	*Naziritor chelynoides*	Cyprinidae
393.	*Nemacheilus anguilla*	Balitoridae
394.	*Nemacheilus arunachalensis*	Balitoridae
395.	*Nemacheilus baluchiorum*	Balitoridae
396.	*Nemacheilus barapaniensis*	Balitoridae
397.	*Nemacheilus carletoni*	Balitoridae

Contd...

Table 1.7–Contd...

Sl.No.	Species	Family
398.	*Nemacheilus chindwinicus*	Balitoridae
399.	*Nemacheilus devdevi*	Balitoridae
400.	*Nemacheilus doonensis*	Balitoridae
401.	*Nemacheilus elongatus*	Balitoridae
402.	*Nemacheilus gangeticus*	Balitoridae
403.	*Nemacheilus himachalensis*	Balitoridae
404.	*Nemacheilus horai*	Balitoridae
405.	*Nemacheilus keralensis*	Balitoridae
406.	*Nemacheilus kodaguensis*	Balitoridae
407.	*Nemacheilus labeosus*	Balitoridae
408.	*Nemacheilus monilis*	Balitoridae
409.	*Nemacheilus mooreh*	Balitoridae
410.	*Nemacheilus multifasciatus*	Balitoridae
411.	*Nemacheilus nagaensis*	Balitoridae
412.	*Nemacheilus nilgiriensis*	Balitoridae
413.	*Nemacheilus pambarensis*	Balitoridae
414.	*Nemacheilus pavonaceus*	Balitoridae
415.	*Nemacheilus petrubanarescui*	Balitoridae
416.	*Nemacheilus reticulofasciatus*	Balitoridae
417.	*Nemacheilus rueppelli*	Balitoridae
418.	*Nemacheilus sikmaiensis*	Balitoridae
419.	*Nemacheilus singhi*	Balitoridae
420.	*Nemacheilus subfuscus*	Balitoridae
421.	*Nemachilichthys shimogensis*	Balitoridae
422.	*Neoeucirrhichthys maydelli*	Cobitidae
423.	*Neolissochilus dukai*	Cyprinidae
424.	*Neolissochilus spinulosus*	Cyprinidae
425.	*Neonoemacheilus peguensis*	Balitoridae
426.	*Neotropius khavalchor*	Schilbeidae
427.	*Notopterus notopterus*	Notopteridae
428.	*Odonteleotris macrodon*	Eleotridae

Contd...

Table 1.7–Contd...

Sl.No.	Species	Family
429.	*Olyra horae*	Olyridae
430.	*Olyra longicaudata*	Olyridae
431.	*Omobranchus ferox*	Blenniidae
432.	*Ompok bimaculatus*	Siluridae
433.	*Ompok goae*	Siluridae
434.	*Ompok malabaricus*	Siluridae
435.	*Ompok pabda*	Siluridae
436.	*Ompok pabo*	Siluridae
437.	*Oncorhynchus mykiss*	Salmonidae
438.	*Ophieleotris aporos*	Eleotridae
439.	*Ophiocara porocephala*	Eleotridae
440.	*Ophisternon bengalense*	Synbranchidae
441.	*Oreichthys cosuatis*	Cyprinidae
442.	*Oreochromis mossambicus*	Cichlidae
443.	*Oreochromis niloticus niloticus*	Cichlidae
444.	*Oryzias carnaticus*	Adrianichthyidae
445.	*Oryzias dancena*	Adrianichthyidae
446.	*Osphronemus goramy*	Osphronemidae
447.	*Osteobrama alfrediana*	Cyprinidae
448.	*Osteobrama bakeri*	Cyprinidae
449.	*Osteobrama belangeri*	Cyprinidae
450.	*Osteobrama bhimensis*	Cyprinidae
451.	*Osteobrama cotio cotio*	Cyprinidae
452.	*Osteobrama cotio cunma*	Cyprinidae
453.	*Osteobrama cotio peninsularis*	Cyprinidae
454.	*Osteobrama dayi*	Cyprinidae
455.	*Osteobrama neilli*	Cyprinidae
456.	*Osteobrama vigorsii*	Cyprinidae
457.	*Osteochilichthys brevidorsalis*	Cyprinidae
458.	*Osteochilus godavariensis*	Cyprinidae
459.	*Osteochilus hasseltii*	Cyprinidae

Contd...

Table 1.7–Contd...

Sl.No.	Species	Family
460.	Osteochilus longidorsalis	Cyprinidae
461.	Osteochilus nashii	Cyprinidae
462.	Osteochilus thomassi	Cyprinidae
463.	Osteogeneiosus militaris	Ariidae
464.	Otolithoides pama	Sciaenidae
465.	Pangasius pangasius	Pangasiidae
466.	Pangio goaensis	Cobitidae
467.	Pangio longipinnis	Cobitidae
468.	Pangio oblonga	Cobitidae
469.	Pangio pangia	Cobitidae
470.	Papillogobius reichei	Gobiidae
471.	Parachiloglanis hodgarti	Sisoridae
472.	Parambassis baculis	Ambassidae
473.	Parambassis dayi	Ambassidae
474.	Parambassis lala	Ambassidae
475.	Parambassis ranga	Ambassidae
476.	Parambassis thomasi	Ambassidae
477.	Parapsilorhynchus discophorus	Cyprinidae
478.	Parapsilorhynchus prateri	Cyprinidae
479.	Parapsilorhynchus tentaculatus	Cyprinidae
480.	Pareuchiloglanis kamengensis	Sisoridae
481.	Pareuchiloglanis macrotrema	Sisoridae
482.	Parluciosoma labiosa	Cyprinidae
483.	Pastinachus sephen	Dasyatidae
484.	Periophthalmodon septemradiatus	Gobiidae
485.	Periophthalmus barbarus	Gobiidae
486.	Periophthalmus chrysospilos	Gobiidae
487.	Pillaia indica	Chaudhuriidae
488.	Pinniwallago kanpurensis	Siluridae
489.	Pisodonophis boro	Ophichthidae
490.	Pisodonophis cancrivorus	Ophichthidae

Contd...

Table 1.7–Contd...

Sl.No.	Species	Family
491.	*Plotosus canius*	Plotosidae
492.	*Plotosus lineatus*	Plotosidae
493.	*Poecilia reticulata*	Poeciliidae
494.	*Polydactylus macrochir*	Polynemidae
495.	*Polynemus paradiseus*	Polynemidae
496.	*Poropuntius burtoni*	Cyprinidae
497.	*Poropuntius clavatus*	Cyprinidae
498.	*Pristis microdon*	Pristidae
499.	*Pristis pectinata*	Pristidae
500.	*Pristis pristis*	Pristidae
501.	*Pristolepis fasciata*	Nandidae
502.	*Pristolepis marginata*	Nandidae
503.	*Proeutropiichthys taakree taakree*	Schilbeidae
504.	*Pseudapocryptes elongatus*	Gobiidae
505.	*Pseudecheneis sulcata*	Sisoridae
506.	*Pseudeutropius atherinoides*	Schilbeidae
507.	*Pseudeutropius mitchelli*	Schilbeidae
508.	*Pseudobagrus chryseus*	Bagridae
509.	*Pseudogobiopsis oligactis*	Gobiidae
510.	*Pseudogobius javanicus*	Gobiidae
511.	*Pseudogobius poicilosomus*	Gobiidae
512.	*Pseudosphromenus cupanus*	Belontiidae
513.	*Pseudosphromenus dayi*	Belontiidae
514.	*Pseudotrypauchen multiradiatus*	Gobiidae
515.	*Psilorhynchus balitora*	Psilorhynchidae
516.	*Psilorhynchus homaloptera*	Psilorhynchidae
517.	*Psilorhynchus microphthalmus*	Psilorhynchidae
518.	*Psilorhynchus sucatio*	Psilorhynchidae
519.	*Pterocryptis afghana*	Siluridae
520.	*Pterocryptis gangelica*	Siluridae
521.	*Pterocryptis indicus*	Siluridae

Contd...

Table 1.7–Contd...

Sl.No.	Species	Family
522.	*Pterocryptis wynaadensis*	Siluridae
523.	*Ptychobarbus conirostris*	Cyprinidae
524.	*Puntius amphibius*	Cyprinidae
525.	*Puntius arenatus*	Cyprinidae
526.	*Puntius arulius*	Cyprinidae
527.	*Puntius bimaculatus*	Cyprinidae
528.	*Puntius cauveriensis*	Cyprinidae
529.	*Puntius chalakkudiensis*	Cyprinidae
530.	*Puntius chola*	Cyprinidae
531.	*Puntius conchonius*	Cyprinidae
532.	*Puntius coorgensis*	Cyprinidae
533.	*Puntius deccanensis*	Cyprinidae
534.	*Puntius denisonii*	Cyprinidae
535.	*Puntius dorsalis*	Cyprinidae
536.	*Puntius fasciatus*	Cyprinidae
537.	*Puntius filamentosus*	Cyprinidae
538.	*Puntius fraseri*	Cyprinidae
539.	*Puntius gelius*	Cyprinidae
540.	*Puntius guganio*	Cyprinidae
541.	*Puntius melanostigma*	Cyprinidae
542.	*Puntius morehensis*	Cyprinidae
543.	*Puntius mudumalaiensis*	Cyprinidae
544.	*Puntius muzaffarpurensis*	Cyprinidae
545.	*Puntius nangalensis*	Cyprinidae
546.	*Puntius narayani*	Cyprinidae
547.	*Puntius ophicephalus*	Cyprinidae
548.	*Puntius orphoides*	Cyprinidae
549.	*Puntius parrah*	Cyprinidae
550.	*Puntius phutunio*	Cyprinidae
551.	*Puntius punctatus*	Cyprinidae
552.	*Puntius puntio*	Cyprinidae

Contd...

Table 1.7–Contd...

Sl.No.	Species	Family
553.	*Puntius roseipinnis*	Cyprinidae
554.	*Puntius sahyadriensis*	Cyprinidae
555.	*Puntius setnai*	Cyprinidae
556.	*Puntius shalynius*	Cyprinidae
557.	*Puntius sharmai*	Cyprinidae
558.	*Puntius sophore*	Cyprinidae
559.	*Puntius terio*	Cyprinidae
560.	*Puntius ticto*	Cyprinidae
561.	*Puntius vittatus*	Cyprinidae
562.	*Racoma labiata*	Cyprinidae
563.	*Raiamas bola*	Cyprinidae
564.	*Raiamas guttatus*	Cyprinidae
565.	*Rama chandramara*	Bagridae
566.	*Rasbora caverii*	Cyprinidae
567.	*Rasbora daniconius*	Cyprinidae
568.	*Rasbora rasbora*	Cyprinidae
569.	*Rhinomugil corsula*	Mugilidae
570.	*Rita chrysea*	Bagridae
571.	*Rita gogra*	Bagridae
572.	*Rita kuturnee*	Bagridae
573.	*Rita rita*	Bagridae
574.	*Rohtee ogilbii*	Cyprinidae
575.	*Salmo trutta fario*	Salmonidae
576.	*Salmo trutta trutta*	Salmonidae
577.	*Salmostoma acinaces*	Cyprinidae
578.	*Salmostoma bacaila*	Cyprinidae
579.	*Salmostoma balookee*	Cyprinidae
580.	*Salmostoma belachi*	Cyprinidae
581.	*Salmostoma boopis*	Cyprinidae
582.	*Salmostoma horai*	Cyprinidae
583.	*Salmostoma kardahiensis*	Cyprinidae

Contd...

Table 1.7–Contd...

Sl.No.	Species	Family
584.	*Salmostoma novacula*	Cyprinidae
585.	*Salmostoma orissaensis*	Cyprinidae
586.	*Salmostoma phulo*	Cyprinidae
587.	*Salmostoma sardinella*	Cyprinidae
588.	*Salmostoma untrahi*	Cyprinidae
589.	*Salvelinus fontinalis*	Salmonidae
590.	*Scatophagus argus*	Scatophagidae
591.	*Schismatogobius deraniyagalai*	Gobiidae
592.	*Schismatorhynchos nukta*	Cyprinidae
593.	*Schistura beavani*	Balitoridae
594.	*Schistura cincticauda*	Balitoridae
595.	*Schistura corica*	Balitoridae
596.	*Schistura dayi*	Balitoridae
597.	*Schistura denisoni*	Balitoridae
598.	*Schistura kangjupkhulensis*	Balitoridae
599.	*Schistura manipurensis*	Balitoridae
600.	*Schistura montana*	Balitoridae
601.	*Schistura prashadi*	Balitoridae
602.	*Schistura punjabensis*	Balitoridae
603.	*Schistura rendahli*	Balitoridae
604.	*Schistura rupecula*	Balitoridae
605.	*Schistura savona*	Balitoridae
606.	*Schistura scaturigina*	Balitoridae
607.	*Schistura semiarmata*	Balitoridae
608.	*Schistura sijuensis*	Balitoridae
609.	*Schistura tirapensis*	Balitoridae
610.	*Schistura vinciguerrae*	Balitoridae
611.	*Schizopygopsis stoliczkae*	Cyprinidae
612.	*Schizothoraichthys curvifrons*	Cyprinidae
613.	*Schizothoraichthys progastus*	Cyprinidae
614.	*Schizothorax esocinus*	Cyprinidae
615.	*Schizothorax kumaonensis*	Cyprinidae

Contd...

Table 1.7–Contd...

Sl.No.	Species	Family
616.	*Schizothorax microcephalus*	Cyprinidae
617.	*Schizothorax molesworthi*	Cyprinidae
618.	*Schizothorax nasus*	Cyprinidae
619.	*Schizothorax niger*	Cyprinidae
620.	*Schizothorax plagiostomus*	Cyprinidae
621.	*Schizothorax richardsonii*	Cyprinidae
622.	*Scoliodon laticaudus*	Carcharhinidae
623.	*Securicula gora*	Cyprinidae
624.	*Semiplotus manipurensis*	Cyprinidae
625.	*Semiplotus modestus*	Cyprinidae
626.	*Setipinna breviceps*	Engraulidae
627.	*Setipinna brevifilis*	Engraulidae
628.	*Setipinna phasa*	Engraulidae
629.	*Sicamugil cascasia*	Mugilidae
630.	*Sicyopterus griseus*	Gobiidae
631.	*Sicyopterus microcephalus*	Gobiidae
632.	*Sillago sihama*	Sillaginidae
633.	*Silonia childreni*	Schilbeidae
634.	*Silonia silondia*	Schilbeidae
635.	*Silurus morehensis*	Siluridae
636.	*Sinilabeo dero*	Cyprinidae
637.	*Sisor rabdophorus*	Sisoridae
638.	*Somileptus gongota*	Cobitidae
639.	*Sperata aor*	Bagridae
640.	*Sperata seenghala*	Bagridae
641.	*Stigmatogobius sadanundio*	Gobiidae
642.	*Taenioides gracilis*	Gobiidae
643.	*Tenualosa ilisha*	Clupeidae
644.	*Tenualosa toli*	Clupeidae
645.	*Terapon jarbua*	Terapontidae
646.	*Tetraodon cutcutia*	Tetraodontidae

Contd...

Table 1.7–Contd...

Sl.No.	Species	Family
647.	*Tetraodon fluviatilis*	Tetraodontidae
648.	*Tetraodon nigroviridis*	Tetraodontidae
649.	*Tetraroge niger*	Tetrarogidae
650.	*Thynnichthys sandkhol*	Cyprinidae
651.	*Tinca tinca*	Cyprinidae
652.	*Tor khudree*	Cyprinidae
653.	*Tor kulkarnii*	Cyprinidae
654.	*Tor mussullah*	Cyprinidae
655.	*Tor progeneius*	Cyprinidae
656.	*Tor putitora*	Cyprinidae
657.	*Tor tor*	Cyprinidae
658.	*Toxotes chatareus*	Toxotidae
659.	*Toxotes jaculatrix*	Toxotidae
660.	*Travancoria elongata*	Balitoridae
661.	*Travancoria jonesi*	Balitoridae
662.	*Trichogaster chuna*	Belontiidae
663.	*Triplophysa gracilis*	Balitoridae
664.	*Triplophysa kashmirensis*	Balitoridae
665.	*Triplophysa ladacensis*	Balitoridae
666.	*Triplophysa marmorata*	Balitoridae
667.	*Triplophysa microps*	Balitoridae
668.	*Triplophysa shehensis*	Balitoridae
669.	*Triplophysa stoliczkae*	Balitoridae
670.	*Triplophysa tenuicauda*	Balitoridae
671.	*Triplophysa yasinensis*	Balitoridae
672.	*Valamugil cunnesius*	Mugilidae
673.	*Valamugil speigleri*	Mugilidae
674.	*Wallago attu*	Siluridae
675.	*Xenentodon cancila*	Belonidae
676.	*Zenarchopterus dispar*	Hemiramphidae
677.	*Zenarchopterus ectuntio*	Hemiramphidae
678.	*Zenarchopterus striga*	Hemiramphidae

Source: NBFGR (ICAR).

Figure 1.2: Some Common Fishes Found in India

Indian Major Carps

Labeo rohita

Catla catla

Cirrhinus mrigala

Contd...

Figure 1.2–Contd...
Exotic Carps

Cyprinus carpio

Ctenopharyngodon idella

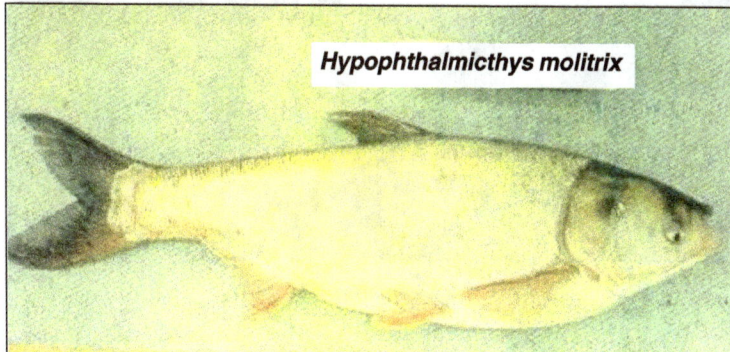

Hypophthalmicthys molitrix

Contd...

Figure 1.2–Contd...

Glossogobius sp.

Ompok pabda

Contd...

Figure 1.2–Contd...

Liza parsia

Gudusia chapra

Contd...

Figure 1.2–Contd...

Ctenualosa ilisha

A Hall of Shrimp Catch

Contd...

Figure 1.2–Contd...

Notopterus chitala

Polynemus sp.

Contd...

Figure 1.2–Contd...

Aorichthys aor

Etroplus suratensis

Contd...

Figure 1.2–Contd...

Pampusargenteus

Table 1.8: Some Important Freshwater Culture Fishes

	Indigenous Exotic
Catla catla (Ham)	*Hypothalmicthys molitrix (val.)*
Labeo rohita (Ham)	*Cyprinus carpio* (Linne)
L.calbasu (Ham)	*Ctenopharyngodon idella* (Val)
L. fimbriatus (Bloch)	*Tinca tinca* (Linne)
L. bata (Ham)	*Carasius carassius* (Linne)
Cirrhinus mrigala (Ham)	*Osphromenus goramy* Lacepede
C. reba (Ham)	*Orcochromis mossambica* (Peters)
Etroplus suratensis (Bloch)	Larvicidal fishes
Anabas testudneus (Bloch)	*Poecilia reticulata* (Peters)
Wallago attu (Bloch)	*Gambusia affinis* (Baird and Girard)
Aorichtlys seenghala (Sykes)	Game fishes
Channa striatus (Bloch)	*Salmo trutta fario*
Clarias batrachus (Linne)	*S.garidneri*
Heteropneustes fossilis (Bloch)	*Salvelinus fontinalis*
	Onchorhynchus nerka

And also suggested several remedial measures which can save the species. Such studies when taken up for the whole of India can provide the status of our freshwater species. The present state of our knowledge about the freshwater fish fauna of India contains several gaps. Many species have not been collected since their original description several hundred years ago. Apart from these several aquarium species have also been introduced. Of the culturable fishes 66 per cent are constituted by Cyprinids, which subsist mostly on vegetable matter, and algae, while the Channids and Silurids depend on a carnivorous diet.

Role of Government and NGO

(*a*) Government

The Ministry of Environment and Forest of the Government of India has been taking active steps in the protection of the Environment and the wealth of India. In recent times the biodiversity conservation has been given top priority and several projects in these lines have been financed with a view to protect the Environment, create awareness among the masses and conserve India's Biodiversity, Biosphere reserves, National parks and Wildlife sanctuaries are declared to protect the endangered, threatened vulnerable and rare species.

The NBFGR are actively involved in building a database on genetic biodiversity and gene banks, developing strategies for conservation of mahseers and other threatened species. Needless to say, the Zoological Survey of India, is the custodian of the faunal wealth of India. The fishery survey of India has been contributing much to our understanding of the demersal fauna and coastal pelagics.

(*b*) NGO

With regard to the conservation our Mahseers and other game of fishes several industrial and corporate sectors have had both positive and negative impact.

At a time when there was a sharp decline in catches of large Mahseers the National Commission on Agriculture recommended a comprehensive survey of rivers and streams known to have Mahseer stocks. In 1976 wildlife Association of South India (WASI) leased about 22 km of Cauvery, 60-80 km from Bangalore for conservation of tor khudree.

During Aug-Sep 1978, Trans World Fishing Expedition (TWFE) collected the species ranging from 13.6-30.8 Kg in weight, the biggest Mahseer captured at Moshelli Halle weighing 32.0 kg. There are recent reports of the successful fishery management of this species in River Coorg. For a number of years WASI has been stocking the leased stretch with Mahseer fingerlings. Today the fishing season is open to licenced fishermen from October to May. Only rod and line fishing is allowed and there is a ban on using dynamites and poison. Fishing is permitted between 6 am to 6 pm only. Records of catches are carefully maintained. Between 1989 and 1996 the Mahseers captured ranged from 21.6 to 48.1 kg. Mahseer stock in this carefully managed stretch of the Cauvery is now protected from overfishing, with large-sized mahseers available to licenced sport fishermen.

The Coorg Wild Life Society, Madikere is another voluntary organization engaged in mahseer protection in the Cauvery. This society has leased a 28 km stretch and has been stocking young mahseers since 1993, organizing sport fishing and maintaining fish catch statistics.

The Tata Electric Companies (TEC) fish seed farm at Lonavala, Maharashtra, has been carrying out artificial propagation, rehabilitation, and conservation since 1970's of *Tor khudree, Tor tor* and *Tor putitora*, the species being successfully bred using hypophysation in 1995 and 1996.

The Department of Fisheries, Government of Karnataka and the Fish Farmer's Development Agency, Yadavagiri, Mysore have rehabilitated the river with fry and fingerlings of *Tor khudree.*, from TEC's fish farm, as part of the implementation of the project launched in 1987 by the Department of Fisheries, Karnataka by "Rehabilitation and Development of Mahseer Fishery in the River's and Reservoirs of Western Ghats".

The Karnataka Power Corporation Limited has initiated in 1996 the stocking and management of the species in 5 hectare hatchery.

The other important species which provides sport fishing is the exotic rainbow trout. It is present in streams of the Bababudan Range in Karnataka, in Nilgris and Palani hills in Tamil Nadu and in High Range in Kerala, by the Nilgiri Game Association, Palani Hills

Game Association and the High Range Angling Association (Grammar problem). A decline in catches in Nilgiris was reported due to the ecological conditions being not conclusive to this trout. No published data are available on the catch statistics in these ranges for the post 1970 period. Thought the introduction has been successful in the ranges, the species having developed self-sustaining populations, there has been a steep decline in its size, indicating heavy fishing pressure. More recently the hatcheries in the High ranges of Kerala were taken over by Tata Tea Company.

Invasive predatory species such as bass and trout which were introduced widely in streams that had a small natural fish community of a few minnows. Not only the indigenous species lack natural defenses against large specialized predators, but also the predators themselves had no natural controls. Consequently, the populations of indigenous species were quickly eliminated from many water bodies and some threatened species now survive only in refuge streams beyond the reach of introduced predators. Similar problems are being caused by the translocation of indigenous species.

Biodiversity Related Initiatives

1. Government Policy and Legal Measures

(a) Though the several restrictions on indiscriminate fish catching has been imposed by the Government, scant response is being given to the regulations and restrictions, by the fishing community. Unless it dawns on each and every group involved in fishing that conservation is very essential for sustained catches, it is impossible to implement any legal measures.

(b) Aquaculturists in their eagerness to increase fish catches and provide sport fishing concentrate on culturing and introducing any species with least concern about its impact on the resident species.

Introduction of species in the lower reaches may not have much of a deleterious effect as it might in the upper reaches that may have few and small sized species. These small hill stream fishes mostly endemics, are the wealth of the country. "There is little evidence of ill-effects of trout on other fish species of streams and lakes of the western Ghats. Waters stocked with trout had no local fish of commercial or sport importance prior to trout introduction. The endemic fish fauna consists of mainly of small-sized cyprinids and coptids and which are prey fish for the introduced species". It is not all the fishing pressure but the fishing pleasure of the sport lovers which have caused havoc to biodiversity.

2. NGO

NGO should be encouraged to take active part in conservation measures. They should interact with the Government departments and take decisions with the approval of concerned authorities.

3. Community and People movements

(*i*) Regeneration and Conservation

It is the local community which will have the final say on the environmental policy applications. Unless co-operation, is there no forceful regulations can *be implemented.*

(*ii*) Protests Against Destruction

The scientists', conservationists', and the people should react in unison when detrimental decisions are taken in the name of development.

Strategies to Conserve Freshwater Biodiversity

Current Strategy and Action Plan

Though the need to conserve animal and forest resources was taught and decreed in parts of China and India as far back as 700 B.C., the fish genetic resource conservation on modern lines is a very young concept and recent venture in the country. This is being implemented through the newly established research organization, National Bureau of Fish Genetic Resources. The identified immediate priorities and thrusts for the task are (*i*) distinct populations with the help of advanced techniques, (*ii*) cataloguing of the genotypes, (*iii*) collection of information on genetic variability for suggesting steps for conservation of genetic diversity, (*iv*) ascertaining the characters that are polymorphic in nature for inclusion in studies of genetic resources, (*v*) development of practicable methodologies for *in situ* and *ex situ* conservation of the exploited and endangered species under Indian conditions.

Sustainable Management and Conservation

Fish conservation as a concept is being encouraged since Ashoka's time (246 BC) when catching of fish was prohibited during certain periods. The Indian Fisheries Act-IV of 1897 is a landmark in the conservation of fishes in India, where by the use of explosives or poison to kill fish in any water is prohibited. The Wild Life Act of 1972 provides the creation of several National Parks and Sanctuaries in India primarily for the protection of birds and mammals. Protection of fish species has not been included in the above Act. IUCN (International Union for Conservation of Nature and Natural Resources) the major global agency for conservation of animals and plants, has published lists of species of fishes, amphibians, reptiles, birds, etc. However, the IUCN Red Data book does not list any Indian fishes as endangered, threatened or rare fish so far. The U.N. Conference on Human Environment, held in Stockholm in 1972 stressed the need for conservation of all genetic resources. FAO (1981) proposed two essential and complementary requirements for genetic resources conservation, *i.e. in situ* and *ex situ*.

A. *In situ* Conservation

In situ conservation for land races and wild relatives is useful where genetic diversity exists and where wild forms are present. This is done through their maintenance within natural or man-made ecosystems in which they occur. The major advantages of *in situ* conservation are:

1. Continued co-evolution where in the wild the species can continue to coevolve with other forms providing the breeder with a dynamic source of resistance that is lost in *ex situ* conservation and

2. Natural parks and biosphere reserves may provide less expensive protection for the wild relatives than *ex situ* measures.

Conservation of endangered fishes in their natural habitat is the best option available for conservation. In India, *in situ* conservation of aquatic resources is still in its infancy. The steps involved in the process of *in situ* conservation has been summarized as under:

(a) Habitat inventory

A preliminary habitat inventory survey of Ladhiya river from downward of Chalthi Bridge in India was conducted in 1995-96 in following the format of U.S. Fish and Wildlife Service. Soil and water samples were also collected from the eroded beds of Ladhiya and Bisouria (a seasonal stream of first order). Sediment load after the confluence was mainly responsible for the mass mortality of fishes in the river during heavy rains.

(b) Utilization of Remote Sensing to Study Endangered Fish Habitat

Conserving biological diversity involves restoring, conserving and enhancing the variety of life in an area that abundance of distribution of species and communities provide for continued existence and normal ecological functioning.

(c) Spectral Behaviour of Various Fish Habitat Features

The technique of remote sensing along with geographical information system (GIS) was used to study fish habitat over a large area. A study was conducted in Ladhiya stream during March 1997 with the objective to determine spectral behaviour of various features and to observe the extent of differentiation between land features, water bodies and whether differentiation of these features is possible based on spectral signature.

☆ Salt encrusted area

☆ The surfing water, sand and boulder bed

☆ The turbid water, moist sand and blackish soil

☆ Clear water bodies irrespective of their depth and substrate.

The study has indicated that at least broadly land and water features can be differentiated based on spectral observation.

(d) Tagging of Mahseer

To study more about breeding and migration of mahseer, tagging was done in two places of Ladhiya river, Chalthi and Ritha Sahib in June 1997. Fishes were collected by using cast net. At Chalthi, 30 fishes (12 to 23 cm) were tagged and released back into the river. Fabricated Floy tags were used for this study. Seven numbers of *Schizothorax* sp. were also tagged. At Rithasahib, after intense fishing, 65 mahseer, (size ranging 10-20 cm) were collected, tagged and released back into the river.

(e) Restoration by Ranching

As a measure of replenishing the declining population of mahseer, ranching programme was organized in June, 1996 at Garampani in Kosi stream in U.P. hills. This was done in collaboration with National Research Centre on Coldwater Fisheries, NBFGR,Central Inland Fisheries Research Institute and States Fisheries Department.

(f) Induced Breeding of Threatened Fishes

Attempts to breed endangered golden mahseer in captivity in Garhwal region by earlier workers were unsuccessful. On a request of U.P. State Fisheries Department, the prospect for breeding golden mahseer (*Tor putitora*) at the State Fish Farm at Dhakrani (near Dehradun) was taken up.

The necessary arrangements for the induced breeding experiments were finalized with the State Fisheries Department. In September 1997, the brood stocks were checked for maturity and three pairs weighing 1.5–2.0 kg were seggregated. Fishes were induced bred with the administration of a single dose of Ovaprim at 0.2 ml/kg body weight. Fertilization was about 80-85 per cent and hatching was 60 per cent at water temperature 22-24°C and pH 6.9. This success opens avenues for *ex–situ* conservation since mahseer has been bred captivity in a geographic area outside the range where natural breeding takes place.

(g) Mass Awareness and Community Participation

To create awareness among the common people for the conservation of fish genetic resources efforts was made by National Bureau of Fish Genetic Resources (NBFGR) in the upland areas of Uttar Pradesh and North-Eastern Region of the country. Joint efforts were made to discourage the use of destructive methods of fishing. During the mass awareness programme exhibitions were organized along with go this, lectures and pamphlets were distributed.

Public awareness on the status of endemic fishes and the need for their conservation is an important part of species recovery programme. To get a feedback on mechanism involved, a series of awareness camps were organised and local conservation committees were set up. Mahseer conservation committee was formed to stop illegal fishing and to create awareness among the local people for the conservation of fish germplasm in the area.

As visual display plays an important role in mass communication, *photo exhibitions* displaying various aspects of Mahseer and importance of its conservation was organized at Govt. Inter College, Tanakpur, U.P. *Handouts* and *pamphlets* were also distributed among the visitors. This effort on the part of Bureau has no doubt aroused consciousness among local inhabitants regarding conservation of the diminishing fish population.

During 1994-95 mass awareness programmes were organised with active participation of the committees constituted earlier along with fishermen, students, teachers, Gram Sabha and Yuvak Mangal Dal members and villagers for the conservation of mahseer populations.

Mass awareness programme was also held at Department of Zoology, Kumaon University, Almora (Uttarakhand) jointly organized by NBFGR and Department of Zoology, Kumaon University, largely attended by the scientists, students, NGO representatives, farmers and local people. People were motivated through participatory approach, photo exhibition and video campaign. *Bulletins* in Hindi were distributed on "Save endangered fishes" and *road side hoarding* were fixed with conservation slogans. The Masheer Bachao Samitis organized at different places involving fishermen, students, teachers, peoples representatives and common men.

In 1996-97 again several gosthis at various places like Tanakpur, Chalthi, Amrohi and Ritha Saheb were organized. These gosthis were largely attended by local people, fishermen, members of different "Mahseer Bachao Samitis" of the three zones of Ladhiya river. Emphasis was given for the declaration of certain parts of the Ladhiya and Sharda rivers as Fish Sanctuary.

The effect of continuous mass awareness programme in a stretch of 200 kms. was quite encouraging. Local people developed an understanding of importance of conservation of fish germplasm in its natural habitat. Illegal fishing was greatly reduced and the *"Mahseer Conservation Committees"* members captured lots of fishing gears. Size and no. of mahseer in the experimental catch also increased markedly.

(h) Declaration of Aquatic Sanctuaries

Efforts is being made to declare large water bodies as fish semotuaning for this potential ascas falling under the wild life semotuaning are being identified. After declaration as aquatic sanctuaries these water-bodies will serve as a repository for endangered fish species.

B. *Ex situ* Conservation

(a) Cryopreservation

Cryopreservation of fish gametes is of great practical value. It offers a number of practical and economic advantages in fish culture. These include the transportation of gametes from one facility to another permitting exchange of sex products and enhancing collaboration among research groups irrespective of distance. It allows for genetic preservation of desirable gene lines and facilitates selective breeding and hybridization. It also allows banking and ready availability of germplasm of known genetic make-up.

(b) Fish Germplasm Repository

The concept is to establish a live gene bank of fishes. In this connection several wild fish species was collected from the nature and transferred to the germplasm repository where its feeding, breeding behaviour, growth is being studied. There after its breeding will be done and the seeds thus produced will be utilized for ranching or for transportation to distant places where ever needed.

1. *Chitala chitala*
2. *Notopterus notopterus*
3. *Tor putitora*
4. *Hypopthalmichthys molitrix*
5. *Ctenopharyngodon idella*
6. *Labeo calbasu*
7. *Labeo rohita*
8. *Cirrhinus mrigala*
9. *Catla catla*
10. *Channa marulius*
11. *Bagarius bagarius*

In future wild stocks of river Ghaghra, Gomti and Ganga will be collected and transferred to the repository. This repository will serve as a model and the technology thus developed will be transferred to other States for implementation.

(c) Strategies for Aquatic Exotic and Quarantine

For increasing aquaculture production, exotic fish stocks and species are being introduced in the country. Therefore, there is a need to safeguard our indigenous fish fauna from infectious exotic diseases and to develop a protocol for fish quarantine. Indian Fisheries Act for protection of fishery resources is very old which needs amendment according to present situation and need. There is no specific Act is available in India to prevent the illegal introduction and spread of illegally or legally introduced exotic fishes. As the fisheries come under the state fisheries and forest department they have made their own acts and a part of Wildlife Protection Act, which differs from state to state. There is no legislation or policy to restrict the movement of infected fish in India, which may result in spread of EUS disease. So the Government must give the adequate power and control to fishery inspectors to inspect the fisheries farms.

Though, in India there is no specific Act, but in other countries like USA has non indigenous aquatic nuisance prevent and control Act of 1990, Australia has Fisheries Management Act 1994, Quarantine Act of 1908 for control of various aquatic organisms introduction etc. Therefore it is imperative to have an Act in India.

(d) Needs for E-situ Conservation of Fishes

Some gaps and deficiencies in the knowledge and practice of conservation of fish gene pools need be asserted before truly rigorous programmes can be developed. These may be as under:

(i) Genetic Information on Fishes

A primary and immediate need is collection of genetic information on fishes judged to be in peril of extinction or significant genetic losses. Of primary importance is determination of how genetic variance is distributed within and among population (Hamrick 1983). Such type of work has recently been initiated at the NBFGR.

(ii) Establishment of Gene Banks

Bureau has developed and standardized the technique for cryopreservation of fish milt and a mini gene bank with milt of *Catla catla, Labeo rohita, Cirrhinus mrigala, Cyprinus carpio, T. putitora, T. khudree* and *Tenualosa ilisha* and many other species found outside the State has been cryopreserved. Viable hatchlings were produced from milt cryopreserved for over five years.

(iii) Experimental Work on Genetics and Fitness

A better understanding of the role that genetic heterozygosity plays in individual fitness of fishes is necessary. More heterozygous individuals are believed to display higher fitness with respect to characters like growth rate survivorship, fecundity, developmental stability etc. In fact, we need to learn more about heterozygosity and fitness to better serve our interests in maintaining endangered populations of fishes.

(iv) Subtle Aspects of Life Histories

Conservation programme should also consider features of fish life histories such as behaviour, mating patterns and dispersal etc. that could affect the genetic structure of population.

The studies, in this direction are being planned at the Bureau.

(v) Educational Emphasis

Emphasis in fishery management programme that recognizes the inherent value of native fish species, regardless of whether they can be harvested in a fishery, is necessary.

Attempts are to be made through seminars, meetings, and publications to bring about the mass awareness on this aspect.

(vi) Protecting Aquatic Habitats

The most universal goals of fish recovery plans is to protect and enhance existing habitats, Johson and Rinne (1982) from their experience visualize at least three methods of specifying protection of existing habitats. These include increased public sector management, prioritization of protected habitats and ecosystem-based recovery teams.

The protection may be through the National Parks, declaring sanctuaries in appropriate places, bringing about/enforcing legislations regulating destructive methods of fishing and ensuring avoiding water pollution etc. The Bureau has been engaged in formulating workable measures for the above purpose.

Given the above, the two bottom lines that are considered pre-requisites in this action plan, are:

Ecological Security of the country or of any region within it, and Livelihood Security of those most critically dependent on biodiversity and its components:

☆ *Ecological security* refers to the maintenance of the diversity of ecosystems and habitats; the diversity of species, subspecies/varieties, populations and communities; the interactions between species, populations, communities and their habitats and ecosystem; their integrity including biological productivity of ecosystems and taxa; the evolutionary potential of natural and agricultural systems; and critical ecosystem services. This refers to both wild and domesticated biodiversity.

☆ *Livelihood security* refers to the security of human communities and individuals critically dependent on biological resources, including guaranteed access to, and control over, such biological resources and related knowledge.

Both ecological and livelihood security have been severely eroded, and continue to be threatened. Therefore there is a need to take urgent and comprehensive measures to reverse this trend.

Three basic goals need to be achieved to reverse this trend:

☆ Conservation of biodiversity, including the integrity and diversity of genes, species and ecosystems and their evolutionary potential.

☆ Sustainable use of biological resources, referring to the use of components of biological diversity in such a manner and at such rates that does not lead to the long-term decline of the biological diversity, thereby maintaining its potential to meet the needs and aspirations of present and future generations.

☆ Equity in conservation and use, including equitable access to and decision-making control over biodiversity as well as equitable distribution of costs and benefits associated with conservation and sustainable use. In particular, it includes creating democratic spaces for the voices of disprivileged women and men in defining conservation and use priorities.

Gaps in Freshwater Biodiversity

(*i*) Gaps in Information

An understanding of fish diversity is not complete, with many new species still being discovered, in recent years. (Good) Although survey of the rivers, the upper reaches, especially of the west flowing rivers, is sure to reveal many more interesting species. Recently, three subterranean species were discovered from Kerala alone.

The biology of many species are not understood. Unless an understanding of its food habits and spawning requirements are known conservation measures cannot be carried out.

(*ii*) Gaps in Vision

There should be more interaction between the biosystematists and the Aquaculturists. Better results with least loss to endemics can be achieved with a broader vision. When entire trophic levels are understood in an ecosystem and importance is given to each species, whether they are small or big, are of commercial importance or not.

(*iii*) Gaps in Policy and Legal Structure

Strict implementation of the rules should be done and those found erring should be seriously dealt with. Any policy decision should be made with a multidisciplinary approach. (Gaps are not recommendation. What you say here are recommendations!)

(*iv*) Gaps in Institution and Human Capacity

The main cause for Centilinean extinction is the lack of biosystematists in various groups. The ZSI and BSI, the Chief custodians surveyors of the faunal and floral wealth should also act in unison. Since more and more is revealed in recent years about the inter-relationships and interdependence of the flora and fauna, better understanding of the ecosystem can be achieved with more interaction between the two institutions and co-ordinated efforts are taken. Infact even survey and study should be a team effort.

The organization should be recognized as a deemed university with various branches of biosystematics offered for study for master's degree and specialization. More importance should be given to the study of biosystematics. If timely action is not taken the biosystematists themselves (non-commerical group) will soon become an endangered lot, with biotechnologists and aquaculturists soon replacing them. The eternal conflict between conservationists and commercialists can never be overlooked.

Major Strategies to Fill these Gaps

As mentioned earlier, a centralized system should exist, which regulates the function of the various disciplines. At least frequent interactions between the major institutions which have anything to do with faunal diversity can work more effectively and economically than isolated futile attempts by various organizations in various directions.

The importance of biodiversity should be implanted in the minds of every individual. It is heartening that schools and colleges and NGO's organize environment awareness programs.

Required Action to Fill Up Gaps

Action to conserve and sustainable use of freshwater ecosystem and animal diversity. Preventive and protective measures form part of the conservation strategies (Menon, 1994)

Preventive measures

1. Nature reserves
2. Habitat protection.
3. Abatement of pollution.
4. Ban on expansion of plantations.
5. Monitoring stations.
6. Augmenting stocks in rivers.
7. Ecological and life history studies.
8. Wetlands as protected areas.
9. Ban on introduction of exotic species.
10. Repository of endangered species.

Protective measures

1. Fish sanctuaries.
   (a) Deep-pools
   (b) Temple tanks and fort moats.
   (c) Temple sanctuaries.

2. River stretch below dam as sanctuaries.

3. Upper reaches of rivers as sanctuaries.

4. Captive breeding.

Follow up to monitor the mechanism, including the evaluation and recommendation.

The strategies formulated should be given wide publicity to various Government organizations and NGO's and their opinion and inputs should also be assessed. These plan makers should finally be given a central stage to voice their formulations and their decisions should have strong backing for the effective materialization of the plan.

Conclusion

Reorientation of the development process, ensuring that ecological and livelihood security become central concerns, and that the conservation of biological diversity receives the highest priority. Community tenurial rights, ensuring rights of women, children and other disprivileged sections within them. Elimination of absolute poverty and preventing deprivation of indigenous people from the natural resources necessary for them to maintain an acceptable living standard. Development of alternative (including community) intellectual rights systems appropriate for indigenous knowledge, which respect that life forms should not be subjected to private and monopolistic IPR regimes. Recognition and integration of the full range of intrinsic as well as direct values of biodiversity into human activities. Recognition, respect, and revitalisation of gender differentiated indigenous and community knowledge systems relating to biodiversity, and synergising these with mainstream knowledge systems. Balancing of local, national, and international interests related to biodiversity, on the basis of principles of ecological sustainability and social equity; within these principles, local interests to get priority over national, and national over international. Respect for cultural diversity, and the diversity of governance systems, customary practices and laws, and other aspects of human society, in so far as these are in consonance with the basic principles of ecological sustainability and social equity. The importance of biodiversity should be implanted in the minds of every individual.

References

Anon., 1996a. Ninth five-year plan for fisheries. *Report of the Working Group*, New Delhi.

Anon., 1996b. *Handbook on Fisheries Statistics*. Ministry of Agriculture, Department of Agriculture and Co-operation, Fisheries Division), Government of India, New Delhi, 217 p.

Anon., 2000. *Handbook on Fisheries Statistics*. Ministry of Agriculture, Department of Agriculture and Co-operation, Fisheries Division, Government of India, New Delhi, 153 p.

Barman, R.P., 1998. *Faunal Diversity in India: Pisces*. ENVIS Centre, Zoological Survey of India, Calcutta, p. 418–426.

Day, F., 1875–78. *The Fishes of India: Being a Natural History of the Fishes Known to Inhabit the Sea and Freshwater of India, Burma, and Ceylon*, London, xx+778 pp., 195 plts.

Datta Munshi, J.S. and Srivastava, M.P., 1988. *Natural History of Fishes and Systematics of Freshwater Fishes of India*. Narendra Publishing House, Delhi, 430p.

FAO/UNEP, 1981. *Tropical Forest Resources Assessment Project*. Technical Report No. 3. FAO, Rome.

Government of India, 1985. *Research and Reference Division, Ministry of Information and Broadcasting*.

Groombridge, B., 1983. Comments on the rain forests of south-west India and their herpetofauna. Paper Prepared for the *Centenary Seminar of the Bombay Natural History Society*, 6–10 December, Revised, January 1984, 18 pp.

Groombridge, B. (ed). 1993. *The 1994 IUCN Red List of Threatened Animals*. IUCN, Gland, Switzerland and Cambridge, UK, lvi+286 pp.

IBWL, 1972. *Project Tiger: A Planning Proposal for Preservation of Tiger (Panthera tigris tigris Linn.) in India*. Indian Board for Wildlife, Government of India, New Delhi, 114 pp.

ICBP, 1992. *Putting Biodiversity on the Map: Priority Areas for Global Conservation*. International Council for Bird Preservation, Cambridge, UK, 90 pp.

IUCN, 1986. *Review of the Protected Areas System in the Indo-Malayan Realm*. IUCN, Gland, Switzerland and Cambridge, U.K., 461 pp.

IUCN, 1987. *Centres of Plant Diversity: A Guide and Strategy for their Conservation*. An outline of a book being prepared by the Joint IUCN–WWF Plants Conservation Programme and IUCN Threatened Plants Unit.

Jayaram, K.C., 1999. *The Freshwater fishes of the Indian Region*, Narendra publishing House,New Delhi, 551 pp.

Jhingran, V.G., 1991. *Fish and Fisheries of India*. Hindustan Publishing Corporation, New Delhi, 727 p.

Jhingran, V.G., 1983. *Fish and Fisheries of India*. Hindustan Publishing Corporation (India), Delhi, pp. 666.

Kar, C.S. and Bhaskar, S., 1981. Status of sea turtles in the Eastern Indian Ocean. In: *Biology and Conservation of Sea Turtles*, (Ed.) K. Bjorndal. *Proc. World Conf. Sea Turtle Cons.*, Smithsonian Institute Press, Washington, pp. 373–383.

MacKinnon, J. and MacKinnon, K., 1986. *Review of the Protected Areas System in the Indo-Malayan Realm*. International Union for the Conservation of Nature and Natural Resources, Gland, Switzerland and Cambridge, U.K., 284 pp.

Menon, A.G.K., 1999. *Checklist of Freshwater Fishes of India*. Zoological Survey of India, Occ. Pap. No. 175, p. 366.

Menon, A.G. K., 1989. *Conservation of Freshwater Fishes of Peninsular India*. Final Report. GRANT NO. 14/24/87–MAB/RE Dt. 12-08-88.

Menon, A.G.K., 1951. Monograph of Cyprinid fishes of the genus Garra Hamilton. *Mem. Indian Mus.*, 14(4): 173–260p.

Menon, A.G.K., 1974. *A Checklist of Fishes of Himalayan and the Indo-Gangetic Plains*. Inland Fishery Society of India, Barrackpore, Special Publication No. 1, 136 p.

Menon, A.G.K., 1994. Criteria for determining the status of threatened categories of Indian freshwater fishes. In: *Threatened Fishes of India*, (Eds.) P.V. Dehadrai, P. Das and S.R. Verma. Nature Conservators, Muzaffarnagar, p. 1–5.

Ponniah, A.G. and Sarkar, U.K., 2000. *Fish Biodiversity of North-East India.* National Bureau of Fish Genetic Resources, 228 p.

Rao, K.L., 1979. *India's Water Wealth: Its Assessment, Uses and Projections.* Orient Longman Limited, New Delhi, p. 55–102.

Seghal, K.L., 1999. Coldwater fish and fisheries in the Western ghats, India. In: *Fish and Fisheries at Higher Altitudes Asia*. FAO Fisheries Technical Paper 385, p. 103–121.

Sinha, M., 1997. Inland Fisheries resources of India and their utilisation. In: *Fisheries Enhancement of Small Reservoirs and Floodplain Lakes in India*, (Eds.) V.V. Sugunan and M. Sinha. Central Inland Capture Fisheries Research Institute Bulletin, Barrackpore, 75: 167–174.

Sinha, M. and Katiha, Pradeep K., 2002. Management of inland fisheries resources under different property regimes. In: *Institutionalizing Common Pool Resources* (Ed.) Dinesh K. Marothia. Concept Publishing Company, New Delhi, pp. 437–460.

Skelton, Paul H., 1993. *A Complete Guide to the Freshwater Fishes of Southern Africa.* Southern Book Publishers, pp. 388.

Sugunan, V.V., 1995. *Reservoir Fisheries in India.* FAO Fisheries Technical Paper No. 345. Food and Agriculture Organisation, Rome, p. 6, 12.

Talwar, P.K. and Jhingran, A., 1990. *Inland Fishes of India and Adjacent Countries.* Oxford and IBH Publishing Co. Pvt Ltd., New Delhi, 1158 p.

Chapter 2

Water Quality Assessment of Bhadra Reservoir of Karnataka with Reference to Physico-chemical Characteristics

B.R. Kiran & E.T. Puttaiah

Department of Environmental Science, Kuvempu University,
Shankaraghatta – 577 451, Karnataka

ABSTRACT

Monthly and seasonal variations of physico-chemical characteristics of the back water of Bhadra reservoir (Station-I) and its downstream stretch of the Bhadra river (Station-II) were studied during 1998-2000. Water samples were collected from the identified stations at an interval of 30 days and analysed following the standard procedure (APHA, 1998). The water temperature fluctuated from a minimum of 23° C to a maximum of 30.5°C. The pH of water samples was on alkaline side in both the stations except during summer at station-II. Turbidity values ranged from a minimum of 1.2 NTU to a maximum of 30 NTU. The electrical conductivity values were found to be high during summer season at station II and low during winter at station-I and ranged from 32 µmhos/cm to 378 µmhos/cm. Total solids and pH are inversely related. Total suspended solids fluctuated from a minimum of 69 mg/l to a maximum of 1432 mg/l. While, total dissolved solids showed a minimum of 27.61

mg/l at station-I to a maximum of 195.5 mg/l at station-II. High chloride content was recorded during summer months in both the stations. Downstream stretch showed higher value of hardness as compared to upstream station. Seasonwise, total alkalinity reached highest concentration during summer in both the stations and lower during rainy season at station-II. However, the concentration of nitrite did not exceed the threshold concentration level fixed for drinking water by WHO at station-I. While station-II showed higher nitrite content and it is unfit for drinking. Seasonwise, it was found to be high during winter at station-II. It is found that the concentration of ammonia is more during summer and vice-versa in rainy season. High BOD and COD values were recorded at station-II both yearly and seasonally. The phosphate content varied from a 0.001 mg/l at station-I to a maximum of 0.96 mg/l at station-II. Higher calcium concentration was observed during summer at station-II. Nevertheless, magnesium content varied from a minimum of 2.9 mg/l to a maximum of 33.06 mg/l. In the present study free carbon dioxide varied from 4 mg/l to 58 mg/l. From the above study, upstream stretch of the reservoir is oligotrophic in nature and the water is fit for drinking and irrigation. While, the downstream stretch of Bhadra river is polluted due to addition of sewage, domestic waste, agricultural runoff, faecal matter and industrial effluents. Hence, the water is not suitable for drinking. In this paper some of the simple remedial measures are suggested to protect the river Bhadra from eutrophication.

Keywords: Bhadra reservoir, River, Eutrophication, Oligotrophic, Water quality.

Introduction

Water is a natural source which is essential human commodity. India is rich in fresh water resources, which have ecological diversity, reflecting their range of Indian climatic conditions and topography. The Bhadra reservoir basin play an important role in conservation of fishes, birds and other aquatic animals. They are of great economic, aesthetic and scientific importance. Bhadra reservoir have enough potential and scope for the development of fishes, which remained neglected earlier. However, rivers are the most important water resource. Unfortunately, the rivers are being eutrophicated by indiscriminate disposal of sewage, industrial wastes and human anthropogenic activities.

Bhadra reservoir is constructed across the Bhadra river near Lakkavalli village, Chikmagalore district of Karnataka state at an elevation of 601 above mean sea level. The reservoir is located at a latitude 13° 45' 00" N and longitude 75° 30' 14" E. The Bhadra river rises from Varaha hills at a place called "Ganga mula" in the Western ghats about 24 kms west of Kalasa in Chikmagalore district. After flowing for about 190 kms, it joins the river Tunga at Kudli, 14.40 kms east of Shimoga city and becomes "Tungabhadra" river which is a major tributary of Krishna river.

No literature is available regarding any systematic study of the water quality of the present water body except some occasional analysis carried out by some workers. Hence, the present study was undertaken with a view to assess the water quality of the Bhadra reservoir and its downstream stretch and also identify the source and extent of pollution.

Materials and Methods

Water samples were collected monthly for a period of two years from July 1998 to June 2000 in black plastic carbuoys of 3 litre capacity from two stations from backwater of Bhadra reservoir (Station-I) to downstream stretch of the river (Station-II). For the estimation of certain factors like dissolved oxygen, water samples were taken in 300 ml BOD bottle and fixed immediately using Winkler's reagent. Water temperature and pH were recorded at the time of collection. For the estimation of remaining parameters the samples were kept in air cooled chamber. In all cases the final results were calculated by following atleast 3 consecutive readings.

Physico-chemical parameters of the water samples were analysed following the standard methods (APHA, 1998).

Results and Discussion

Figure 2.1 represents monthly variation in physico-chemical characteristics of two sampling stations. During the present investigation the following physico-chemical characteristics have been studied and discussed.

Water Temperature

Water temperature plays an important role in either decreasing or increasing the concentration of certain chemical characteristics

**Figure 2.1: Monthly Variations in Physico-chemical
Characteristics of Stations–I and II**

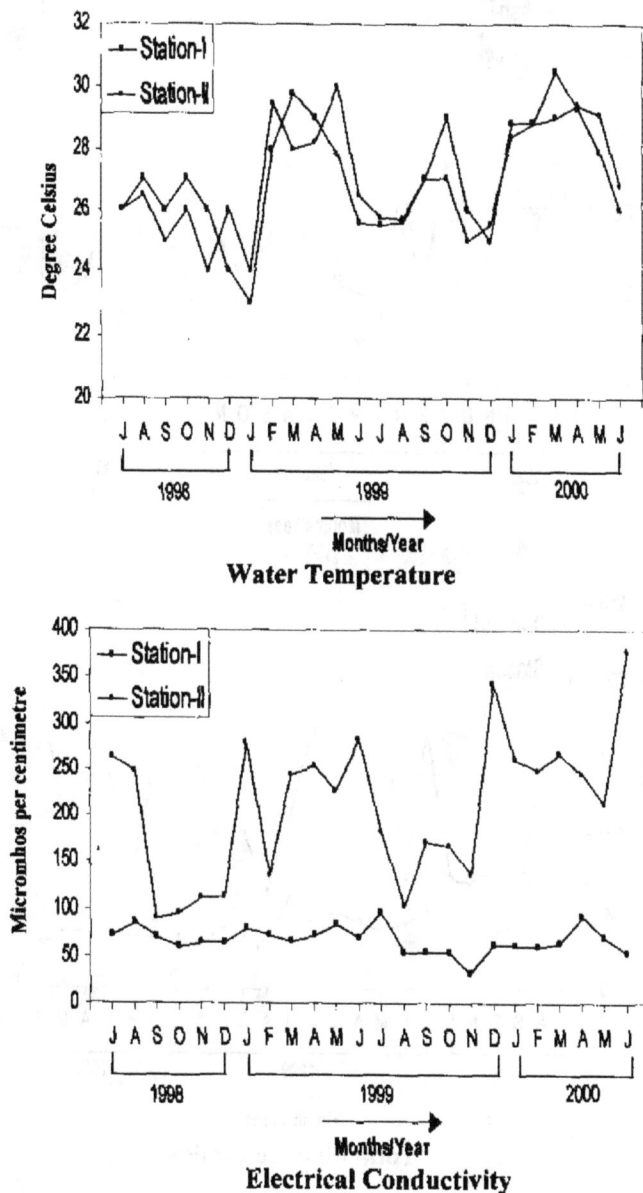

Water Temperature

Electrical Conductivity

Contd...

Figure 2.1—Contd...

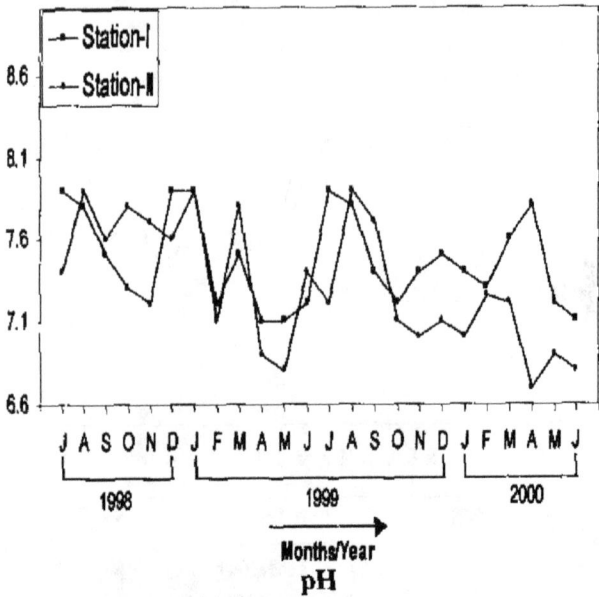

pH

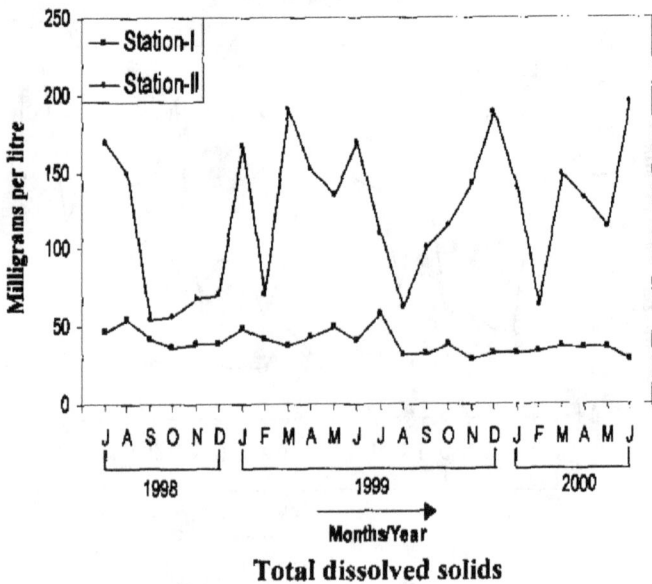

Total dissolved solids

Contd...

Figure 2.1–Contd...

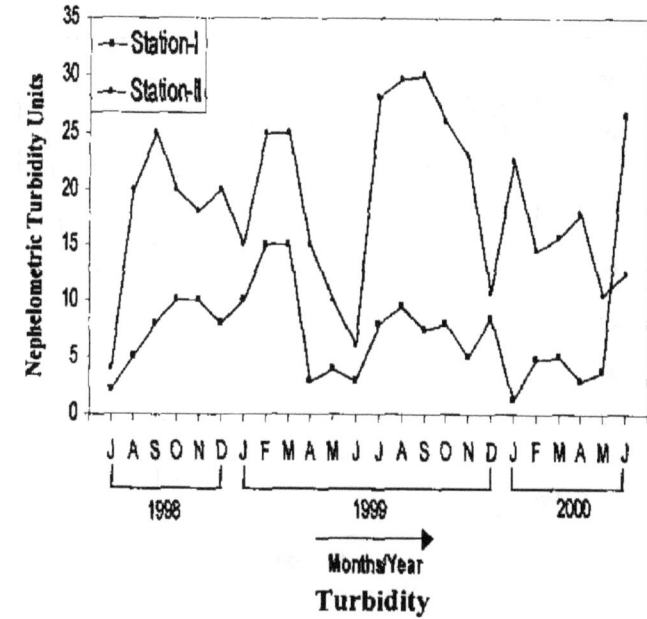

Turbidity

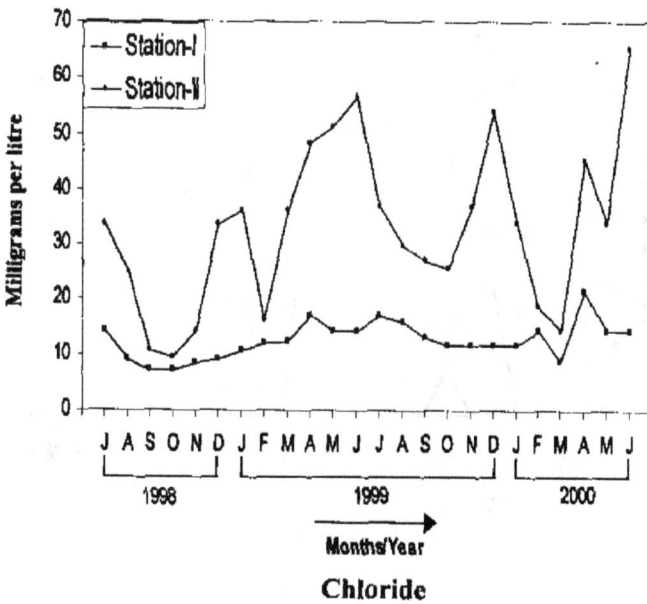

Chloride

Contd...

Figure 2.1–Contd...

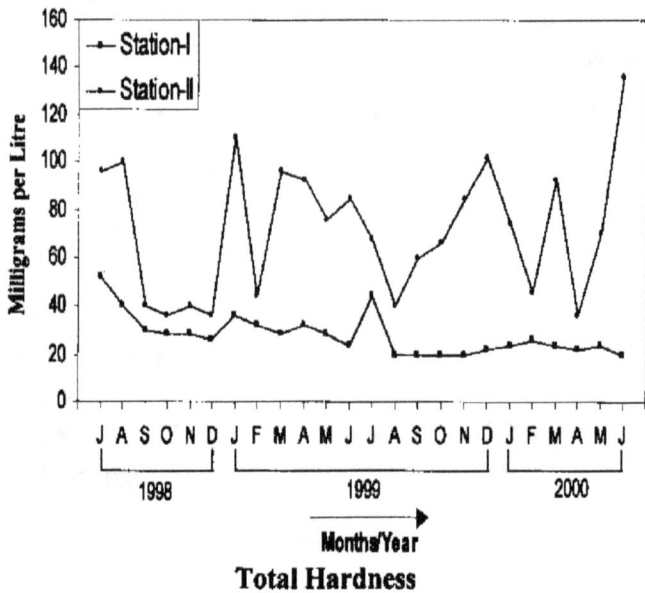

Total Hardness

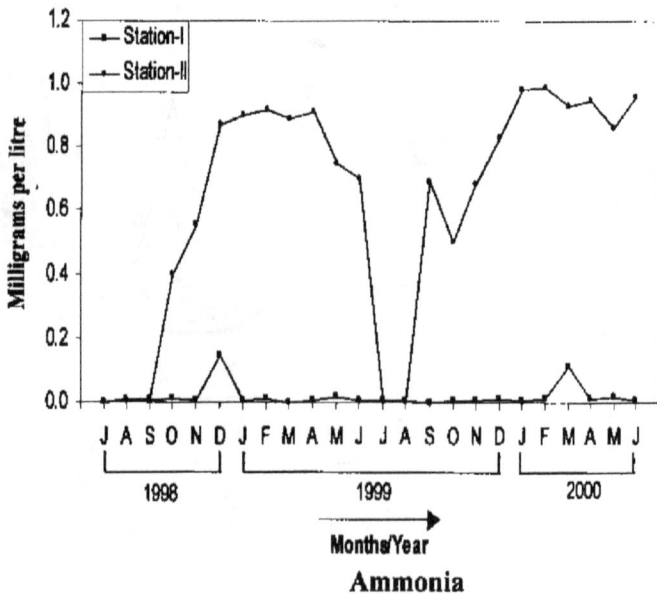

Ammonia

Contd...

Figure 2.1–Contd...

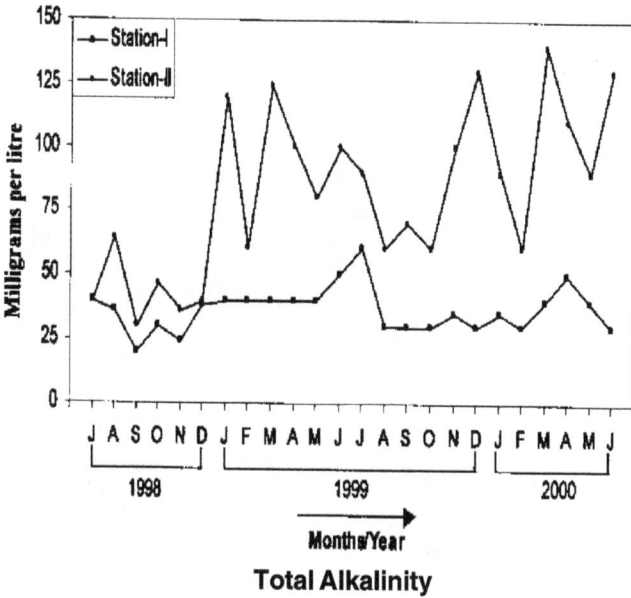

Total Alkalinity

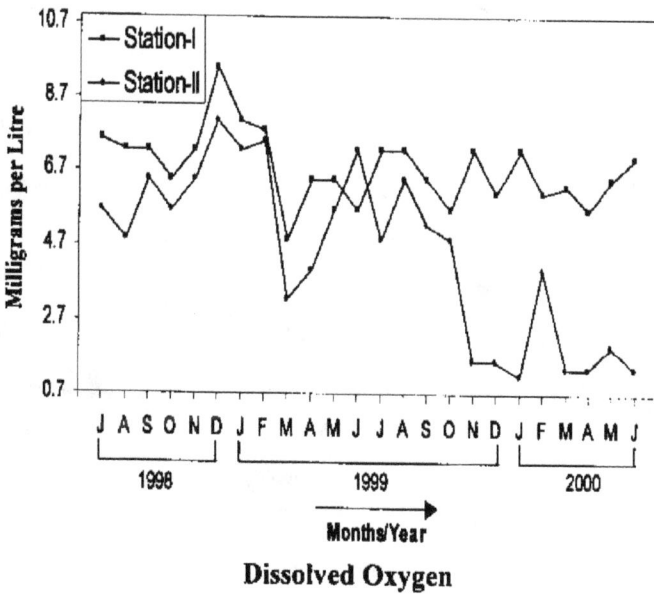

Dissolved Oxygen

Contd...

Figure 2.1–Contd...

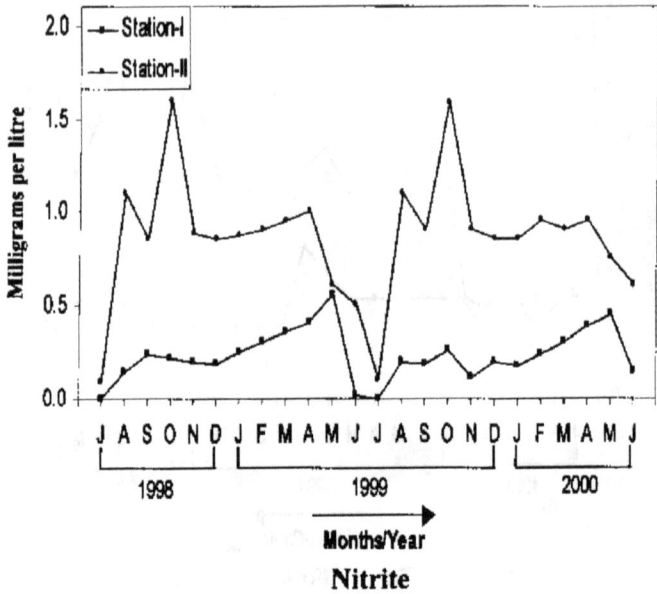

Nitrite

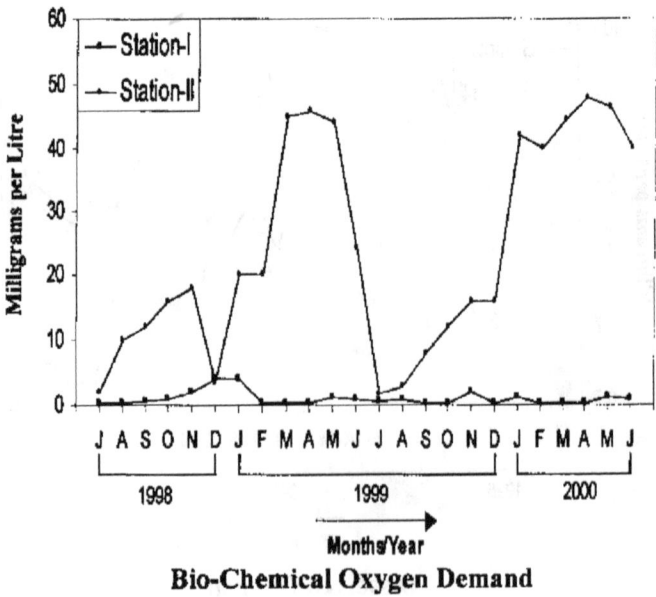

Bio-Chemical Oxygen Demand

Contd...

Figure 2.1–Contd...

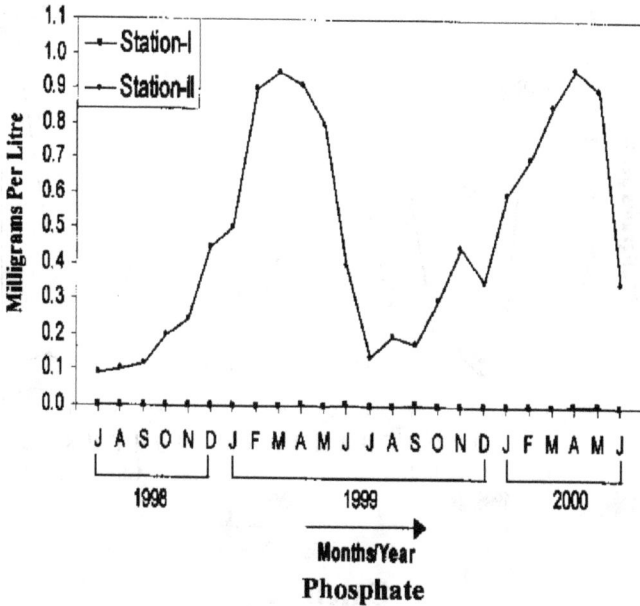

Phosphate

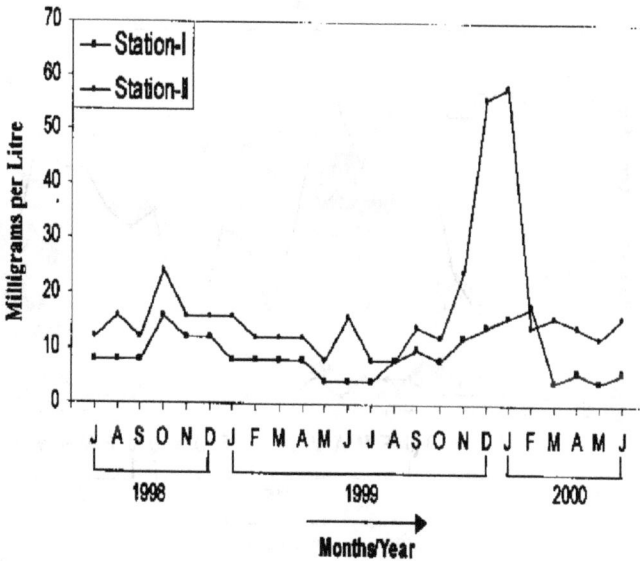

Free Carbon Dioxide

Contd...

Figure 2.1–Contd...

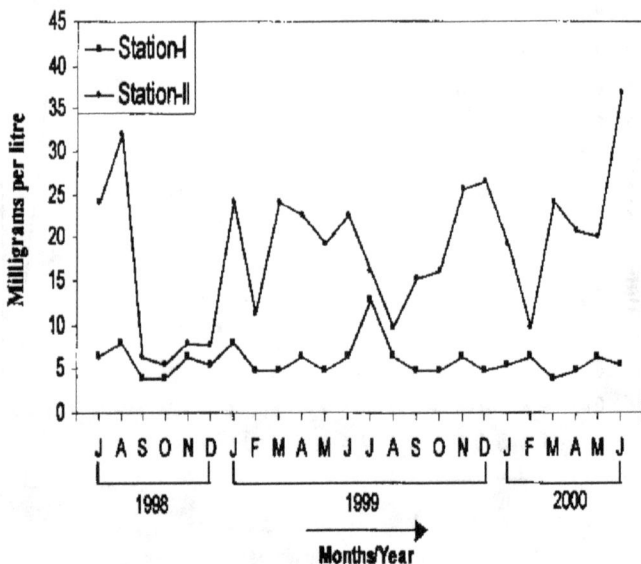

Calcium

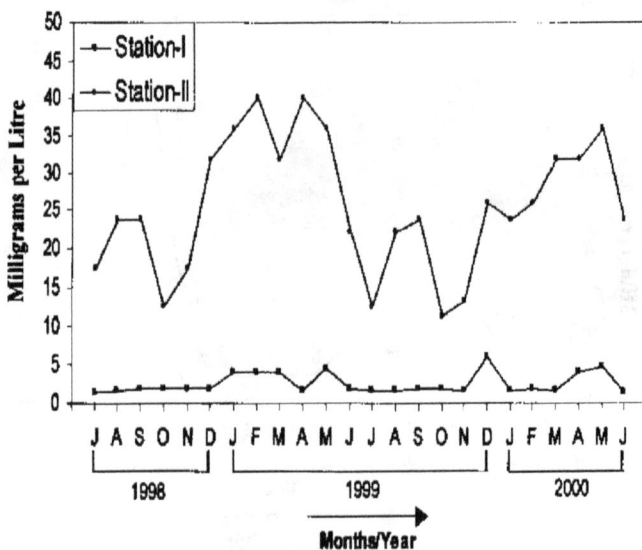

Chemical Oxygen Demand

Contd...

Figure 2.1–Contd...

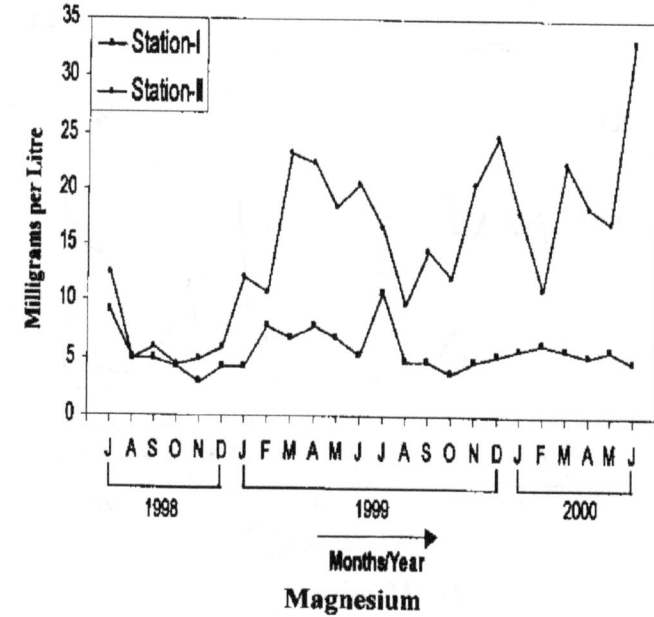

Magnesium

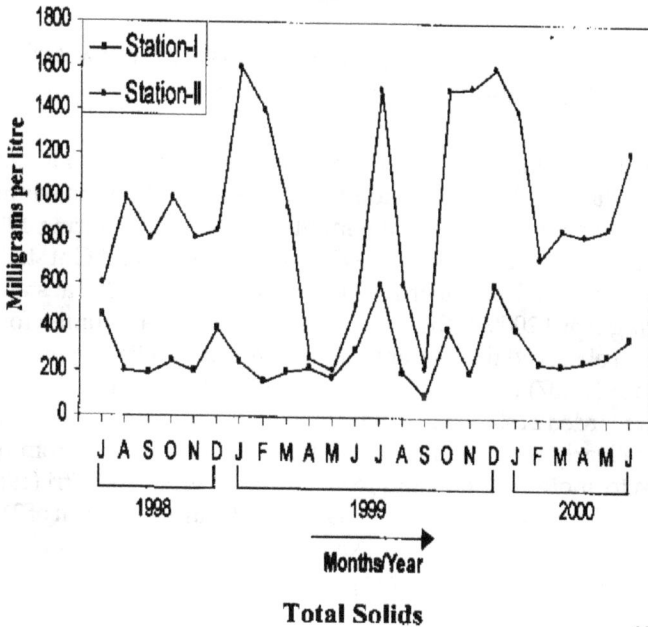

Total Solids

Contd...

Figure 2.1–Contd...

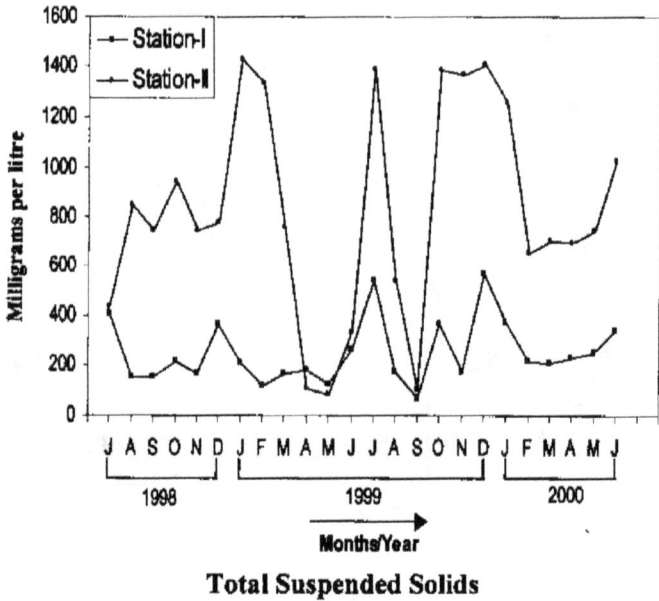

Total Suspended Solids

of water. The occurrence of aquatic organisms of a given water body is also directly or indirectly linked with the variation in temperature ranges.

In the present study, water temperature ranged from a minimum 23° C at station I during the month of January (1999) to a maximum of 30° C at station II during the month of May 1999. During 1999-2000, the water temperature varied from minimum of 25° C at station I in the month of December 1999 to a maximum of 29.4° C at station II during April 2000. If the seasonal variations are accounted for, it has been observed that water temperature reached its maximum at station II (29.07) in summer season (1999-2000) and the minimum value recorded at station I (25.0°C) in winter season (1998-99). Above results clearly indicate that the water temperature vary from one station to another station and from season to season. Khatri (1985) recorded the water temperature that varied from a minimum of 21.8° C to 26.3° C in Idukki reservoir (Kerala state), which is almost similar to the observations made in the present investigation.

However, during the present investigation, the water temperature tends to increase in downstream stations during the summer months. A similar observation has also been made by Khare and Unni (1986) on Pollar river. The record of relatively high temperature at downstream station of the river stretch (Station II) is due to the discharge of industrial effluents from Mysore Paper Mills Ltd., and Visveswaraya Iron and Steel Ltd. factories located on the bank of the Bhadra river near Bhadravathi town. Unni *et al.* (1992) have pointed out that the river systems receiving industrial effluents record relatively high temperature as compared to river stretch not receiving any kind of effluents.

pH

The pH values ranged from 6.7 in station II during April 2000 to 7.9 at station I during the month of December (1998), January (1999) and July (1999). Alkaline pH was recorded in all the months in station I and II except in summer months (April, May and June) during which acidic pH was observed in station II. When seasonal and yearly averages are accounted for, the water is alkaline in both the stations. A similar observation has been made by Venkateswarlu (1986) and Swarnalatha and Narasingarao (1993) in the riverine systems that they have investigated.

The reason for acidic pH at station II during summer season is due to the discharge of industrial effluent containing acidic wastes. Shukla *et al.* (1992) and Rana (1995) have noted acidic pH in the riverine system during their investigations and they concluded that the effluent containing acidic wastes is responsible for the decrease in pH of water. This observation is very much true in case of station II of the present investigation.

Turbidity

The turbidity values ranged between 1.2 NTU to 30 NTU. The lowest value (1.2 NTU) of turbidity was observed at station I and the highest value (30 NTU) of turbidity at station II. Season wise, station I showed lower values (4.15 NTU) of turbidity during summer (Table 2.1) while, station II showed higher values (23.4 NTU) of turbidity during rainy season (Table 2.2). However, the average values of two years is considered, it has been observed that station I showed lowest value (7.56 NTU) of turbidity and the highest value (20.08 NTU) of turbidity was recorded at station II. Vaishya and Adoni (1992) opined

Table 2.1: Seasonal Variations of Physico-chemical Parameters of Backwater of Bhadra Reservoir (Station I)

Sl.No.	Parameters	1998–99			1999–2000			1998–2000		
		Rainy	Winter	Summer	Rainy	Winter	Summer	Rainy	Winter	Summer
1.	Water temperature (°C)	26.33	25.00	28.65	25.92	27.20	29.15	26.12	26.10	28.90
2.	pH	7.73	7.57	7.22	7.57	7.37	7.47	7.65	7.47	7.34
3.	Turbidity (NTU)	5.00	9.50	9.25	7.00	5.67	4.15	6.00	7.33	6.70
4.	Electrical conductivity (µm hos/cm)	76.00	66.95	72.88	67.20	51.82	70.27	71.60	59.38	71.57
5.	Total solids (mg/l)	281.66	274.00	189.00	300.00	400.00	257.00	290.83	337.00	223.00
6.	Total suspended solids (mg/l)	234.33	234.00	119.36	259.97	367.10	220.96	247.15	300.55	170.16
7.	Total dissolved solids (mg/l)	47.33	40.00	42.88	40.02	32.88	36.03	43.67	36.44	39.45
8.	Chloride (mg/l)	10.09	6.75	13.54	14.88	11.35	14.53	12.89	9.05	14.03
9.	Total hardness (mg/l)	40.66	29.50	30.00	27.00	21.50	24.00	33.83	25.50	27.00
10.	Total alkalinity (mg/l)	32.00	33.00	40.00	42.50	32.50	40.00	37.25	32.75	40.00
11.	Nitrite (mg/l)	0.13	0.21	0.40	0.10	0.18	0.34	0.11	0.19	0.37
12.	Ammonia (mg/l)	0.006	0.008	0.009	0.004	0.005	0.037	0.0050	0.0065	0.0230
13.	Dissolved Oxygen (mg/l)	7.37	7.82	6.38	6.65	6.55	6.10	7.01	7.18	6.24
14.	BOD (mg/l)	0.46	2.75	0.55	0.65	1.00	0.55	0.55	1.87	0.55
15.	Phosphate (mg/l)	0.001	0.001	0.002	0.003	0.001	0.003	0.0020	0.0010	0.0025
16.	Calcium (mg/l)	6.13	6.00	5.20	7.60	5.40	5.40	6.86	5.70	5.30
17.	Magnesium (mg/l)	6.30	3.95	7.30	6.43	4.93	5.83	6.36	4.44	6.56
18.	Free carbon dioxide (mg/l)	8.00	12.00	7.00	6.50	12.50	8.00	7.25	12.25	7.50
19.	COD (mg/l)	1.70	2.50	3.55	1.85	2.80	3.10	1.77	2.65	3.32

Table 2.2: Seasonal Variations of Physico-chemical Parameters of Downstream Stretch of Bhadra River (Station II)

Sl.No.	Parameters	1998–99			1999–2000			1998–2000		
		Rainy	Winter	Summer	Rainy	Winter	Summer	Rainy	Winter	Summer
1.	Water temperature (°C)	25.83	25.00	28.92	26.23	26.47	29.07	26.03	25.73	28.99
2.	pH	7.63	7.75	7.15	7.55	7.05	7.01	7.59	7.40	7.08
3.	Turbidity (NTU)	16.33	18.25	18.75	23.4	20.57	14.65	19.86	19.41	16.70
4.	Electrical conductivity (μm hos/cm)	200.00	149.25	213.87	184.37	226.80	242.60	192.18	188.02	228.23
5.	Total solids (mg/l)	800.00	1065.00	707.50	705.00	1502.50	816.50	752.50	1283.75	762.00
6.	Total suspended solids (mg/l)	675.33	974.50	570.25	590.32	1355.62	701.76	632.82	1165.06	636.00
7.	Total dissolved solids (mg/l)	124.66	90.50	137.25	110.17	146.80	114.74	117.41	118.65	125.99
8.	Chloride (mg/l)	22.93	23.17	37.79	37.58	37.58	28.01	30.25	30.37	32.90
9.	Total hardness (mg/l)	78.66	55.50	77.00	63.00	81.50	71.00	70.83	68.50	74.00
10.	Total alkalinity (mg/l)	44.66	60.50	91.00	80.00	95.00	100.00	62.33	77.75	95.50
11.	Nitrite (mg/l)	0.68	1.05	0.86	0.650	1.047	0.88	0.66	1.045	0.87
12.	Ammonia (mg/l)	0.004	0.68	0.86	0.34	0.74	0.93	0.172	0.71	0.89
13.	Dissolved Oxygen (mg/l)	5.64	6.85	5.09	5.94	2.32	2.21	5.79	4.58	3.65
14.	BOD (mg/l)	8.00	14.4	38.75	9.15	21.50	44.75	8.57	17.95	41.75
15.	Phosphate (mg/l)	0.10	0.34	0.89	0.23	0.42	0.85	0.16	0.38	0.87
16.	Calcium (mg/l)	20.80	11.35	19.22	15.81	21.84	18.63	18.30	16.59	18.92
17.	Magnesium (mg/l)	7.80	6.87	18.70	15.31	18.84	17.25	11.55	12.85	17.97
18.	Free carbon dioxide (mg/l)	13.33	18.00	11.00	11.50	37.50	14.00	12.41	27.75	12.50
19.	COD (mg/l)	21.8	24.46	37.00	20.27	18.68	31.50	21.03	21.57	34.25

that the turbidity values increase due to the inflow of rainwater carrying suspended particles and the discharge of industrial effluents. In the present study, the reason for the highest value of turbidity at station II is due to the discharge of industrial effluents and domestic sewage.

Electrical Conductivity

In the present work, the value of electrical conductivity ranged from 32 μmhos/cm (station I) to 378 μmhos/cm (station II). It reached its maximum value at station II (228 μmhos/cm) during summer season (Table 2.2) and minimum value (59.38 μmhos/cm) during winter at station I (Table 2.1). Nevertheless, the average value of two years is considered, it has been observed that station I showed minimum value (64.06 μmhos/cm) and maximum of it was recorded at station II (210.23 μmhos/cm). Natarajan (1983) working on Nagarjunasagar reservoir recorded electrical conductivity which ranged between 172.53 and 1114.7 μmhos/cm. He opined that lotic sector of the reservoir showed maximum value of conductivity. Which is true in the present investigation also particularly at station II.

Total Solids

The total solids in this study varied between a minimum of 100 mg/l at station I and a maximum of 1600 mg/l at station II. Season wise the station II showed highest value of 1502 mg/l (1999-2000) during winter (Table 2.2) and while station I showed lowest value of (189 mg/l) total solids. If the two yearly averages accounted for, station II showed a highest value of 950.25 mg/l and lowest value of 286.87 mg/l at station I (Table 2.3). Singh and Singh (1990) and Tripathy and Adhikari (1990) have recorded higher amounts of total solids in the water bodies that they have investigated and pointed out that the discharge of industrial effluent is the main reason for increase in the total solids.

Total Suspended Solids

Total suspended solids are the cause of suspended particles inside the water body influencing its turbidity and transparency. The amount of total suspended solids in the present study varied between a minimum of 69 mg/l at station I and a maximum of 1432 mg/l at station II. If the two yearly averages accounted for, station II

Table 2.3: Averages of the Physico-chemical Parameters of Back Water of Bhadra Reservoir (Station I) and Downstream Stretch of the Bhadra River (Station II)

Sl.No.	Parameters	Station I			Station II		
		Average 1998–99	Average 1999–2000	Total 1998–2000	Average 1998–99	Average 1999–2000	Total 1998–2000
1.	Water temperature (°C)	26.6	27.45	27.02	26.64	27.28	26.96
2.	pH	7.46	7.46	7.46	7.43	7.15	7.29
3.	Turbidity (NTU)	7.75	7.56	7.65	16.91	20.08	18.49
4.	Electrical conductivity (μm hos/cm)	66.32	61.81	64.06	194.62	225.84	210.23
5.	Total solids (mg/l)	249.75	324.00	286.87	832.50	1068.00	950.25
6.	Total suspended solids (mg/l)	206.87	288.80	247.83	711.25	940.44	825.84
7.	Total dissolved parameters (mg/l)	42.88	35.19	39.03	121.25	126.05	123.65
8.	Chloride (mg/l)	11.21	13.59	12.40	30.78	35.10	32.94
9.	Total hardness (mg/l)	32.00	23.83	27.91	70.83	76.16	73.49
10.	Total alkalinity (mg/l)	36.50	36.66	36.58	70.00	94.16	82.08
11.	Nitrite (mg/l)	0.24	0.22	0.23	0.85	0.87	0.86
12.	Ammonia (mg/l)	0.008	0.015	0.011	0.57	0.69	0.63
13.	Dissolved Oxygen (mg/l)	7.05	6.55	6.80	5.99	3.00	4.49
14.	BOD (mg/l)	1.28	0.75	1.015	21.72	26.46	24.09
15.	Phosphate (mg/l)	0.002	0.002	0.002	0.47	0.49	0.48
16.	Calcium (mg/l)	5.80	6.07	5.93	17.25	19.97	18.61
17.	Magnesium (mg/l)	5.40	5.69	5.54	12.18	18.19	15.18
18.	Free carbon dioxide (mg/l)	8.66	9.16	8.91	14.33	21.00	17.66
19.	COD (mg/l)	2.60	2.53	2.56	27.79	23.62	25.70

showed a highest value of 825.84 mg/l at station I (Table 2.3). Seasonwise the station II showed highest value of 1165.06 mg/l during winter and while station I showed lowest value of 170.16 mg/l. Verma and Shukla (1969) and Sharma *et al.* (1981) have recorded higher amounts of total suspended solids in the water bodies that they have investigated. In the present study, station II showed higher amount of total suspended solids. Therefore, our observations are an essential agreement with that of the above researchers.

Total Dissolved Solids

Total dissolved solids recorded during the present investigation fall under the range between 27.61 mg/l at station I and 195.5 mg/l at station II. If the seasonal variations are accounted for, it has been observed that the total dissolved solids reached its maximum at station II (125.99 mg/l) in summer season and the minimum value at station I (36.44 mg/l) during winter season. Manikya Reddy (1984) correlated the richness of total dissolved solids to the surface run-off during rainy season and concluded that rainy water adds more solids to the water. In the present investigation also, the station I showed maximum value of total dissolved solids during rainy season indicating that the rain water through run-off increases the total dissolved solids in the water bodies. Further, Rajkumar (1984) and Balakrista Reddy (1989) are of the opinion that the polluted water bodies contain more of total dissolved solids when compared to unpolluted water. Which is also very much true in respect of station II.

Chloride

The values recorded ranged between a minimum of 7 mg/l at station I and maximum of 65.24mg/l at station II. It is interesting to note that the content of chloride has been more in station II, which receives domestic waste and industrial effluent. While, station I does not receive any domestic waste. Munawar (1970) based on the limnological work carried out by him concluded that higher chloride content indicates higher degree of pollution in waters. However, Rohini and Puttaiah (1992-93) recorded higher amounts of chloride in the water bodies they have studied and concluded that such water bodies are organically polluted. Similarly, in the present study also station II is organically polluted and hence higher values of chlorides recorded.

When two yearly averages are accounted for, the content of chloride was found to be high (32.94mg/l) at station II and low (12.40mg/l) at station I (Table 2.3). Season wise, the content of chloride was found to be high during summer season at both the stations. This is in close agreement with that of the observations made by Patil *et al.* (1986) and Sinha (1995). The lower content of chloride during rainy season when compared to summer may be due to dilution by rainwater. A similar opinion has been given by Oborne *et al.* (1980) and Lowe and Gale (1980).

Total Hardness

In the present investigation, the total hardness value varied from 20mg/l at station I to a 136mg/l at station II. If the two yearly averages are taken into consideration, the station II shows maximum hardness of 74mg/l during summer season (Table 2.2). A similar observation has also been made by Shukla *et al.* (1992) and are of the opinion that the hardness increases towards the summer season due to the low level of water.

Yogesh Shastri (2000) noticed total hardness content in river Mosam that varied from 160 to 374 mg/l and conclude that the water is very hard. In the present study water is soft at station I and hard at station II as per Kannan (1991) classification. Birsal *et al.* (1985) recorded the maximum total hardness value during rainy season which is also true in the present investigation at station II.

Total Alkalinity

Both the stations showed the absence of phenolphthalein alkalinity. Methyl orange alkalinity, however, exhibited wide range of fluctuation. The total alkalinity values ranged from 20 mg/l at station I to a maximum of 140mg/l at station II. Seasonwise it was found to be high (95.50 mg/l) during summer season at station II (Table 2.2) and low (32.75 mg/l) during winter season at station I (Table 2.1). However, when two yearly average value is accounted for, the content of total alkalinity was 36.58 mg/l at station I and 82.08mg/l at station II. Yogesh Shastri (2000) recorded 40 to 125 mg/l of total alkalinity in river Mosam (Maharashtra) and opines that the Mosam river seems to be rich in nutritive contents. Philipose

(1960) suggested that a water body with alkalinity value greater than 100 mg/l seems to be rich in nutrients.

Nitrite

The nitrate content varied between 0.001 mg/l at station I and 1.60 mg/l at station II. If the seasonal variations are accounted for, nitrite was observed to be more during winter (1.045 mg/l) and less during rainy (0.66 mg/l) season at station II (Table 2.2). While, station I showed more nitrite during summer (0.37 mg/l) and less during rainy (0.11 mg/l) season. However, if two yearly average is taken in to consideration, station I showed 0.23 mg/l of nitrite and station II recorded higher nitrite content of 0.86 mg/l. Nevertheless, Sabater and Sabater (1987) observed very low concentration of nitrite in river Ter and Elobregate (Spain).

Ammonia

The content of ammonia in the present investigation fluctuated between a minimum of 0.001 mg/l at station I to a maximum of 0.99 mg/l at station II. If the two yearly averages are accounted for, station II showed a highest value of 0.63 mg/l and the lowest value of 0.011 mg/l at station I (Table 2.3). Season-wise the station II recorded highest value of 0.89 mg/l during summer and decline towards rainy season and while station I showed lowest value of 0.005 mg/l during rainy season. Kaul *et al.* (1980) observed an increase in the concentration of ammonia during summer months and opines that the change in the climatic condition and type of biological composition is also responsible. Prasad (1990) opined that the waterbody receiving domestic waste, by and large, shows fairly high amount of ammonia reaching up to 13.4 mg/l. In the present investigation, the concentration of ammonia at station II was found to be low although it receives domestic wastes and industrial effluent may be attributed to the microbial activity.

Dissolved Oxygen

The dissolved oxygen content of water sample in the present study varied from a minimum of 1.40 mg/l at station II to a maximum of 9.52 mg/l at station I. Nonetheless, the average value of two years is considered, it has been observed that station I showed maximum of 6.80 mg/l and minimum of 4.49 mg/l (Table 2.3) was recorded at station II. Season-wise, station I showed the dissolved oxygen content to a maximum of 7.18 mg/l during summer season. Sahib and Azis (1989) recorded higher dissolved oxygen content in Parappar reservoir during winter. This is also true in the present investigation at station I.

Croome and Tyler (1988) and Chatterjee (1992) have pointed out that polluted water contains less amount of dissolved oxygen, which is also true in the present investigation at station II as a consequence of discharge of effluents. Therefore, the observations made in the present study at station II are in essential agreement with of the above researchers. A reduction in the content of dissolved oxygen during summer months in both stations could be attributed to higher temperature and increased process of microbial decomposition. A similar observation has been made by Shukla *et al.* (1992) while studying on the river Ganga.

Biochemical Oxygen Demand

Unni *et al.* (1992) recorded the biochemical oxygen demand values which ranged between 0.8mg/l to 4.8mg/l in river Narmada and concluded that these values were well below the threshold concentration level prescribed by WHO and ISI for drinking water. In the present study, biochemical oxygen demand values ranged between a minimum of 0.2mg/l at station I and a maximum of 48mg/l at station II. Station II receives a huge amount of industrial waste and sewage and hence contains as high as 48mg/l of biochemical oxygen demand indicating that the water at this station is not fit for drinking purpose. Season wise, biochemical oxygen demand values were found to be very high (41.75mg/l) during summer season at station II and low (0.55mg/l) during rainy and summer season at station I (Table 2.1). Dakshini and Gupta (1974) and Chatterjee (1992) have observed an increase in biochemical oxygen demand values during summer months and concluded that high temperature enhances the biological activities resulting in the increase of biochemical oxygen demand. In the present study, such a relationship was observed at station II.

Biochemical oxygen demand values correspondingly increase from upstream to downstream station. Similar observation was made by Khare and Unni (1986) in Pollar river (Madya Pradesh). Naidu *et al.* (1990) opined that the increase of biochemical oxygen demand value was due to the addition of organic matter due to bottling activities are noted but even though the content of biochemical oxygen demand is high and it could be attributed to the discharge of industrial effluents along with organic matter. It has been noted that the content of dissolved oxygen and biochemical oxygen demand are inversely related to each other. Wherever dissolved oxygen was low there was always corresponding increase in the content of biochemical oxygen demand. Such observations have also been made by Unni *et al.* (1992).

Phosphate

The amount of phosphate ranges from a minimum of 0.001 mg/l at station I to a maximum of 0.96mg/l at station II. If two yearly averages are accounted for, it varies from 0.002mg/l at station I to 0.48mg/l at station II.

Many workers such as Venkateshwarlu (1969a), Sampath Kumar (1977) and Nirmala Kumari (1984) have observed an increase in phosphate concentration in such of the water bodies that receive domestic waste. In the present study also the water at station II showed richness in phosphate content.

Seasonally, the amount of phosphate was found to be high during summer season and low during winter season at station I. While in station II phosphate was high during summer season (Table 2) and low during rainy season. Zutshi and Khan (1988) have also observed similar situations in the water bodies studied by them.

Jhingran (1989b) and Sugunan and Yadava (1991b) have recorded lower phosphate concentration in natural surface water bodies. Verma and Shukla (1968) while working on water bodies have recorded higher phosphate content during summer months. In the present study, phosphate concentration rises appreciably during summer months at station II as it receives large amount of domestic waste and industrial effluent. Other reason could be the use of phosphate fertilizers in the farming operations in the areas adjoining to the river bank, which ultimately find their way into the river.

It is observed that whenever the dissolved oxygen is high there is a corresponding decrease in the content of phosphate, when two yearly seasonal average is taken into consideration. Thus dissolved oxygen and phosphate show a close relationship between each other.

Zutshi and Khan (1988), Kaushik and Saksena (1994) have made a similar observations.

Calcium

In the present investigation, the content of calcium varied between a minimum of 4.0mg/l at station I and maximum of 36.87mg/l at station II. Kaushik *et al.* (1992) have recorded calcium content in the rivers of Madhya Pradesh which ranged between 23.24 to 35.27mg/l. In the present study also, the values were well within the limits as prescribed by ISI standards. Seasonally, high values (18.92mg/l) of calcium were recorded during summer at station II and low values (5.30mg/l) at station I (Table 2.1) during summer season. Dakshini and Gupta (1974) have stated that calcium concentration increases during rainy season with corresponding decrease in winter and followed by summer. Such observation was noticed at station I.

Nevertheless, in the present investigation, the behaviour of calcium during all the seasons at station II is rather irregular due to the addition of industrial effluents along with organic wastes (sewage). Hence, the observation of the present investigations are in partial agreement with that of Dakshini and Gupta (1974). Nonetheless, the observations made by Zutshi and Khan (1988) are in total agreement with our observation.

Magnesium

The concentration of magnesium varied from a minimum of 2.9mg/l at station I to a maximum of 33.06mg/l at station II.

Sugunan and Yadava (1991b), Kaushik *et al.* (1992) and Kaushik and Saksena (1994) have recorded magnesium content which is lower than the concentration of calcium. In the present study also, the content of magnesium was found to be less than calcium. This observation is in conformity with the findings of Janardhan Rao (1982), Manikya Reddy (1984), Sudhakar (1989) and Balakrista Reddy (1989).

Season-wise, minimum (4.44mg/l) magnesium concentration was observed at station I during winter and a higher (17.97mg/l) concentration of magnesium was observed during summer at station II. This higher concentration is due to continuous discharge of industrial effluents and sewage. When two yearly average is accounted for, the content of magnesium was found to be high (15.18mg/l) at station II and low (5.54mg/l) at station I (Table 2.3). Similarly, Sugunan and Yadava (1991b) have recorded average (8.62mg/l) magnesium content in Kyrdemkulai reservoir, Meghalaya. Therefore, the observations made in the present study at station I are in essential agreement with of the above researchers.

Free Carbon Dioxide

The content of free carbon dioxide ranged from a maximum of 58mg/l at station II to a minimum of 4 mg/l at station I. When two yearly averages are taken into consideration, it is noted to be high at station II (17.66 mg/l) and low (8.91mg/l) at station I. Seasonally, station II recorded highest content of free carbon dioxide (27.75 mg/l) during winter and lowest (7.25mg/l) during rainy season at station I (Table 2.1). Sioli (1972); Adhikary and Sahu (1988) have noted an inverse relationship between free carbon dioxide and dissolved oxygen. Such a relationship has not been observed in the present investigation. Shardendu and Ambasht (1988), Ghosh and George (1989) have opined that free carbon dioxide has an inverse relationship with pH. Such a relationship has not been observed during the present study.

Chemical Oxygen Demand

Monthly variations of chemical oxygen demand values of two stations are presented in the Tables 2.1–2.4, which depict that it was minimum at station I with 1.4mg/l and maximum at station II with 40mg/l. Season-wise, it was found to be high (34.25mg/l) during summer season at station II and low (1.77mg/l) during rainy season at station I. However, when the two yearly average value is accounted for, the content of chemical oxygen demand value was found to be very high (25.70mg/l) at station II and low (2.56mg/l) at station I (Table 2.3). Dakshini and Gupta (1974); Shardendu and Ambasht (1988) and Venkateshwaran et al. (1993) have also noted higher concentration of chemical oxygen demand during summer months, which is also in agreement with our observation.

Kaushik et al. (1992) working on riverine water in Chambal command area, Madhya Pradesh recorded chemical oxygen demand values in the range of 8mg/l to 20mg/l. In the present study, the chemical oxygen demand value was low at station I. Tripathy and

Adhikari (1990) and Shukla *et al.* (1992) could record very high chemical oxygen demand load in river Kali and Ganga respectively, which received domestic and industrial wastes. In the present investigation only station II receives industrial wastes and sewage and hence the chemical oxygen demand values were recorded to be very high at this station.

Remedial Measures

The following remedial measures are made to protect these water bodies from eutrophication.

☆ The reservoir has to be utilized for productive purposes by introducing Indian major and Chinese exotic carps that can be harvested periodically.

☆ It is obvious that, to maintain healthy conditions of the reservoir the catchment area should be protected from the adverse effects of human activities.

☆ Dumping of garbage and religious refuse should be prohibited.

☆ Size of immersion idols should be reduced to minimum, idols should be of clay, people should be motivated for not to immerse idols, instead have permanent one.

☆ Washing of vehicles, cloths and bathing should be strictly prohibited in the river stretch.

☆ Harvesting of macrophytes in river should be done to reduce nutrients and they should be properly disposed off.

☆ Construction of proper sewer system in entire city.

☆ Regular water quality monitoring by authentic agencies.

☆ Establishing educational centers for creating awareness about the importance of an eco-friendly heritage and also as a source of income.

☆ The farmers of the adjoining areas of the river system should be informed not to use the chemical fertilizers thereby rendering the river system useful for public. The people living in the areas of the river bank should be kept informed about the impact of environmental degradation and sustainable conservation programmes.

Conclusion

Thus, physico-chemical characteristics of reservoir varied from season to season depending upon the volume of water entering the water body from different rocky regions. However, downstream stretch of Bhadra river is polluted due to the discharge of industrial effluents and domestic wastewater. Therefore, it is suggested that a proper care must be taken to avoid the pollution of the water body through periodic monitoring of the water quality. Going through the above observations, it can be concluded that the Bhadra reservoir can be classified as oligotrophic while, downstream stretch of the Bhadra river is eutrophic.

References

Adhikary, S.P. and Sahu, J.K., 1988. Limnology of the thermal springs of Orissa (India). *J. Bombay Nat. Hist. Soc.*, 84(2): 497–503.

APHA, 1998. *Standard Methods for the Examination of Water and Wastewater*, 20th Edition. APHA, AWWA and WEF, N.W. Washington D.C.

Balakrista Reddy, G., 1989. Ecological studies in the river Kagna near Tandur (A.P) with special reference to water quality and pollution. *Ph.D. Thesis*, Osmania University, Hyderabad.

Birsal, N.R., Deshpande, V.K., Gouder, B.Y.M. and Nadakarni, V.B., 1985. Chemical changes in the surface waters during the filling phase of a tropical reservoir on Kali river, North Canara district, Karnataka state. In: *proceedings of Environmental Impact Analysis of Water Resources Project*, Roorkee, India, 2: 801–810.

Chatterjee, A.A.K., 1992. Water quality of Nandankanan lake. *Indian J. Environ. Hlth.*, 34(4): 329–333.

Croome, R.L. and Tyler, P.A., 1988. Phytoflagellates and their ecology in Tasmanian polyhumic lake. *Hydrobiologia*, 161: 254–263.

Dakshini, K.M.M. and Gupta, S.K., 1974. Physiography and limnology of three lakes in the environs of the union territory of Delhi, India. In: *Proc. Indian Natn. Sci. Acad.*, 4: 417–430.

Ghosh, A. and George, J.P., 1989. Studies on the abiotic factors and zooplankton in a polluted urban reservoir: Hussain sagar. Impact on water quality and embryonic development of fishes. *Indian J. Environ. Hlth.*, 31(1): 49–59.

Janardhan Rao, C.H., 1982. Ecology of algae in polluted channel entering the river Moosi at Hyderabad. *Ph.D. Thesis,* Osmania University, Hyderabad.

Jhingran, A.G., 1989b. Limnology and production biology of two man–made lakes in Rajasthan (India) with management strategies for their fish yield optimisation. *Final Report, IDA Fisheries Management in Rajasthan,* Central Inland Capture Fisheries Research Institute, Barrackpore, India, p. 1–63.

Kannan, K., 1991. *Fundamentals of Environmental Pollution.* S. Chand and Co. Ltd., New Delhi.

Kaul, V., Handoo, J.K. and Raina, R., 1980. Physico-chemical characteristics of Nilnang: A high altitude forest lake in Kashmir and its comparison with the valley lakes. In: *Proc. Indian Nat. Sci. Acad.,* B. 46: 528–541.

Kaushik, S. and Saksena, D.N., 1994. Dispersal pattern of planktonic cladocera in three water bodies of Gwalior (Central India). *J. Aqu. Biol. Fish.,* 1(2): 15–24.

Kaushik, S., Agarkar, M.S. and Saksena, D.N., 1992. Distribution of phytoplankton in Riverine waters in Chambal command area, Madhya Pradesh. *Bionature,* 12(1 and 2): 1–7.

Khare, S.K. and Unni, K.S., 1986. Changes in physico-chemical factors and distribution of periphyton in Pollar river. In: *Proc. All India Seminar on Water Quality Around Urban Ecosystems,* (Ed.) K.S. Unni.

Khatri, T.C., 1985. A note on the limnological characters of the Idukki reservoir. *Indian J. Fish.,* 32(2): 267–269.

Lowe, R.L. and Gale, W.F., 1980. Monitoring river periphyton with artificial benthic substrates. *Hydrobiol.,* 69(3): 235–244.

Manikya Reddy, P., 1984. Ecological studies in the river Tungabhadra (A.P) with special reference to the effect of paper mill effluents on the River. *Ph.D. Thesis,* Osmania University, Hyderabad.

Munawar, M., 1970. Limnological studies on freshwater certain polluted and unpolluted environments. *Hydrobiol.,* 39(1): 105–128.

Naidu, N.V.S., Naidu, D.V., Babu, D.R. and Naidu, P.R., 1990. Water quality of reservoir and temple tanks in Tirupati and Tirumala. *Indian J. Environ. Hlth.*, 32: 413–415.

Natarajan, A.V., 1983. Final Report-All India Co-ordinated Research Project on Ecology and Fisheries of Freshwater Reservoirs, Nagarjunasagar. Research Information Series, 3: 30–31.

Nirmal Kumari, J., 1984. Studies on certain biochemical aspects of Hydrobiology. *Ph.D. Thesis*, Osmania University, Hyderabad.

Oborne, A.C., Brooker, M.P. and Edward, R.W., 1980. The chemistry of the river Wye. *J. Hydrol.*, 45: 233–252.

Patil, M.K., Usha Namboodiri and Unni, K.S., 1986. Changes after sediment removal in a sewage pond physical-chemical characteristics. In: *All India Seminar on Water Quality Around Urban Ecosystem*, (Ed.) K.S. Unni, p. 222.

Philipose, M.T., 1960. Freshwater phytoplankton of inland fisheries. *Proc. Sympom. Algology*, ICAR, New Delhi, p. 272–291.

Prasad, D.Y., 1990. Preliminary productivity and energy flow in upper lake Bhopal. *Indian J. Environ. Hlth.*, 32(2): 132–139.

Rajkumar, B., 1984. Ecological studies in the river Manjira with special reference to its lithophytic flora. *Ph.D. Thesis*, Osmania University, Hyderabad.

Rana, B.C., 1995. Evaluation of pollution stuatus of river khari near Kheda region of Gujarat. *Geobios New Report*, 14(2): 146–148.

Rohini, R. and Puttaiah, E.T., 1992–93. Preliminary observations on the ecological characteristics of Dalvai lake, Mysore, Karnataka. *J. Mysore Univ.*, Section–B. 33: 125–130.

Sabater, S. and Sabater, F., 1987. Diatom assemblage in the river Ter. *Arch. Hydrobiol.*, 111: 397–408.

Sahib, S.S. and Azis, P.K.A., 1989. Post Impoundment water quality of the Kallada river a preliminary report. In: *Proceedings of the Kerala Science Congress*, February, Cochin, p. 153–160.

Sampath Kumar, P.T., 1977. Further studies on the ecology of algae in the river Moosi, Hyderabad (India) with special reference to pollution and potential fertility of the water. *Ph.D. Thesis*. Osmania University, Hyderabad.

Shardendu and Ambasht, R.S., 1988. Limnological studies of a rural pond and an urban tropical aquatic ecosystem: Oxygen enforms and ionic strength. *Tropical Ecology*, 29(2): 98–109.

Sharma, K.D., Lal, N. and Phatak, K.P.D., 1981. Water quality of sewage drains entering Yamuna at Agra. *Indian J. Environ. Hlth.*, 23:118–122.

Shukla, S. C., Tripathi, B. D., Mishra, B. P. and Chaturvedi, S. S., 1992. Physico-chemical and Bacteriological properties of the water of river Ganga at Ghazipur. *Comp. Physiol. Ecol.*, 17: 92–96.

Singh, D.K. and Singh, C.P., 1990. Pollution studies on river Subernarekha around industrial belt of Ranchi (Bihar). *Indian J. Environ. Hlth.*, 32: 26–83.

Sinha, S.K., 1995. Potability of some rural pond's water at Muzaffarpur (Bihar): A note on water quality index. *Poll. Res.*, 14(1): 135–140.

Sioli, H., 1972. Tropical river. In: *The Amazon, Vol. 2: River Ecology*, (Ed.) Whitton. p. 461–488.

Sudhakar, G., 1989. Studies on the impact of paper mill effluents on the ecology of the river Godavari. *Ph.D. Thesis*, Osmania University, Hyderabad.

Sugunan, V.V. and Yadava, Y.S., 1991b. Feasibility studies for fisheries development of Nongmahir reservoir. Central Inland Capture Fisheries Research Institute, Barrackpore, p. 30.

Swarnalatha, N. and Narasingarao, A., 1993. Ecological investigation of two lentic environments with reference to cyanobacteria and water Pollution. *Indian J. Microbial. Ecol.*, 3: 41–48.

Tripathy, P.K. and Adhikari, S.P., 1990. Preliminary studies on the water pollution of river Nandira. *Indian J. Envriron. Hlth.*, 32: 363–368.

Unni, K.S., Anil Chauhan, Varghese, M. and Naik, L.P., 1992. Preliminary hydrobiological studies river Narmada from Amarkantak to Jabalpur. In: *Aquatic Ecology*, (Eds.) S.R. Mishra and D.N. Saksena.

Vaishya, A.K. and Adoni, A.D., 1992. Phytoplankton seasonality and their relationships with physical and chemical properties in a hyper-eutrophic central Indian lake. *Proc. Indian Natn. Sci. Acad.*, B59(2): 153–160.

Venkateswaran, K., Shimada, A., Sakou, H. and Maruyama, T., 1993. Microbial characteristics of palau jelly fish lake. *Canadian J. of Microbiology*, 39(5): 506–512.

Venkateswarlu, V., 1969a. An ecological study of the algae of the river Moosi, Hyderabad (India) with special reference to water pollution. I. Physico-chemical complexes. *Hydrobiol.*, 33: 352–363.

Venkateswarlu, V., 1986. Ecological studies on the rivers of Andhra Pradesh with special reference to water quality and pollution. *Proc. Indian Acad. Sci. Plant Sci.*, 96: 495–508.

Verma, S.R. and Shukla, G.R., 1968. Hydrobiological studies of a temple tank "Devikund" in Deoband (U.P.), India. *Environmental Hlth.*, 10: 177–188.

Verma, S.R. and Shukla, G.R., 1969. Pollution in a perennial stream "Khala" by the sugar factory effluent near Laskar, Dist. Saharanpur, U.P. *Indian J. Environ. Hlth.*, 11: 145–162.

Yogesh Shastri, 2000. Physico-chemical characteristics of river Mosam. *Geobios*, 27: 194–196.

Zutshi, D.P. and Khan, A.U., 1988. Eutrophic gradient in the Dal lake, Kashmir. *Indian J. Environ. Hlth.*, 30: 348–354.

Chapter 3

Analysis of Zooplankton Population in Garhwal Himalayan Lake, Deoria Tal, Uttarakhand

M.S. Rawat[1] & Ramesh C. Sharma[2]

[1]Department of Zoology, P.O. Box No. 8,
Government P.G. College, Uttarkashi
[2]Department of Environmental Science, P.O. Box No. 67,
R.N.B. Garhwal University, Srinagar Garhwal – 246 174

ABSTRACT

The article deals with diversity, community structure and seasonal fluctuation of zooplankton population. Fifteen species belonging to 14 genera were collected and identified under rotifera, copepoda and cladocera. A maximum of 10 species occurred in September and the species count decreased upto 5 in April and May. It was observed that the density of zooplankton was maximum during autumn (1510.50±46.85 unit l^{-1}) and minimum (428.25±37.76 unit l^{-1}) in winter. On the mean annual basis rotifers (52.14 per cent) were the largest contributor followed by copepods (32.82 per cent) and cladocerans (15.04 per cent) to the total population. The rotifers were represented by maximum number of species and formed the most dominant group among the zooplanktionic population during the study period. *keratella tropica, Daphnia* sps. *Cyalopa* sp. and *Anuraeopsis fissa* sp. were the dominant zooplankters of lake Deoria Tal.

Keywords: Zooplankton, Population, Density, Deoria Tal, Rotifera.

Introduction

The zooplankton constitute an important component of secondary production in aquatic ecosystem and play key role in the energy allocation in different trophic levels. The seasonal changes in zooplankton numbers and their abundance are closely related to the physico-chemical environment and trophic status of aquatic system. Inspite of their great importance not much information is available on their diversity, dynamics and productivity in the lacustrine environment. Studies on various aspect of zooplankton community in fresh water bodies of India and abroad were carried out by Arora (1964,65), Burgis (1969), Bricher *et al.* (1973), Nasar (1973), Fernando (1980), Lewis (1979), Jyoti and Sehgal (1980), Ganapati (1943), Pasha (1961), Gannon and Sternberger (1978), Negi and Pant (1983), Pant *et al.* (1985), Chauhan (1988), Khan (2002) etc. Garhwal Himalaya have thick net work of lentic water bodies. But, unfortunately very few scientific studies were carried out on Garhwal lakes by Rawat *et al.* (1993, 95), Rawat and Sharma (1997, 04), Prasad (2004). Studies on zooplankton community of lake Deoria Tal remain unknown. Therefore, the present investigation were carried out.

Material and Methods

For the analysis of zooplankton the samples were collected monthly from three sampling sites selected in the lake. Zooplankton were collected by a silk net (0.5 m diameter) of 25 mm^{-2}. Surface water were directly taken in polyethylene bottles, whereas a nansen's water samplers was also used to collect water from 1 m, 3 m, and 5 m depths. The samples (10 liters of water) were filtered through a silk cloth net (mesh size 25). The organisms collected were fixed in 4 per cent formalin solution. 1 ml aliquots of these samples were then counted in Segdwick Rafter counting cell. Five slides of each samples were counted and the censeus was computed to unit l^{-1}.

Results

A total of 15 species comprising 10 rotifers, 2 copepods and 3 cladocerans were collected and identified during the study period (Table 3.1). A monthly variation in the species of zooplankton was recorded round the year. There occurred a maximum number of 10 species in September of which 8 species rotifers, 1 species copepoda and cladocerans each, to the total species count in this mounth. The species number was minimum of 5 in April and May of which 3 species belong to rotifers and 1 species copepoda and cladocerans (Table 3.2).

Table 3.1: List of Zooplankton Species and their Mean

Species	Density
Cyclops sp.	26.25
Nauolius sp.	6.70
Daphnia sp.	14.11
Alona rectengula	0.70
Chydorus sp.	0.18
Keratella tropica	41.03
K. yulga	0.57
Anuraeopsis fissa	7.93
Monostyla bulla	0.19
Philodina sp.	0.32
Asplanchna sp.	0.09
Collurella sp.	0.07
Trichocerca sp.	0.12
Cephalodella sp.	0.86
Polyarthra sp.	0.77

Table 3.2: Temporal Variation in Total Species Count and Individual Groups of Zooplankton in Lake Deoria Tal

Months	Cladocera	Copepoda	Rotifera	Total
September	1		8	10
October	0	1	7	8
November	1	1	5	7
December	1	1	5	7
January	1	1	5	7
February	1	1	4	6
March	1	1	5	7
April	1	1	3	5
May	1	1	3	5
June	1	1	4	6
July	1	1	5	7
August	1	1	7	9

Numerically, zooplankton population (combining all water strata) exhibited marked monthly variation (Figure 3.1). In general, the population was higher during autumn (1510.50":1: 46.85 unit l^{-1} and lower during winter (428.25 ± 37.76 unit l^{-1}). The population density ranged from 238 unit l^{-1} to 1881 unit l^{-1} during the year with a mean of 770.57 ± 125.36 unit l^{-1}. On the mean annual basis, rotifers were the largest contributor (52.14 per cent) followed by copepods (32.82 per cent) and cladocerans (15.04 per cent) to the total population (Figure 3.2). The zooplankton population in lake Deoria Tal exhibited two peaks of abundance being first in October (1881

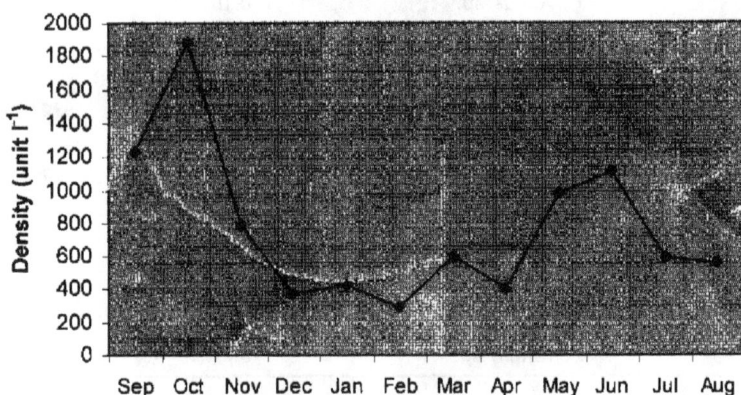

Figure 3.1: Monthly Variation in the Zooplankton Population

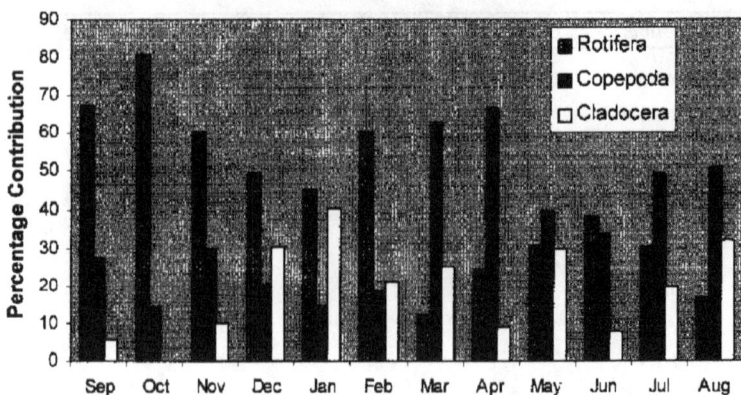

Figure 3.2: Percentage Contribution of Different Zooplanktonic Group

unit l⁻¹), the peak of greater magnitude and second in June (1113 unit l⁻¹). The peak of greater magnitude were largely contributed by rotifers (81.03 per cent) where as the second peak of low magnitude was equally shared by rotifers (38.25 per cent) and copepods (33.78 per cent).

The zooplankton which contributed on mean annual basis $\geq$ 5.0 per cent and $\geq$ 1.0 per cent in terms of number of their respective groups and total population (combining all water strata) are *Daphnia* sp., *Cyclops* sp., *Nauplius* sp., *Keratella tropica* and *Anuraeopis* sp. have been considered dominant.

Discussion

Zooplankton, a heterogeneous assemblage may be used as an index to assess the trophic status of an aquatic environment. The fresh water zooplankton diversity of the India are comprised of 330 species of rotifera, 110 of cladocera and 88 of copepoda which constituted nearly 13.2 per cent, 27.2 per cent and 8.8 per cent of the world fauna respectively (Khan, 2002). Ruttner (1952) listed 23 rotifers, 13 cladocerans and 6 copepods from 15 tropical lakes. Pennak (1957) from 27 Colorado lakes recorded 1–3 copepods, 1–5 cladocerans and 1–10 rotifers. Lewis (1979) recorded 7 rotifers, 3 cladocerans and 2 copepods from lake Lanano. In Kumaun lake Khurpa Tal Negi and Pant (1983) identified 7 rotifers, 4 cladocerans and 2 copepods while in lake Sat Tal Pant *et al.* (1985) recorded 13 rotifers, 6 cladocerans and 3 copepods. The zooplankton diversity of lake Deoria Tal comprised of 10 species of rotifers, 3 species of cladocerans and 2 species of copepods.

Qualitatively, rotifers make a major contribution to the zooplanktonic community. Nayar and Nayar (1969) reported 15 species of rotifers from Kerala. Zutshi and Khan (1977) reported 5 species of rotifers from subtropical lake in Jammu region. Negi and Pant (1983) reported 7 species of rotifers from Khurpa Tal. In the present study 10 species of rotifers were collected from lake Deoria Tal. Rotifers were represented by *Keratella tropica, K. vulge, Anuraeopsis fissa, Monstyla bulla, Philodina* sp., *Collurella* sp., *Trichocera* sp., *Chephalodella* sp., *Polyartgra* sp., *Asplanchna* sp. In present studied lake Deoria Tal two peaks of zooplanktonic density were found at all the sampling sites. The bimodal pattern in zooplankton population has also been reported by Ganapati (1943), Das and

Shrivastava (1959), Alikunhi *et al.* (1955), Negi and Pant (1983) and Pant *et al.* (1985). During present study major contribution was made by rotifers followed by copepods and cladocerans in the total zooplanktonic community. The annual contribution was 52.14 per cent, 32.82 per cent and 15.04 per cent respectively. Zutshi *et al.* (1980) in 9 lakes of Jammu Kashmir reported that the zooplankton population was dominated by rotifers which is followed by copepods and cladocerans. Pant *et at.* (1985) also reported predominance of rotifers over other zooplanktonic components. In lake Deoria Tal higher growth of copepods were recorded when the temperature of the lake ranged from 13.0°C to 24.0°C. Negi and Pant (1983) observed the peak growth of *Mesocyclops leueksrti* at the temperature ranging from 14.0°C to 28.0°C in the lake Khurpa Tal while Pant *et al.* (1985) reported its peak at the temperature 14°C to 29°C in lake Sat Tal.

Acknowledgement

We express our gratitude to Professor H.R. Singh, Department of Zoology, University of Allahabad for providing valuable suggestion and encouragement.

References

Alikhunhi, K.H., Chaudhury, H., and Rama Chandran, V., 1955. On the mortality of carp fry in nursery ponds and role of plankton in their survival and growth. *Indian J. Fish.*, 2: 257–313.

Arora, H.C., 1964. Studies on Indian Rotifera. Part III. On *Brachionus calyciflorus* and some 4 varieties of the species. *Zool. Sci., India*, 16: 1–6.

Arora, H.C., 1965. Studies on Indian Rotifera. Part IV. On a collection of rotifera from Nagpur, India with four new species and a new variety. *Hydrobiol.*, 26: 444–456.

Burgis, M.J., 1969. A preliminary study of the ecology of zooplankton in lake George Uganda. *Verh. Int. Verein. Limnol.*, 17: 297–302.

Bricher, K.S., Bricker, F.J. and Gannon, J.E., 1973. Distribution and abundance of zooplankton in the U.S. water of lake St. Clair. *J. great Lakes Res.*, 2(2): 256–271.

Chauhan, R., 1988. Seasonal abundance of zooplankton in Rewalsar Lake, H.P. *Geobios New Report*, 7(2): 117–121.

Das, S.M. and Srivastava, V.K., 1959. Studies on fresh water plankton III. Quantitative composition and seasonal fluctuations in plankton components. *Ibid*, 26(4): 243–254.

Farnando, C.H., 1980. The fresh water zooplankton of Sri Lanka with a discussion of tropical fresh water zooplankton composition. *Int. Revue. Ges. Hydrobiol.*, 65: 85–125.

Ganapati, S.V., 1943. An ecological study of a garden pond containing abundant zooplankton. *Proc. Ind. Acad. Sci.*, 18: 41–58.

Gannon, J.E. and Sternberger, R.S., 1978. Zooplankton (especially crustaceans and rotifers) as indicator of water quality. *Trans. Amer. Micros. Soc.*, 97(1): 16–35.

Jyoti, M.K. and Sehgal, H., 1980. Ecology of Rotifers of Surinsar: A sub-tropical freshwater lake in Jammu (J&K), India. *Hydrobiol.*, 65: 23–32.

Khan, R.A., 2002. Fresh water zooplankton and their role in aquatic productivity. In: *Proc. 89th Ind. Sci. Cong. Part III*, 100–101.

Lewis, W.M., 1979. *Zooplankton: A Community Analysis Approach*. Srinagar-Verlag, New York.

Nasar, S.A.K., 1973. The zooplankton fauna (Rotifera) of Bhagalpur. *Bhagalpur Univ. J.*, 6: 55–62.

Nayar, C.K.G. and Nayar, K.K.N., 1969. A collection of Brachionid rotifers from Kerala. In: *Proc. Ind. Acad. Sci.*, 69: 223–233.

Negi, V. and Pant, M.C., 1983. Analysis of zooplankton community of lake Khurpa Tal (Kumaun Himalaya). *Tropical Ecology*, 24(2): 271–282.

Pant, M.C., Joshi, A. and Sharma, P.C., 1985. Species composition, temporal abundance and community structure of zooplankton in lake Sat Tal (U.P.), India. *Arch. Hydrobiol.*, 102(4): 519–602.

Pasha, S.M.K., 1961. On a collection of fresh water rotifers from Madras. *J. Zool. Soc. India*, 13: 50–55.

Pennak, R.W., 1957. Species composition of limnetic zooplankton communities. *Limnol. Oceanogr.*, 2: 222–232.

Prasad, S., 2004. Ecology of high altitude lake Nachiketa Tal, Uttarkashi. *Master Degree Dissertation*, H.N.B. Garhwal University, Srinagar Garhwal.

Rawat, M.S., Gusain, O.P., Juyal C.P. and Sharma, R.C., 1993. First report on limnology: Abiotic profile of Garhwal Himalayan lake Deoria Tal. In: *Advances in Limnology*, (Ed.) H.R. Singh, p. 87–92.

Rawat, M.S., Juyal C.P. and Sharma, R.C., 1995. Morphomentry and physico-chemical limnology of a high altitude lake Deoria Tal of Garhwal Himalaya. *J. Freshwater Biol.*, 7(1): 1–6.

Rawat, M.S. and Sharma, R.C., 1997. Macrozoobenthos in the high altitude lentic environment of Deoria Tal of Garhwal Himalaya. *Indian J. Fish*, 44(2): 217–220.

Rawat, M.S. and Sharma, R.C., 2004. Phytoplankton population of Garhwal Himalayan Lake Deoria Tal, Uttarakhand. *J. Ecophysiol. Occup. Hlth.* (Communicated).

Ruttner, F., 1952. Limnological studies in einigen Seen der ostalpen. *Arch. Hydrobiol.*, 32: 167–319.

Zutshi, P. and Khan, M.A., 1977. Limnological investigations of two sub tropical lakes. *Geobios*, 4: 45–48.

Zutshi, D.P., Subla, B.A. and Khan, M.A., 1980. Comparative limnology of nine lakes of Jammu and Kashmir Himalaya. *Hydrobiologia*, 72: 101–112.

Chapter 4

Physico-Chemical Limnology of Yeldari Reservoir (Maharashtra) in Relation to Fisheries

V.B. Sakhare[1] & P.K. Joshi[2]

[1]Department of Fishery Science, Yeshwantrao Chavan College,
Tuljapur – 413 601, M.S.
[2]Department of Zoology, Dnyanopasak College,
Parbhani – 431 401, M.S.

ABSTRACT

A systematic analysis of water quality from Yeldari reservoir in Parbhani district of Maharashtra was conducted during October 2000 to September 2001. Present paper deals with physico-chemical aspects of reservoir and its significance to fisheries. The water temperature ranged from 21 to 34°C. pH was found to be 7.7 to 8.3.Total dissolved solids ranged from 96 to 262 mg/l. Transpoarency was found to be 36.5 to 95 cm. The dissolved oxygen differed from 6.4 to 13.9 mg/l. Phenolphthalein and total alkalinity ranged from 20.5 to 47.5 and 189 to 394 mg/l. Free carbon dioxide was totally absent during the entire study period. Chlorides and total hardness of water varied from 14.18 to 49.63 and 68 to 226 mg/l respectively. The study revealed that the water of reservoir is suitable for fisheries.

Keywords: Reservoir, Physico-chemical characters, Fisheries.

Introduction

Considerable limnological investigations are carried out on the reservoirs in India (Kulshrestha *et al.*, 1992; Thomas and Aziz, 2000; Das *et al.*, 2001 and Thirumathal *et al.*, 2002). Similarly number of studies has been conducted on limnology from different regions of Maharashtra (Shastri and Pendse, 2001 and Sakhare and Joshi, 2002). However, no such work was carried out on Yeldari reservoir in Parbhani district of Maharshtra. Therefore the present work was undertaken to study the physico-chemical limnology of Yeldari reservoir in relation to fisheries.

The Yeldari reservoir, a purely hydro-electric project was constructed across river Puma in year 1962 in hilly area of Jintur tahsil. The reservoir is bounded by latitudes 19°43'N and longitudes 76°45'E and is included in the survey of India toposheet map No. 56A/10. The water of reservoir is also used for irrigation through another reservoir and for supplying drinking water to Parbhani, Vasmat, Jintur and other towns. No canal for irrigation takes off directly from this reservoir. However, a few miles down stream, there is the Siddeshwar reservoir at Siddeshwar village, which is built on the same river. From the Siddeshwar reservoir, a system of irrigation canals originates. As it is mainly a hydro-electric project, all rights of reservoir management are belonging to the irrigation Department and the State Electricity Board. The area around reservoir comprises forest-covered hills. The reservoir is having catchment area of 7,330.00 sq.km. The maximum level of reservoir is of 462.380 m. The reservoir consists of three outlets, water through which goes to the Siddeshwar reservoir after passing through the turbines and ten spillway gates. The salient morphometric features of Yeldari reservoir in Table 4.1. The normal annual rainfall over the reservoir area varies from about 850 mm to about 980 mm. The air is generally dry from February to May, the relative humidity during the afternoons being 20 per cent. During the south-east monsoon season, the humidity is as high as 80 per cent in the mornings and 60per cent in the afternoons. The climate of the reservoir area is semiarid sub-tropical, characterized by a hot summer. Except for the monsoon period the weather is generally dry. The summer outsets in the area generally in March and continues up to May. This is followed by the south-west monsoon from June till September. However, there are periods of dry spells even within this monsoon period. The winter season stretches from December to the end of February.

Table 4.1: The Salient Features of Yeldari Reservoir

1.	Catchment area	733.00 km²
2.	Water spread area	106.84 km²
3.	Full reservoir level	461.772 m.
4.	Maximum water level	462.30 m.
5.	Top of the dam	465.90 m.
6.	Crest level	450.190 m.
7.	M.D.D.L.	
	Without carry over	477.751 m.
	With carry over	454.54 m.
8.	Length of masonry	350.215 m.
9.	Total length of earthen dam	4430.00 m.
10.	Length of spillway	149.65 m.
11.	Height of dam above the reservoir bed	51.35 m.
12.	Gross storage capacity	934.44 m.ha.m.
13.	Power generation	22500 kw
14.	Design spillway	
	At FRL	9740.80 Cumecs
	At MWL	10874 Cumecs
15.	Maximum Flood	10478 Cumecs
16.	No. of irrigation outlets	3 Nos. (1.92 × 2.44 m.)
17.	No. of spillway gates	10 Nos. (Radial gate)
18.	Size of radial gates	12.5 × 11.65 m.
19.	Power outlets	
	Service gates	3 Nos.
	Stop log gates	9 Nos.

In the study area, winds are generally moderate with appreciable increase in force during south-west monsoon season. During the monsoon season the wind is mainly from west and south-west directions. During the rest of the year, wind blows from north and east directions. The maximum-recorded wind speed during the month of May and August is 11.4 to 14.2 km/hour with annual average wind speed being of 8.6 km/hour.

Table 4.2: Physico-chemical Profile of Yeldari Reservoir (from October 2000 to September 2001)

Sl.No.	Parameters	Range
1.	Water temperature (°C)	21–34
2.	pH	7.7–8.3
3.	Transparency (cm)	36.5–95
4.	Total Dissolved Solids (mg/l)	96–262
5.	Dissolved Oxygen (mg/l)	6.4–14
6.	Phenolphthalein alkalinity (mg/l)	20.5–48
7.	Total alkalinity (mg/l)	189–394
8.	Free carbon dioxide (mg/l)	Nil
9.	Chlorides (mg/l)	14.18–56.72
10.	Total hardness (mg/l)	68–226

Materials and Methods

Monthly samples were collected from the three stations for a period of two years from October 2000 to September 2001. The water samples (at a depth of one meter) were collected with the help of sampler in the morning hours. Water samples were brought in one-litre plastic containers to the laboratory for analysis. Parameters like temperature, transparency, pH, dissolved oxygen, free carbon dioxide, and alkalinity were analyzed at the study sites, whereas parameters *viz.,* the total dissolved solids, total hardness, and chlorides were analyzed in the laboratory.

The methods used for the analysis of various physico-chemical parameters except pH are as given in methodology for water analysis (Trivedy and Goel, 1984; APHA, 1980 and Kodarkar *et al.,* 1998). The hydrogen ion concentration (pH) values were recorded at the water sample collection sites with the help of Hanna made pH meter.

Results

During the present study period water temperature ranged between 21 to 34°C. It minimum value (21°C) was recorded in January, while its maximum (34°C) was recorded in April.

pH in the reservoir varied from 7.7 to 8.3. The minimum pH was 7.7 was recorded in months of December and April, while the

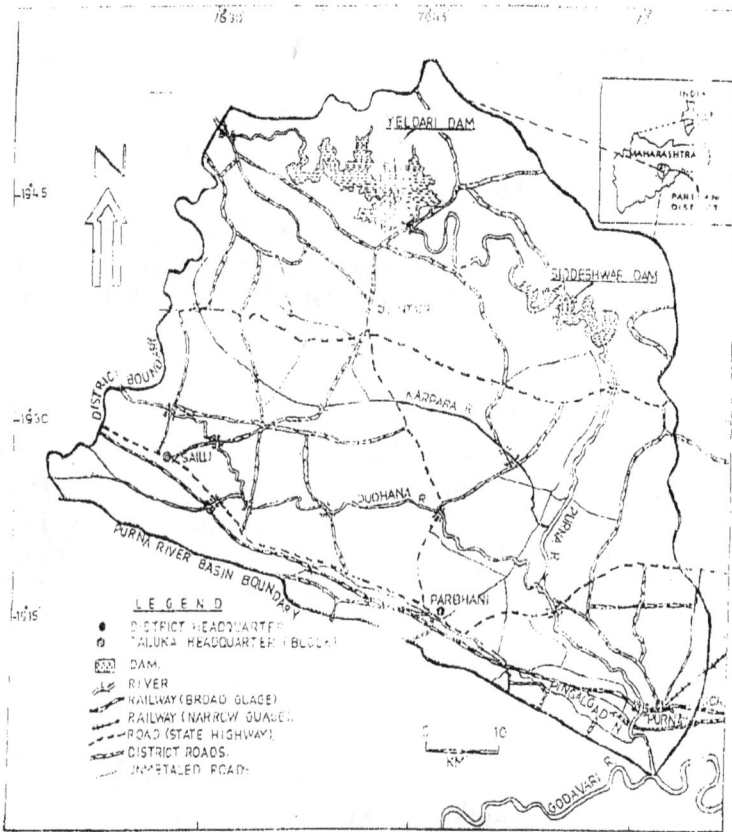

**Figure 4.1: Location and Road Map of
Purna River Basin Parbhani District**

maximum recorded was 8.3 in February. The observations indicate
that the water was alkaline through out the study period.

By the Secchi disk, the transparency of 65.5 was recoded in
September and maximum 95 cm was observed during February. The
less transparency might may be due to the silt brought into the
reservoir during rainy season.

Total Dissolved Solids (TDS) irrespective of the seasons ranged
from 96 to 262 mg/l. The maximum value of 262 mg/l in April and
minimum of 96 mg/l in August. The observation on TDS clearly
indicate that TDS values were high in summer followed by winter
and monsoon.

The dissolved oxygen ranged from 6.4 to 14mg/l. Its minimum (6.4) was found in March and maximum of 14 mg/l was in August. It is seen that, the high dissolved oxygen was found during monsoon, while lower values were recorded in summer.

Phenolphthalein alkalinity ranged between 20.5 mg/l in August to 48 mg/l in December.

The minimum value of total alkalinity (189 mg/l) was noticed in August, while the maximum value (394) was in December. During the study period higher values were observed in winter followed by summer and monsoon.

Free carbon dioxide was totally absent in reservoir water.

The minimum and maximum values of chloride were 14.18 mg/l and 56.22 mg/l in month of February and April respectively.

During present investigation the maximum total hardness (226) was found in May, it minimum (68) was recorded in January. The higher values of total hardness were recorded during summer and lower values were recorded in the winter season.

Discussion

The temperature of air and water is an important factor indicating the water quality which influences the aquatic life and concentration of dissolved gases (such as O_2, CO_2) and chemical solutes. Changes in the temperature produce characteristic pattern of circulation and stratification (Kulshrestha *et al.*, 1992). The air and water temperatures depend on geographical location and meteorological conditions such as rainfall, humidity, cloud cover, wind velocity at a particular place. Generally, the rise in temperature accelerates the rate of chemical reactions, reduces the solubility of gases, amplifies taste and odour and elevates overall metabolic rate. Reservoirs having water temperature more than 22°C are the highly productive reservoirs (Jhingran and Sugunan, 1990; Sugunan, 1995). The average water temperature was recorded at 28.58°C, which reveal that Yeldari reservoir is highly productive. Sugunan (1995) further mentions the wide seasonal variations in air temperature is the predisposing factor in the thermal features of the north Indian and peninsular reservoirs. According to Sharma and Jain (2000) the fluctuations in water temperature have relationship with the air temperature. Same results were indicated by Rao (1955) and Saha and Pandit (1986). The highest temperature recorded in summer months can be attributed to the direct relationship between bright sunshine, its duration and air temperature in the tropical countries (Lakshminarayana, 1965; Hussain, 1977).

The pH is affected not only by the reaction of carbon dioxide but also by the organic and inorganic solutes present in water. Any alteration in water pH is accompanied by the changes in other physico-chemical parameters (Kulshrestha *et al.*, 1992). In the present study the water pH ranged between 7.7 to 8.3. The water was alkaline throughout the study period with no definite seasonal variation. This observation is similar to the investigation made by Chandrasekhar (1996) on Saroornagar lake of Hyderabad. The present findings also corroborate with findings by Kulshrestha *et al.* (1992), which mentions the pH values of 7.2 to 9.5 in Mansarovar reservoir of Bhopal. The pH range of 6 to 9 is most suitable for pond fish culture, while pH more than 9 is unsuitable (Swingle, 1967). Thus, the pH range indicate that Yeldari reservoir which is having a pH range of 7.7 to 8.3 is suitable for survival of fish. Optimal growth and survival of aquatic plants required that their environment pH should be confined with a very short range below or above which they will be subjected to various kinds of stresses, and the diurnal fluctuation of pH of a water body should remain in the range of 6.4 to 8.5 in order to support the optimum fish growth (Das, 1996). The average pH value (7.9) of Yeldari is suitable for the optimum fish growth. According to Jhingran and Sugunan (1990), the pH range between 6 to 8.5 was medium productive reservoirs, more than 8.5 were highly productive and less than 6 were less productive reservoirs. In case of Yeldari reservoir the average value during the study period was observed to be 7.9. This indicated that the reservoir is a medium productive type. Generally the pH values increased in summer and decreased in monsoon and winter. The decrease in pH during winter could be due to decrease in photosynthesis while during monsoon it may be due to greater inflow of water. Maximum values during summer were probably due to increased photosynthesis in the algal blooms resulting into the precipitation of carbonates of calcium and magnesium from bicarbonates causing higher alkalinity (Kulshrestha *et al.*, 1992).

Water transparency is a physical measurable variable and is quite significant for production. Transparency is inversely proportional to turbidity created by suspended inorganic and organic matter. During the present study period transparency expressed in cms ranged between 36.5 to 95. The water was less transparent during monsoon as compared to winter or summer. Present findings corroborate with findings of Singh *et al.* (1993). They have mentioned the lowest transparency values during rainy season. The low values recorded in rainy season were stated to be due to the heavy rains and winds of high velocity and high values recorded during winter and summer months were probably due to low and moderate velocity of winds.

During present study the total dissolved solids were high in summer followed by winter and monsoon. Devi (1997) also reported the maximum total dissolved solids during pre-monsoon season and lowest during monsoon in Shathamraj and Ibrahimbagh reservoirs of Hyderabad. Tripathi and Pandey (1990) reported the seasonal fluctuations in total dissolved solids from two ponds of Uttar Pradesh, where the maximum concentration of total dissolved solids was recorded during summer which decreased during rainy season due to dilution of pond water.

In the present investigation the dissolved oxygen was found to range between 6.4–14 mg/l. Dissolved oxygen in some medium sized Indian reservoirs like Gularia, Baccharia, Baghla, Aliyar, Chapparwara and Kyrdemkulai were in the range of 4.9 to 9.0, 2.5 to 8.6, 2.4 to 12.8, 4.2 to 11.6, 6.10 to 10 and 6.7 to 7.10 mg/l respectively (Sinha, 1998). The dissolved oxygen range (6.4–14 mg/l) of Yeldari reservoir is quiet more than that of the above mentioned reservoirs.

Total alkalinity is the measure of the capacity of water to neutralize a strong acid. The alkalinity in the waters is generally imparted by the salts of carbonates, bicarbonates, phosphates, nitrates, borates, silicates etc. together with the hydroxyl ions in free state. However most of the waters are rich in carbonates and bicarbonates with little concentration of other alkalinity imparting ions (Trivedy and Goel, 1984). According to Lagler (1978) total alkalinity in water depends on the geology of the region. The fish may be affected by total alkalinity because waters with low values are generally biologically less productive than those with high values of alkalinity. Waters having 40 mg/l or more total alkalinities are considered to be more productive than waters of lower alkalinities (Moyle, 1945). The low alkalinity is not conducive for good productivity since highly productive water have alkalinity over 100 mg/l $CaCO_3$ (Jhingran, 1983). During present investigation the total alkalinity was more than 100 mg/I. Hence the reservoir water IS highly productive. During the present study higher total alkalinity values were observed in winter followed by summer and monsoon. The present findings are in accordance with the findings of workers like Kulshrestha *et al.* (1992) and Deshmukh (2001).

Lagler (1978) mentioned that the amount of free carbon dioxide in water is important in fish management because it is perhaps the best single criterion of environmental suitability for fishes. High concentrations of free carbon dioxide which are in themselves toxic to fish are usually accompanied by low values for dissolved oxygen. Free carbon dioxide in excess of 20 ppm may be regarded as harmful to fishes, although lower values may be equally harmful in waters of low oxygen content (less than 3 to 5 ppm). During the present study free carbon dioxide was totally absent. The absence of the free CO_2 may be due to its complete utilization in photosynthetic activity (Sreenivasan, 1971) or its inhibition by the presence of appreciable amount of carbon dioxide in water (Sahai and Sinha, 1969).

In the present investigation values of chlorides ranged between 14.18 to 56.72 mg/l. The range was similar to one (33.25 to 97.93) reported by Piska *et al.* (2000). In the present investigation higher values were recorded in winter and lower in monsoon. Lower values in monsoon could be attributed to effect and renewal of water mass after summer stagnation. Many authors (Subamma and Rama Sarma, 1992 and Devi, 1997) have discussed the fluctuation in chloride values of various water bodies and have recorded minimum concentration in monsoon period.

The hardness of water is mainly due to the presence of calcium and magnesium. Maximum values were found during summer and lowest values were found during winter months. Higher hardness values in summer were also reported by Devi (1985) and Devi (1997).Productive waters should have hardness value above 20 mg/1, calcium content above 5 mg/1 and magnesium content above 2 mg/l. Fishes have been found to be susceptible to diseases in water with hardness below 20 mg/l. Very hard water (> 300 mg/l) also becomes uncongenial for fish production because of higher pH. Optimum hardness for fish culture has been observed to be around 75 to 150 mg/l (Das, 1996).

References

APHA, 1980. *Standard Methods for the Examination of Water and Wastewater,* 15th edn. New York, pp. 1134.

Chandrasekhar, S.V.A., 1996. Ecological studies on Saroornagar Lake, Hyderabad. *Ph.D. Thesis,* Osmania University, Hyderabad.

Das, R.K., 1996. Monitoring of water quality, its importance in disease control. Paper presented in *Nat. Workshop on Fish and Prawn Disease, Epizootics and Quarantine Adoption in India,* October 9, CIFRI, pp. 51–55.

Das, A.K., Gopalakrishnayya, Ch. and Ramakrishiah, M., 2001. Guidelines for management of Andhra Pradesh reservoirs. *Fishing Chimes,* 21(5): 25–28.

Deshmukh, U.S., 2001. Ecological studies of Chhatri lake, Amravathi with special reference to planktons and productivity. *Ph.D. Thesis,* Amravathi University, Amravathi.

Devi, M.J., 1985. Ecological studies of the limnoplankton of three freshwater bodies of Hyderabad. *Ph.D. Thesis,* Osmania University, Hyderabad.

Devi, Sarla, B., 1997. Present status, potentialities, management and economics of fisheries of two minor reservoirs of Hyderabad. *Ph.D. Thesis,* Osmania University, Hyderabad.

Hussain, M., 1977. Ecobiology of freshwater protozoa. *Ph.D. Thesis,* Osmania University, Hyderabad.

Jhingran, Arun G., 1988. Reservoir fisheries in India. *Jr. of the Indian Fisheries Association,* 18: 261–273.

Jhingran Arun G. and Sugunan, V.V., 1990. General guidelines and planning criteria for small reservoir fisheries management. In: *Reservoir Fisheries in India,* (Eds.) A.G. Jhingran and V.K. Unnithan. Asian Fisheries Society, Indian Branch, Mangalore, India, pp. 1–8.

Kodarkar, M.S., Diwan, D.D., Murugan, N., Kulkarni, K.M. and Ramesh, Anuradha, 1998. Methodology for water analysis: Physico–chemical, biological and microbiological. *Indian Asso. of Aqua. Biologists,* Hyderabad, pp. 102.

Kulshrestha, S.K., George, M.P., Saxena, Rashmi, Johri, Malini and Shrivastava, Manish, 1992. Seasonal variations in the limnochemical characteristics of Mansarovar reservoir of Bhopal. In: *Aquatic Ecology,* (Eds.) S.R. Mishra and D.N. Saksena. Ashish Publishing House, New Delhi, India, pp. 275- 292.

Lagler, Karl F., 1978. *Freshwater Fishery Biology.* WmC Brown Company Publishers, Dubuque, Iowa, pp. 259.

Lakshminarayana, J.S.S., 1965. Studies on the phytoplankton of the river Ganges, Varanasi, India, Part I: The physico–chemical characteristics of river Ganges. *Hydrobiol.,* 25: 119–137.

Lakshminarayana, J.S.S., 1965. Studies on phytoplankton of the river Ganges, Varanasi, India, Part I–IV: Phytoplankton in relation to fish population. *Hydrobiol.,* 25: 171–175.

Mani Bharat and Gaikwad, S.A., 1998. Physico–chemical characteristics of lake Pokhran. *Indian J. Environ. and Toxicol.,* 8(2): 56–58.

Moyle, J.B., 1945. Some chemical factors influencing the distribution of aquatic plants in Minnesota. *Amer. Midl. Natur.,* 34: 402–420.

Piska Ravi Shankar, Sarla Devi B. and Chary Divakara, K., 2000. The present status of Ibrahimbagh: A minor reservoir of Hyderabad. *Fishing Chimes,* 20(2): 41–43.

Rao, C.B., 1955. On the distribution of algae in a group of six small ponds. II: Algae periodicity. *J. Ecol.,* 43: 291–308.

Saha, L.C. and Pandit, B., 1986. Comparative limnology of Bhagalpur ponds. *Comp. Physiol. Ecol.,* 1(14): 213–216.

Sahai, R. and Sinha, A.B., 1969. Investigations on bioecology of Inland waters of Gorakhpur (U.P.), India. I. Limnology of Ramgarh lake. *Hydrobiol.,* 34(3–4): 433–447.

Sakhare, V.B. and Joshi, P.K., 2002. Ecology of Palas–Nilegaon reservoir in Osmanabad district, Maharashtra. *J. Aqua. Biol.,* 17(2): 17–22.

Sharma, Dushyant and Jain, Renu, 2000. Physico–chemical analysis of Gopalpura tank of Guna District (M.P.). *Ecol. Env. and Cons.,* 6(4): 441–445.

Shastri, Yogesh and Pendse, D.C., 2001. Hydrobiological study of Dahikhuta reservoir. *J. Environ. Biol.,* 22(1): 67–70.

Singh, C.S., Sharma, A.P. and Deorari, B.P., 1993. Limnological studies for bioenergetic transformation in a Tarai reservoir,

Nanak Sagar (U.P.). In: *Advances in Limnology,* (Ed.) H.R. Singh. Narendra Publishing House, Delhi, India, pp. 29–44.

Singh, C.S., Sharma, A.P. and Deorari, B.P., 1993. Plankton population in relation to fisheries in Nanaksagar reservoir, Nainital. In: *Recent Advances in Freshwater Biology,* 1: 66–79.

Sinha, M., 1998. Policy options for integrated development of reservoir fisheries: From production to marketing. *Fishing Chimes,* 18(1): 54–59.

Sreenivasan, A., 1971. Recent trends of limnological investigations in Indian reservoirs. In: *Workshop on All India Coordinated Research Project on Ecology of Freshwater Reservoirs,* CIFRI, Barrackpore, India, August 30–31.

Subbamma, D.V. and Rama Sarma, D.V., 1992. Studies on the water quality characteristics of a temple pond near Machillipatnam, Andhra Pradesh. *J. Aqua. Biol.,* 7(1 and 2): 22–27.

Sugunan, V.V. 1995. *Reservoir Fisheries of India.* FAO Fisheries Tech. Report 345, Daya Publishing House, Delhi, India.

Swingle, H.S., 1967. Standardization of chemical analysis of water and ponds muds. *FAO Fish. Rep.,* 44: 397–342.

Thirumathal, K., Sivakumar, A.A., Chandrakantha, J. and Suseela, K.P., 2002. Physico–chemical studies of Amaravathy reservoir, Coimbatore district, Tamil Nadu. *J. Ecbiol.,* 14(1) 13–17.

Thomas, Sabu and Abdul Azis, P.K., 2000. Physico–chemical limnology of a tropical reservoir in Kerala, South India. *Ecol. Env. and Cons.,* 6(2): 159–162.

Tripathi, A.K. and Pandey, S.N., 1990. *Water Pollution.* Ashish Publishing House, New Delhi.

Trivedy, R.K. and Goel, P.K., 1984. *Chemical and Biological Methods for Water Pollution Studies.* Environmental Publications, Karad, pp. 244.

Chapter 5

Application and Role of Remote Sensing in Coastal Zone Management

N.V. Prasad

Division of Marine Biology, Department of Zoology,
Andhra University, Visakhapatnam – 530 003

ABSTRACT

The costal zones throughout the world are very precious and delicate ecological environment, both man and for nature. Since they have often fertile soils, and are in favor by man through their location near the sea, the pressure on the yet disturbed coastal zones is great. The concentration of human settlements, economic activity and resource mobilization is exceeding the capacity of natural systems and thus, there arise the need for coastal zone management. To achieve this, an understanding of the coastal processes that influence the coastal environments and the ways in which they interact is necessary. This requires crucial knowledge about the natural variability of marine ecosystem and dynamics of costal environment in time and space. Remote sensing technology in recent years has proved to be of great importance in acquiring data for effective resources management and hence could also be applied to coastal environment monitoring and management. An attempt was made here to provide detail information on the state of art coastal habitats using remote sensing data. The basic aim of this article is to have realistic and reliable information on the potential of coastal habitats and for sustainable development and management.

Keywords: *Coastal zone, Critical habitats, Critical issues, Management, Remote sensing.*

Introduction

Costal zone is the interface where the land meets the ocean, encompassing shoreline environment as well as adjacent costal water. Its components can include river deltas, coastal plains, wetlands, beaches and dunes, reefs, mangroves forests, lagoons, and other coastal features. The limits of coastal zone is often arbitrarily defined, differing widely among nations, and are often based on jurisdictional limits or demarcated by reasons of administrative ease. It has often been argued the costal zone should include the land area from the watershed to the sea, which theoretically would make sense, as this is the zone where biophysical interactions are strongest.

The coastal zone of world is under increasing pressure to high rates of human population growth due to development of various industries (tourism, chemical, petrochemical, fishing, aquaculture, shipping, mining, etc.), municipal sewage, industrial waste effluents and offshore petroleum exploration activities. This industrial development on coast has resulted in degradation of coastal ecosystems and diminishing the living resources of Exclusive Economic Zone in form of coastal and marine biodiversity and productivity. Apart from these the coastal processes of erosion, deposition, sediment transport as well as sea level rise continuously modify the shoreline and affect costal ecosystem. Thus there is an urgent need to conserve the coastal ecosystems and habitats including individual plant species and communities so that their current and potential usefulness to people is not impaired.

Ecologically sustainable development of the costal habitats requires that management and use is compatible with the attributes of exploited resources (Pearce, 1999). This requires crucial knowledge about the natural variability of marine ecosystem and dynamics of costal environment in time and space. Remote sensing, which can be defined as the acquisition of information from an object or an event without physical contact, helps in managing the vast area of coastal zone in space and time. The advantages of remote sensing is that it provides near real time information on various application so that strategies can be prepared to exploit the natural resources particularly in the costal realm. So many oceanographic satellites are available now, which provides lot of information on various aspects of costal habitats. Availability of sun synchronous satellites like NOAA, INSAT etc., for weather monitoring and polar-orbit satellites like MODIS, SeaWiFS, SPOT, ERS-I, IRS 1A–1D and IRS-P3, IRS-P4, IRS-P6 etc., which takes care of physical, chemical and biological aspects of coastal habitats. These satellites have different sensor to monitor the coastal and its associated environment.

The present article deals with the review of various coastal habitats that can be monitored by remote sensing. The objective of this article is to discuss major issues concerning to the coastal zone through inputs provided by orbital remote sensing data.

Coastal Zone Management

The Value of Coastal Resources

Estuaries, mangroves, coral reefs, sea grass beds, mudflats and sand beaches are the diverse habitats that make the ocean and costal environments and provide valuable benefits to human and marine life. In nature, coastal system maintains an ecological balance that account for shoreline stability, beach replenishment, and nutrient generation and recycling, all of which are of great ecological and socio-economic importance. In coastal rural areas fishing of nearshore water and farming of coastal lowlands are the major economic activities supplying fish and agricultural products for subsistence of the inhabitants and urban centers. Activities that add further value to coastal resources include recreation and tourism, which have become major resources of domestic and foreign earnings in many coastal nations. In recent years, coastal habitats are rapidly being cleared for urban, industrial, and recreation growth as well as for aquaculture ponds. In tropical countries, the loss of mangroves averages well over 50 percent of the pre-agricultural area (Pearce, 1999). In India, at places such as Kochi, wetlands have been drained for development and prevention of malaria. Concern is growing in particular about the destruction of natural coastal ecosystems by the demands placed upon them by population and economic growth.

These natural ecosystems have considerable value for sustainable extractive and non-extractive use which is often undervalued in comparison with other often non-sustainable uses.

The Importance of Coastal Zone Management

More than half of the world's population lives within 60 km of a coastline. The world's coastal areas are thus exposed to rapid urban growth, increasing population pressure, expansion of major industries and extensive exploitation of marine resources. The results include the pollution of marine and freshwater resources, air pollution, loss of marine and land resources, the loss of cultural resources, loss of public access, soil degradation, and increasing levels of noise and congestion (Figure 5.1).

Although there have been many attempts to protect coastal areas and to encourage sustainable forms of coastal development, few have been successful. The main reason is that they have largely been sectoral, and there is fierce competition for coastal resources in many areas. For example, there is often conflict in coastal areas over access to the coastline, which is required for tourist beaches, marinaes, aquaculture and cooling for power generation. As a result, in several

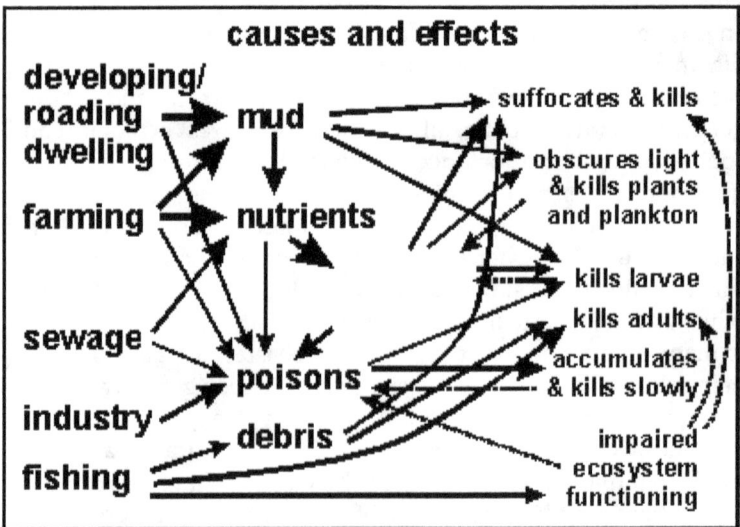

Figure 5.1: Various Factors Influencing Coastal Zone

areas of the coastal water is polluted, the sea is over-fished, natural wetlands are drying up, and disposal of waste and sewage presents difficult problems.

Almost all coastal and marine areas produce or support multiple products and services. Sectoral solutions usually 'transfer' the problem between resources, products and services. But coastal resources cannot be sustainably used by any interest group as their exclusive right. Most importantly, the sea cannot be regarded as a common basin for effluent disposal. Tourism will not flourish if the area loses its attraction to visitors; fisheries are usually on the receiving end of everyone else's problems. Industry and energy facilities can degrade the environment for all other activities. Integrated coastal zone management ensures sustainable use of coastal nature resources, maintenance of high levels of biodiversity and conservation of critical habitats by allowing policy orientation and development of management strategies to control impact of human intervention on the environment; addressing resources use conflicts through multi-sector planning process. An integrated approach to coastal and marine management is needed to resolve the conflicting demands of society for products and services, taking into account both current and future interests (UNEP, 1995).

Critical Issues of Coastal Zone Management

Population growth in the coastal zone is a major concern. The world population is expected to grow at an exponential rate from 5.8 billion in 1995 to 8.5 billion by the year 2025. Nearly 60 per cent of the world population is already concentrated within 60 km of the coast while there is considerable migration of population to the coast from the inland areas. In developing countries, by the turn of this century two-third of population (3.7 billion) is expected to live along the coast. This growth will exacerbate already severe coastal-use conflicts in terms of land and water space and resource utilization. The negative impact of the increased human settlements and industrial development are also more acutely felt in the coastal zone since it is at the receiving end of land–water-based pollution. Compounding the problem, the coastal zone is often subject to overlapping governance of local, provincial and central governments resulting in inter-agency conflicts and unclear policy concerning resource development and management and environmental protection

The following issues are critical in context of coastal zone management (Sudarshan *et al.*, 2000):

1. Coastal Habitat Conservation Related

- ☆ Availability of base line data.
- ☆ Preservation, conservation and monitoring of critical habitats (coral reefs, mangroves etc.).
- ☆ Appropriate site selection for industries, landfall points, aquaculture, recreation activities etc.
- ☆ Assessment of conditions in regulation zones, areas under construction set-back-lines and mega cities.
- ☆ Reclamation of wetland for agriculture and industrial purpose.

2. Coastal Processes Related

- ☆ Planning and implementation of coastal protection work (erosion, flood protection, salt water intrusion, etc.)
- ☆ Interaction between developmental activities and modification of coastal processes.
- ☆ Impact of dam construction on shoreline equilibrium.
- ☆ Suspended sediment dynamics.
- ☆ Change in bottom topography.

3. Coastal Hazards

- ☆ Cyclones
- ☆ Coastal erosion
- ☆ Sea level rise and possible effect
- ☆ Non-point and point pollution
- ☆ Toxic phytoplankton blooms.

4. Availability of Resources and its Utilization (Sand Mining and Fisheries)

Role of Remote Sensing and its Major Applications

Environmentally effective coastal zone management depends upon accurate and comprehensive scientific data on which policy decision can be based. A basic problem confronting our country is limited availability of geographical data on coastal zone (Nayak *et al.*, 1989). Remote sensing data because of its repetitive, multi-spectral and synaptic nature have proved to be extremely useful in providing information on various components of the coastal environment *viz.* tidal wetland conditions, mangrove degradation, coastal land forms, shoreline changes, tidal boundaries, brackish water areas, suspended sediment dynamics, coastal currents, oil pollution etc. The repetitive nature of the data has helped in monitoring vital and critical areas, periodically. Important of remote sensing data for inventorying, mapping, monitoring and recently for the management and development planning the optimum sustainable utilization of natural resources has been well established. Remote sensing offers solution to get wide range of information over inaccessible areas in space and time intervals. The major applications of remote sensing for coastal ecological studies as follows:

Mangroves

Mangroves are very important as they help, in the production of detritus organic matter and recycling of nutrients in coastal waters (Balasco, 1975). The mangroves of the world spread over 30 countries in tropical and sub-tropical regions, spread an area of 99,330 sq. km. In India, the earlier estimation of mangroves was 6,740 sq. km (GOI, 1997). The present satellite based estimates of mangroves was 4,460 sq. km, which indicates loss of about 2,280 sq. km. At many places, mangroves are degraded (~25 per cent) and destroyed due to their use as fuel, fodder and convention of these areas for aquaculture purpose.

Mangroves have different optical properties in the visible, near infrared (NIR) and middle infrared (MIR) region. Based on remote sensing data, mangroves have been assessed in terms of extent, density of community, condition, diversity, identifying potential habitats and heterogeneity in ecologically rich areas. Dense mangroves have been mapped in Sundarbans, deltanic regions of Mahanadi and Bhitarkanika (Orrisa), Krishna and Godavari delta (Andhra Pradesh), fringing the coast in Andaman Islands, Deltanic region of Kori creek in Gujarat coast and Pichavaram and Vedaranyam mangrove forest in Tamil Nadu (Nayak, 1994). Delineation between old and new mangroves can be possible using combination of Landsat TM bands (MIR, NIR and red). It has been seen that these combination of Landsat TM is best for differentiation mangroves from evergreen, semi-evergreen, and moist/dry deciduous forests. Merging images of IRS 1C/1D LISS III and PAN can be also used for mangrove delineation.

There are few studies highlighting the species wise identification of mangroves using satellite data. Mangroves shows distinct zonation characterized by the presence of particular species, having specific physico-chemical environment and related dominant genus, being depending on the extent and frequency of inundation under tidal waves, salinity and soil characteristics. Information regarding different mangrove community zonation is a vital remote sensing based input for biodiversity assessment and for preparing management plans for conservation. Mangrove zonations based on species association, tidal inundation and density has been carried out in Bhitarkanika forest, Orrisa, by using IRS-1C LISS III data (Bhahuguna and Nayak, 1997).

Coral Reefs

Coral reef around the world is deteriorating rapidly due to human activities. In the whole world, south-east Asia contains 30 per cent of the world coral reefs. It has been found here that 60 per cent are already destroyed or on the verse of destruction. In India, these are mostly fringing reefs in the gulf of Kachchh, Gulf of Mannar and Andaman and Nicobar Islands, with a few platform, patch and atoll reefs and coral pinnacles. Lakshadweep Islands are mostly atolls with few coral heads. Coral reefs of the Gulf of Kachchh and few reefs in Andaman and Nicobar are in degraded conditions as indicated by mud depositions (Desai *et al.*, 1991). The felling of mangroves and clearing of forests have increased sedimentation and affected live coral and species diversity. Unless drastic action is taken immediately most of the reef will be eradicated during next 40-50 years (Desai *et al.*, 1991).

Remote sensing offers solution to get wide range of information over inaccessible coral reef area in space and time intervals. The latest Indian satellites IRS–1C, 1D, IRS-P4 and IRS-P6 with their improved spatial resolution extended special resolution and increased repetitive (5 days for WiFS) has opened up new vista for application in coastal zone (Choudary *et al.*, 2002). The PAN data combine with LISS-III are extremely useful in providing detailed spatial information about reclamation, construction activity and ecological sensitive areas which are vital for coastal zone regulatory activities. The information available from merge PAN and LISS III data about coral reef zonation especially for atolls, patch reef is valuable for coral reef conservation (Choudary *et al.*, 2002). The majority of remote sensing provides the geographical information on reef zones. These zones are primarily of a geographical, rather than ecological one and include fore reef, reef crest, algal rim, spur and groove. In general, sensors with higher resolution offer greater details. Landsat TM allows better discriminations of varies coral zones than Landsat MSS and SPOT XS helps to find out the density of coral reef. IRS-1C data have been found to be extremely useful for coral reef zonation study (Nayak *et al.*, 1997; Bahuguna *et al.*, 1992).

Sea Grass

Remote sensing data has useful in monitoring sea grass location and extent. Sea grass species identification appears to be beyond the capability of remote sensing (Monaghan, 1988). Landsat TM is probably the best for sea grass mapping. Sensors like MSS, TM, XS, API and CASI also useful for the mapping of sea grass. Arial photographs and most satellite sensors are poor at mapping of sea grass beds of low medium standing crop but their performance was higher of better biomass bed.

Monitoring Protected Areas

In India, 3 National parks and 13 wildlife sanctuaries have been established for protecting critical/vital habitats, such as mangroves, coral reefs and wetlands. 4 National parks and 16 wildlife sanctuaries have been identified for conservation measures. Apart from this, 17 parks and 28 wildlife sanctuaries have been proposed on the Islands territories of India (FAO, 1991). It is necessary to monitor these areas to assess impact of conservation measures. The repetitive nature of satellite data helps in monitoring vital ecosystem areas as well as in assessing the effect of conservation measures. In one such study, in the Marine National Park, Jamnagar, on the Gujarat coast revealed that there is a significant change in the mangrove vegetation and coral reef during 1975 to 1991 (Nayak *et al.*, 1989).

Ecosystem Diversity

India has important examples of all the main ecosystems found in the region. Government of India has surveyed and prepared maps and created baseline information on coastal habitats both critical and vital with associated shore land features along the entire Indian coast on 1:250,000/1:50,000, scale for macro level planning. The major finding of the inventor of Indian carried out through visual interpretation of multi spectral IRS LISS II and Landsat TM data.

Brackish Water Aquaculture

Brackish water aquaculture has tremendous potential due to ever increasing demand of prawns. In India, the aquaculture development started with essentially to provide employment in rural coastal areas as well as to increase the export to developing countries (Mitchell, *et al.*, 1986). Intensive commercial aquaculture practiced both in developed and developing countries, growing in popularity among many Asian and Latin America countries as an export industry, could have harmful consequences as it replaces coastal mangrove habitats, reducing breeding grounds of wild stocks (FAO, 1991). In order to provide information related to environments, IRS LISS II data was utilized to prepare coastal land use maps on 1: 50,000 scale along the Indian coast showing wetland features between high and low water lines and land use features of the adjoining shore as a major element in the site selection for brackish water aquaculture. These maps along with other engineering, biological, meteorological, socio-economic and infrastructure related parameters were integrated for evaluating site suitability for brackish water aquaculture (Gupta *et al.*, 1995).

The high resolution data sets with some image enhance technique (MSS Landsat, IKNOS and IRS-PAN) are useful in monitoring the spacing between ponds, ponds and other land forms and also infrastructure facilities in aquaculture projects. The destruction of mangrove forests due to construction of aquaculture ponds is also clearly seen on the IRS-1D LISS-III + PAN merged images. The high resolution data from varies satellites will be useful in mapping and monitoring of aquaculture surrounding environment which provide strong proof on the encroachment in the absence of *in situ* information (Congalton, 1991).

Coastal Regulation Zone Mapping

The increasing pressure on the coastal zone due to concentration of population, development of industries, discharge of water effluents and municipal sewage and spurt in recreational activities, has adversely affected the coastal environment. In view of this, coastal stretches of bays, estuaries, backwaters, seas, creeks, which are influenced by tidal action up to 500m high tide line has been declared as the Coastal Regulation Zone (CRZ). One of the main objectives of this act was to prepare maps showing wetland features up to 500m from HTL on 1:25000 scale for the entire Indian coast, Using IRS LISS II and SPOT data.

Shoreline Change Mapping

Orbital data have been useful for detecting long term and seasonal changes in entire coastline. Shoreline-change mapping (1967-68, 1985-89, 1990-92 periods) for entire Indian coast has been carried out by using Landsat MSS/TM and IRS LISS II data on 1:250,000 and 1:50,000 scale. The satellite data revealed that the most area on the east coast was affected by depositional activities. Shifting of river mouths, formation of shoals, growth of spits, have been noticed along the Maharashtra and Goa coasts (Nair and Sankar, 1994).

Primary Production

The water colour remote sensing determine concentration of chlorophyll-α in water as a parameter for estimation of the phytoplankton biomass and hence in primary productivity. Satellite remote sensing of ocean colour is only tool that can provide information on marine primary production on a global scale. Attempts have been made to estimate and mapping the primary production from Landsat 5 TM data over west coast of India. SeaWiFS/IRS-P4 OCM and other ocean colour sensors provides more efficient means to determine primary production from satellite derived surface chlorophyll concentration. Available IRS-P4 OCM data is useful in monitor and identification algal blooms in the ocean.

Potential Fishing Zones

Ocean is a cornucopia of resources. The full potential of ocean resources is being appreciated during the past 2-3 decades. Even though food resources from maritime sources are explored, fishery resources are not exploited to its full potential. Barton (1974) opined that effective utilization of fishery resources could be achieved by using remote tracking devices to locate shoals of fish more quickly. At present, instruments like SONARR, RADAR and satellites *i.e.*

NOAA are used for locating sites of fish aggregation. The information about important oceanographic conditions and processes affecting fish population, such as surface temperature isotherms, ocean frontal boundaries, currents, circulation patterns and coastal upwelling may often be deduced by using ocean surface measurements made by satellites. Ocean temperature measurements made by satellite remote sensing can be extremely useful in defining the distribution of marine fish habitat conditions. As a part of National programme of Department of Ocean Development, the NRSA utilizing SST information for identifying Potential Fishing Zones for dissemination of such information for the benefits of fishermen.

Coastal Salinity

The agriculture land around coastal region is being affected by ingress of seawater. The coastal belt around the Gulf of Khambhat is severally affected by salinity and this had resulted in low productivity in this area. Using of pre and post monsoon images of Landsat TM and employing condition of vegetation, colour, association, location as criteria, salt-encrustations, saline and slightly areas and non-saline areas, were delineated (Nayak *et al.*, 1987). This study proved that remote sensing data is useful for monitoring coastal salinity effectively.

Coastal Water Quality

Due its repetitive, multi-spectral and synoptic nature, remote sensing data has provide to be extremely useful in providing information on varies aspects of oceanographic parameters and coastal landforms. Remote sensing can detect turbidity (suspended sediments) and colour (chlorophyll) which are indicators of water quality (Vinu Chandra, 2004). Chlorophyll indicates trophic status, nutrients load and possibly presence of man made pollutants in coastal waters. Excess nutrients over fertilize a marine area, causing algal blooms, including those of toxic algae, can lead to mass kill of fish and invertebrates. In view of this it is necessary to monitor chlorophyll in coastal waters. The present IRS-P3 MOS data allows measurements of suspended sediments, chlorophyll, yellow substances etc (Sarangi *et al.*, 2001).

Oil Pollution

The numerous methods used for oil detection at the sea surface include visual detection by eye, aerial camera, MSS and CZCS;

microwave detection by SMMR and SAR; fluorescence detection by lidar; and thermal detection by IR scanner (Klemas, 2001). The visual method images the change in colour and brightness due to the presence of oil. Other visible-light phenomena used to detect oil slicks include EMR interference effects (colour banding) and the suppression of solar speckle by slicks. The microwave method, when passive techniques are used, is based on the difference of emissivity between the sea surface and the oil slick. Active radar sensors depend on small capillary wave backscatter to be dampened by the oil slick as a means of oil detection. Fluorescent properties of hydrocarbons may be detected and discriminated by appropriate lidars. These laser fluorosensors can also identify the basic types of oil (heavy, light, etc.) and provide a measurement of oil slick thickness. Thermal sensors identify oil by means of the difference in solar absorption and thermal emissivity between oil and water and they also provide a basic measurement of oil thickness.

Other Applications

☆ Lagoon and lake studies

☆ Flood zone mapping

☆ Critical habitat analysis mapping

☆ Estuarine river morphological studies

☆ Thermal pollution studies

☆ Sea weeds

☆ Suspended sediment dynamics

☆ Coastal land form studies

☆ Sea level rise

☆ Waves and current pattern

☆ Sea state forecasting

☆ Sea surface temperature

Status on Utilization of Remote Sensing Data in India

Remote sensing data is useful for wide variety of management and scientific issues in coastal zone. The present status on utilization of remote sensing data for marine ecological studies in India is summarized in the Table 5.1.

Table 5.1

Parameter	Remote Sensing Compliances	Status
Mangroves, Coral reefs, Salt pans, Aquaculture, Wetlands and other coastal inland resources	Mapping and monitoring in different scale	Operational using high resolutions multispectral sensors data from IRS series.
Fisheries	Forecasting and monitoring	Semi operational with NOAA and IRS-P4.
Mineral and Energy	Exploration and monitoring	R&D stage with existing RS data.
Coastal geomorphology and shoreline changes	Mapping and monitoring in different scale	Operational using high resolutions multi-spectral sensors data from IRS series.
SST, Wind, Waves, Water vapour content etc.	Fisheries forecasting, monsoons, Ocean and atmospheric studies	Operational with IRS-P4 and other foreign satellites.
Upwelling, Eddies, Gyres etc.	Fishery and Ocean dynamics studies	Operational with IRS-P4 and other sensors.
Coastal Regulation Zone	Mapping and monitoring in 1:50,000 and	Operational using IRS 1C and IRS 1D. 1: 25,000 scale
Suspend Sediment Concentration	Mapping and monitoring	Operational with IRS-P4.
Oil slicks	Mapping and monitoring	Semi operational with IRS series and other foreign satellites.
Chlorophyll concentration	Mapping and monitoring	Semi operational with IRS-P4.
Current and Surface circulation pattern	Mapping and monitoring	Semi operational with IRS series and other foreign satellites.

Conclusions

Integrated Coastal Zone Management (ICZM) has become a concept, which coastal nations around the world are adopting to wisely plan and manage the use of coastal resources. "Integrated" is used to describe the bringing together of participant, initiatives and government sectors (Kay and Alder, 1999). Achieving ICZM will need to address several dimensions of integration *i.e.*, interaction among sectors (inter-sector integration); integration between the land and water sides of coastal zone (spatial integration); integration among scientific aspects of coasts, levels of government and among agencies within each level of government (inter-agency integration); and integration among disciplines, policymaking and implementation (National Research Council, 1995).

The ICZM of critical coastal habitats required comprehensive and integrated set of data with continuous monitoring (Klemas, 2001). At present, remote sensing is the only technology available which can be effectively used for continuous monitoring of coastal zone ecosystem. The role of remote sensing will increasingly become an indispensable tool for coastal planning and management. Satellite imagery in visible, infrared and microwave bands and side-looking radar images are found to be a reliable source of scientific and managerial information for the environmental monitoring of coastal zones. The remote sensing based management plan helps sustainable development of coastal zone without endangering the environment.

References

Balasco, B., 1975. *The Mangroves of India*. Institute Francais de Pondichery, Podicheri Pub, pp. 160.

Bhahuguna, A. and Nayak, S., 1997. Mangrove community discrimination using IRS-1C data. In: *Proceedings of National Symposium on Remote Sensing for Natural Resources with Special Emphasis on Water Management*, ISRS, Pune, December 4–6, 1996, p. 12–18.

Bhahuguna, A., Ghosh, A., Nayak, S., Patel, A. and Agarwal, J.P., 1992. Ecological status of the coral reefs of Gulf of Kachchh and Lakshadweep. In: *Proceedings of National Symposium on Remote Sensing for Sustainable Development*, ISRS, Lucknow, p. 57–61.

Choudary, S.B., Rao, K.H. and Rao, M.V., 2002. Satellite remote sensing for marine resources assessment. *Tropical Ecology*, 43(1): 187–202.

Congalton, F., 1991. A review of assessing the accuracy of classification of remote sensing data. *Remote Sensing of Environment*, 37: 35–46.

Desai, P.S., Narain, A., Nayak, S.R., Manikiam, B., Adiaga, S. and Nath, A.N., 1991. IRS-1A application for coastal and marine resources. *Current Science*, 61: 204–208.

FAO, 1991. *Environment and Sustainability of Fisheries.* Committee on Fisheries, FAO, Rome, p. 1–7.

GOI, 1987. *Mangroves in India: Status Report.* Ministry of Environment and Forest, Govt. of India, pp. 156.

Gupta, M.C., 1995. *Brackishwater Aquaculture Site Selection Using Techniques of GIS.* Scientific Note, SAC, Ahmedabad, August, p. 56.

Kay, R. and Alder, K., 1999. *Coastal Planning and Management.* Routledge Pub., London, pp. 189.

Klemas, V.V., 2001. Remote sensing of landscape-level coastal environmental indicators. *Environmental Management*, 27: 47–57.

Mitchell, J.R., Ckefors, H.A., Bardach, J. and Egidius, E., 1986. Statement committee. In: *Realism in Aquaculture: Achievements, Constrains, Perspectives.* European Aquaculture Society Pub., Belgium, pp. 585.

National Research Council, 1995. *Science and Policy and the Coast: Improving Decisionmaking.* National Academy Press, Washington D.C.

Nayak, S., 1994. Application of remote sensing in study of mangroves ecosystem. In: *Conservation of Mangrove Forest Genetic Resources: A Training Manual.* Swaminathan Foundation, Chennai and International Timber Organization, Japan, p. 203–220.

Nayak, S., Bhahuguna, A., Shaikh, M.G. and Gupta, M.C., 1989. Coastal environmental studies of the Kavaratti and the Agatti coral Islands. In: *Proc. of National Symposium on Engineering*

Application of Remote Sensing and Recent Advances, ISRS, Indore, p. 184–188.

Nayak, S., Bhahuguna, A., Chauhan, P., Chauhan, S.B. and Rao, R.S., 1997. Remote sensing application for coastal environment management in India. *Maeer's Pune Journal*, 4: 113–125.

Nayak, S., 1996. *Monitoring of Coastal Environment of India Using Satellite Data*. Frankcass and Co. Ltd., Essex Pub., pp. 175.

Pearce, F., 1999. An unnatural disaster-clearing Indian's mangrove forests has left the coast defenseless. *New Scientist*, 164: 12–19.

Sarangi, R.K., Chandran, P. and Nayak, S.R., 2001. Phytoplankton bloom monitoring in the offshore waters of northern Arabian sea using IRS-P4 OC satellite data. *Indian Journal of Marine Science*, 30(4): 214–221.

Smith R.C., Baker, M. and Baker, K.S., 1989. Estimation of a photon budget for the upper ocean in the Sargasso Sea. *Limnology and Oceanography*, 34: 1673–1693.

Sudarshan, R., Mitra, D. and Mishra, A.K., 2000. Rapidly changing environment: Hidden risks in Gulf of Cambay, Gujarat, India. In: *Subtle Issues in Coastal Management*, IIRS Pub., pp. 176.

UNEP, 1995. *Guidelines for Integrated Management of Coastal and Marine Areas*, pp. 165.

Vinu Chandra, R., Jeraram, S. and Adiga, S., 2004. Studies on the drift of ocean colour features using satellite derived sea surface wind for updating potential fishing zone. *Indian Journal of Marine Science*, 33(2): 122–128.

Chapter 6

Conceptual Aquatic Ecosystems and M.I.S.

D. Das

Central Inland Fisheries Research Institute, ICAR

ABSTRACT

Aquatic resource assessment and database on water-quality parameters, dynamics and depth of aquatic system, volume of water are the key to aquatic productivity. Parameters like water temperature, dissolved nutrient elements and gases, transparency at many extent could be moderated through aquatic management.

Keywords: Hydrophyte, Hydrophonic, Aquatic database.

Introduction

Fisheries are performed in nature by the grace of fishermen and unlike agriculture probably very little is there in between researchers and fishermen hence aquatic resource management remains critical. Entrepreneur intervention in aquatic ecosystem may be desired for enhanced aquatic production by minimizing gap of aquatic resource potential and actual output obtained. At the same time an integrated approach may be necessary in aquatic ecosystem for better sustainability.

Dynamics and Stratification

Aquatic resource is dynamic in nature. Seasonal variation in aquatic area and its depth are found due to geographic, climatic parameters. Geographic parameters mainly like topology, elevation, bottom soils, structure and texture, water holding capacity, contours. Climatic parameters like atmospheric humidity, variation of temperature, solar eclipse, rainfall etc. As well as water used for irrigation, drinking purposes from the aquatic sources. Overall in Indian topography variation of areas and depth of aquatic resources are observed more in plain and desert zone than in plateau. Seasonal variation also in plankton population has been found and maximum attains during summer (Jackson and Meir, 1965).

Aquatic strata determine the nature and quality of aquatic bodies. Hence the management need of aquatic resources remains within the aquatic layers. The strata may be named as:

1. Bottom soil
2. Lower aquatic
3. Middle aquatic
4. Upper aquatic and
5. Floating stratum.

Floating aquatic flora often takes crucial role in aquatic production. The reasons are indicated.

Floating Hydrophyte

Nutrient Loss from Aquatic System

Floating hydrophytes absorb major nutrients like nitrogen (N/NH_3^+), phosphorus(P/PO_4^{---}), potassium(K/K^+), calcium (Ca/Ca^{++}), magnesium (Mg/Mg^{++}), sulphur(S/SO_4^-), iron (Fe/Fe^{+++}) and as well as few trace elements from aquatic medium dissolved within. Depleted nutrients in aquatic medium is limiting factors for growth and multiplication of immerged and submerged hydrophyte, algae. Thus faster growth rate of floating hydrophytes greatly inhibits the growth of submerged or immerged plankton. As a result floating hydrohyte often dominates and management is critical in aquatic production.

Regulating Water Transparency

Floating hydrophytes controls water transparency at a great extent in aquatic system. It has the negative effect on water transparency, occurrence of more floating hydrophyte resulted lesser water transparency. The main reason is simply light could not penetrate freely within the water body due to presence of floating vegetation. This kind of less water transparency does not mean there is more algal biomass persists. Since existence of more algal biomass also reduces water transparency. The main reason is due prevailing of darkness shadows under the floating hydrohytes.

Controlling Water Temperature

Water temperature has many roles in aquatic system. Most basic is enzyme kinetics and habitat of living organisms *i.e.,* flora and fauna. Water temperature regulated by amount of heat-wave absorbed and by the earth surface temperature through heat conduction. Water has unique feature by acting as heat capacitor. During the sunshine hours floating hydrophytes reflect solar waves in the air. Amount of solar rays reflected depend on the thickness and leaf area indexes of floating hydrophyte. At the same time the floating vegetation act as non conductor of heat waves. Natural process of heat conduction is significantly reduced on emergence of floating hydrophyte. As a result under this situation more differentiation in air and water temperature prevails as compare to aquatic system where there is less or no floating weeds.

Root Exudation and Decomposition

Root secretion might have little effects on increasing alkalinity in aquatic system. Microbial more often bacterial decomposition of floating plant tissues on the surface of aquatic bodies, like cellulose decomposing bacteria, found in many areas of Ganges plains and plains of north east regions rottening of Jute added tannin and more bacterial growth and activities. These grasses has little effect on water pH. Methane from wetlands and SO_2 generated from aquatic system otherwise could be managed for mankind.

Surface Aquatic Fauna

Aquatic weed mass create a habitat zone of aquatic fauna and some of them are deleterious towards managing aquatic resources. Aquatic fauna may sometime act as predators of many fish species within the aquatic system. Floating hydrophytes can encourage migratory birds, reptiles, primates and may be active ecological part of a natural sanctuary where dimension of aquatic body is determining factor.

Repellent Effect on Domestic/Migratory Ducks

Ducks, either domestic or migratory, prefer to swim in waters where there is no excessive floating weeds in aquatic surface. It is found that swimming Duck has a role in changing water quality parameters like dissolved oxygen, and organic matter. Swimming bird adds extra oxygen through aeration in aquatic medium thus dissolve oxygen in aquatic medium become increases. This dissolved oxygen helps in growth of plankton and fisheries within the aquatic system.

Water Stagnation

Water stagnation may occur if there is heavy infestation of aquatic weeds. Stagnation is more in closed water-bodies in such situation even a little tide could not occurs resulting water stagnation. The reason is where there is excess aquatic weed prevails in aquatic condition turbidity of water reduced at a great extent. It is known that less water turbidity indicates lesser aeration in aquatic body. As a result low dissolved oxygen persists under stagnant water.

Inhibition in Fish Breeding

Instances, when aquatic weeds prevalent in excess, it is hypothesized that floating hydrophytes might have inhibitory role during the process of natural fish breeding. Since during breeding time some fishes need to migrate some extend for its' physiological demand prior to ovlution or in some cases requiring a glimpse of flow water within the aquatic system. Although most fishes complete natural egg laying and hatching activities during/just after rainy season when there persisting a less rain-water current, which is some extend controlled by aquatic floating weeds.

Reducing Fish Run Depth

Unlike top feeders in aquatic system, middle and bottom feeders swim in certain water depth, may be called as fish run-depth, once aquatic weed is heavily infested on aquatic body effective fish run depth become minimized due to formation of ' root zone of floating hydrophyte. Reduced water height often restricts the normal fish swim resulting normal habitude and growth of fishes restricted. This situation is severe once there is aquatic body whose depth is shallow to medium. Reduction of water depth by root zone is less significant under the aquatic condition where the depth of water is more than normal depth of around of six fits.

Estimation and Identification of Hydrophyte

Aquatic floating weeds of vast water areas could be estimated with the help of remote sensing techniques. Since fisheries are performed in surface water, visual identification of aquatic floating weeds and converting into organic manure within the aquatic system may also be advisable for sustainable aquatic system. Although, nature and occurrence of hydrophytes and their intensities depends on aquatic environments and climate exposed. Database of most commonly observed weeds are enlisted. Among them rate of propagation and surface multiplication is maximum in *Eichhornea crassipes* along with other species of *Cyperus, Paragrass, Saccharum* and considered to be aquatic weeds in wetland.

Hydrophyte of Importance

Need of aquatic database of hydrophytes, mode of propagation and possible hydrophonic culture in aquatic system may require attention.

Ipomoea bonanox L. / *I. alba* L.	Moon flower
Ipomoea hederacea Jaq.	Indian jalap
Ipomoea reptans Poir. / *I. aquatica*	Waterbind weed
Hygrophila spinosa T.	Vegetable
Hydrocotyle asiatica L.	Vegetable
Nelumbium speciosum	Wild Lotus
Nelumbo montana Small.	Lotus
Euryale ferox Salisb.	Foxnut/Makhna
Herpestis monnieria H. B. and K.	Vegetable
Isoetes spp.	Algae/Isoetes
Anabena spp.	Bio-fertilizer
Nostoc spp.	Bio-fertilizer
Cucumis sativus L.	Fruits/Vegetable

Conventional Usage

1. Mechanization in aquatic system by converting into organic form and adding them within the aquatic medium for sustainable aquatic production
2. Using as animal feed *i.e.*, buffalo by integrating animal husbandry along with aquatic production cycle.
3. Bio-gas

Aquatic System and Linkages

Outside the floating hydrophytes, we may consider the possible floating linkages while managing a aquatic system. An aquatic system and probable integration could return more production than their cumulative effect. Type of integration depends on dimension, distribution of aquatic resource, type of tenureship/ownership, scientific and technical personnel, human resources and overall the community needs. Linkage with other systems needs to be sustainable in a long run.

1. Integrated glass house or green house in aquatic system, specially on wetlands.
2. Generation of organic fertilizers from the aquatic resources through culturing and multiplication of floating fern with symbiotic Anabenae azollae, nostoc or BlueGreenAlgae based on the environment and suitability of aquatic system to make economic use of solar energies and atmospheric nitrogen supplementary to fisheries.
3. Growing vegetables, fruits in aquatic system in a integrated approach and hydrophonic culture giving without inhibiting major aquatic production.
4. Establishment of animal husbandry in close proximity to aquatic resource areas, on embankments to add organic matter to maintain water fertility for growth of primary feed. In flood prone areas elevated settlement of animal husbandry could be managed and found that duckery are suitable if frequency of birds is adjusted.
5. Fish breeding habitats, fish seed immunology and point database of hatcheries and potentialities for non-mortality

during transportation and fish recruitment in safer aquatic regions of existing lesser fish stock.

6. Enhancement of seasonal fisheries based on the availability of seasonal waters especially for smaller and medium size fishes of ecological and ethical importance.

7. Floating gardening either for recreational or economic usage for making greenery on the aquatic surfaces helps in maintaining hygienic system surrounding aquatic bodies.

8. Illumination or photo directed trapping of aerial insects *i.e.*, 'fauna' for frog culture, air breaths and cat fishes helps in fisheries at the same time countries of tropics where rice cultivation is more, reduces epidemics.

9. Capital intensive aquatic resource utilization through crabs, prawns, shrimps, turtle/tortoise in long run aquatic resource development process.

10. Aquatic resource usage through establishing tank aquatic system of fisheries and olericulture in the regions of humid tropics where evaporational water loss is minimum.

11. Wave management–Possibilities of solar wave management on aquatic ecosystem with suitable bands for plankton growth and photo sensitive multiplication.

12. Subsidiaries along with capture fisheries in aquatic bodies like utilization through fish processing, value addition in fish and fisheries and research need in amino acids or lipids extraction for medicinal purposes by avoiding lipid rancidation or protein denaturation or intoxication. Also fish meal, dry fish and packaging to mitigate nutritional demands during the fish breeding seasons.

Conclusion

It has been found that aquatic systems, spatially situated in limited geographic location, differs in aquatic production due to differential management followed as primary factor. Once values of aquatic parameters remain almost same for consecutive three years in a geographic region of North 24 Parganas, aquatic production differs due to differential management methods followed by aquatic/ fishermen communities.

Name of Society	First Year	Second Year	Third Year*
Kola Baor	36.3	28.3	28.5
Sindhrani	21.9	26.1	24.7
Hariarpur	16.0	13.1	12.6
Berbari	10.6	10.1	4.6
Akaipur	35.4	57.0	61.3

*: Aquatic Yield 'Q/Ha' of aquatic resource, during the year of 1997-1999.

Different management policies followed by different fishermen societies, observed in columns. Differential annual management observed within a society observed in rows.

Acknowledgement

Author is acknowledged to Shri R A Gupta, Head, Resource Assessment Division and Dr. D.Nath, Director, CIFRI.

Reference

Jackson, D.F. and Meir, H.F.A., 1965. Variation in summer phytoplankton populations of Skaeateles Lake, New York. Abstract, XVI. *Limnologorum Conventus in Polaria*, 77 p.

Chapter 7

Fluoride Toxicity and Management in Community Ecosystem

M.K. Mahapatra

Department of Zoology Khariar College,
Rajkhariar, Orissa – 766 107

ABSTRACT

Fluoride is present in earths crust. It is an essential compound for mankind. Its deficiency and efficiency has adverse effects on mankind. When it is present in excessive amount (more than 1PPM) it causes residual fluoride toxicity. In a community where it is found to be 11.73 ppm in water, passes to the producer (35.5), consumer (0.42-2.03) and decomposer (population) tropic levels and affecting all of them. It ultimately results primary, secondary and tertiary fluorosis in the human population and became lethal to the animal husbandry. It can be Preventable but not curable. The appropriate technologies are discussed and it needs public awareness, Government attention and scientist inventions.

Keywords: Fluoride, Fluorosis, Residual, Toxicity, Community.

Introduction

The water is an elixir of life for the survivility and sustainability of mankind in the biosphere. The demand of water increasing tremendously with the increasing urbanization, industrialization, pasteurization, population explosion and enhancement of agriculture, aqua farming etc. The water is a renewable natural resource gradually becoming degraded to a great extent and evoke a serious concern. The availability of safe drinking water to the common people (rural) is still a dream. The water is polluted artificially as well as naturally. Athvale (2003) classified the source of pollution as follows:

1. Pollution due to anthropology activities of man.
2. Pollution due to organic contamination.
3. Pollution due to presence of toxic chemicals above their permissible limit.

The first and second category is regarded as artificial pollution whereas the last category is regarded as natural pollution. The rural community are mostly dependant on their traditional water bodies (lentic) for their day to day activities. The presence of heavy metal such as calcium, magnesium, copper, iron, manganese, chloride, fluoride, mercury, cadmium, arsenic lead, zinc etc. in excessive amount provoke natural pollution in community ecosystem. These heavy metals are non-biodegradable and toxicated community ecosystem.

Fluoride is the common element in the earths crust as a component of rocks and minerals. (Nagarajan *et al*., 2003). According to WHO, the permissible limit of fluoride ion in drinking water is fixed as 0.5 PPM and ICMR has suggested it as 1 PPM. Deficiency of fluoride leads to dental carries and higher concentration leads to dental and skeletal fluorosis (Short *et al*., 1937). Fluoride an excess of 1.5 Mg/l leads to mottled enamel of teeth, concentration beyond 3.0 mg/l would result skeletal fluorosis and consuming water with an excess of fluoride form 6-10 mg/l and above leads to crippling fluorosis (Hand, 1988).Fluoride loss are directly toxic to aquatic live and accumulate in its tissues at concentration where absorption rates exceeds excretion rates. Some accumulation occurs in all tissues, but in most tissues subsequent losses may occur when ambient fluoride levels decrease. However, in bone, tooth and scales accumulation is permanent and cumulative. Fluoride is toxic to tissues, animal husbandary and the human it self (Mahapatra *et al*., 2004). Fluoride is also phytotoxic is nature (Halwachs, 1963). In vice the high accumulation of fluoride usually shows extensive chlorosis. Patel et. al 2004 observed residual fluoride toxicity deteriorate the quality and quantity of rice. The fluoride toxicity deteriorates the reproduction capacity of cattle (Rosenberger, 1964). In domestic animal it shows dental and skeletal changes, lethargy, emaciation, poor health (Pedini, 1967) under extreme toxicity the animal die (Cass, 1961). Keeping all these factor in mind the present work of the author is concocted.

Materials and Methods

The fluoride is estimated spectrophotometrically by following the procedure from APHA (1985). Determination of fluoride content in plant (rice) sample were determined by standardizing Megregian (1958). Fluoride in urine of animal husbandry (sheep, goat, cow and bufallow)were estimated by standardizing the methods of Allen *et al.* (1974). Fluoride toxicity in blood serum, milk was determined by applying the procedure of Armstrong and Singer (1959). Fluoride in flesh was also estimated by applying the procedure of Megregion (1958).

In the present work Karlakote community was taken as the study site. Karlakote is situated between 20° to 5' north latitude and 82°5' to 88° East longitude. The fluorosis was reported in this region by Mahapatra *et al.* (2004).

Results

Table 7.1: Fluoride Concentration of Fluoride Reported Villages of Nuapada District

Sl.No.	Name of the Village	Dug Well	Bore Well	WHS	Average
1.	Karlakote	7.59	9.12	1.24	6.15
2.	Dargaon	3.51	9.21	2.03	4.25
3.	Kotmal	4.12	4.87	1.34	3.44
4.	Tangripada	3.59	4.16	0.96	2.90
5.	Kureikela	7.17	5.87	1.92	4.98
6.	Kureijhola	4.39	2.98	1.79	3.05
7.	Sajumund	3.51	3.17	1.84	2.84
8.	Dudkaibahal	6.92	8.51	3.12	5.51
9.	Mandobirli	2.58	3.41	1.32	2.43

Table 7.2: Fluoride Accumulation and Impact in Producer i.e. (*Oryza sativa*)

Age of Plant in Days	Conc of NaF. in ppm		Accumulation of Fluoride in ppm		Average Root Length in cm		Average Shoot Length in cm		Biomass Gm/Plant	
	C	V	C	V	C	V	C	V	C	V
15	1.5	2	1.20	4.7	75.72	69.0	56.7	40.3	0.16	0.12
30	1.5	4	1.23	9.7	114.97	89.1	64.8	54.8	0.23	0.19
45	1.5	6	1.26	13.2	121.04	103.8	82.3	72.3	0.27	0.23
60	1.5	8	1.40	18.3	100.23	80.5	87.6	77.6	0.49	0.47
75	1.5	10	1.35	29.7	95.11	77.4	96.3	89.1	0.81	0.72
90	1.5	12	1.70	35.5	89.5	69.5	97.1	92.8	1.50	1.23

Table 7.3: Dental Fluorosis in Cattle Foraging

Animal	Age	Nos. of Animal Examined	0	1	2	3	Nos.	%
			Dental Score				*Fluorosis Teeth*	
Sheep	3–5	180	52	71	39	18	128	71.11
	5–7	62	23	23	08	04	39	62.8
	> 7	27	12	07	06	02	15	55.5
Goat	3–5	235	68	46	82	35	163	69.2
	3–7	52	14	19	11	08	38	73.1
	>7	21	07	04	07	03	14	66.6
Cow and	3–5	312	105	48	78	87	207	67.4
Bullock	5–7	102	39	31	27	15	63	61.7
	> 7	56	19	08	23	06	37	66.1
Bufallow	3–5	28	07	13	06	02	21	75.0
	5–7	14	03	04	02	03	09	64.3
	> 7	06	01	03	01	01	05	83.2

0: Normal Teeth; 1: Teeth with yellow glistering and patches; 2: Chalky white teeth; 3: Reputed Teeth.

Table 7.4: Concentration of Fluoride in Urine and Blood of Animal Husbandry

Animal		Control	Age Group		
			3–5	5–7	> 7
Sheep	U	0.21–1.28	4.39	8.37	10.92
	B	0.02–0.18	0.61	1.23	1.96
Goat	U	0.18–1.12	5.06	6.52	11.46
	B	0.04–0.16	0.32	1.09	2.12
Cow	U	0.27–1.42	5.79	10.18	13.14
	B	0.03–0.17	0.52	1.57	1.92
Bufallow	U	0.35–1.27	6.39	9.14	12.18
	B	0.04–0.21	0.48	1.96	2.13

Table 7.5: Concentration of Fluoride in Milk and Flesh of Animal Husbandry

Animal		Control	Age Group		
			3–5	5–7	> 7
Sheep	M	0.06	0.42	0.69	0.83
	F	0.08	0.56	0.89	1.13
Goat	M	0.04	0.75	0.82	1.12
	F	0.12	0.68	0.79	1.46
Cow	M	0.07	0.82	0.87	1.49
	F	0.17	–	–	–
Bufallow	M	0.09	0.78	0.96	2.03
	F	0.19	–	–	–

N.B.U,B,M,F are Concentration of Fluoride in urine, blood, milk and flesh of Animal husbandry respectively.

The flesh of cow and bufallow were not studied. The breast milk of human studied only from two samples and fluoride content is reported to 1.67 mg/l.

Discussion

Fluoride in Nature (Abiotic)

Fluoride, a member of a group of elements called halogenes,is the thirteenth most common element in earth crust. Elemental fluoride is a pale yellow gas and a deadly poison. The most reactive of all halogens, fluorine will bind with almost any other element to form a fluoride. Elemental fluorine does not occur in nature,it is found only in compounds with other elements. It has a particularly strong affinity calcium with which it forms a strong bond.In its natural states it commonly found as calcifluoride. The chief fluorine containing material are fluorite (CaF_2) cryolite ($Na_3\ AlF_6$) and fluroapatite ($Ca_{10}F_2(Po_4)_6$) (Sahu, 2004). The high incidence of fluoride is owing to a grate variety of factors such as geological, climatological, social and economical factors. The distribution of fluoride is dependent on a variety of factors such as amount of soluble and insoluble fluoride in source rocks, the duration of contact waiter with rocks and soil, temperature, rainfall, oxidation-reduction reactions and cation exchange softening (Reddy *et al.*, 1999; Selvaraj,

2001). Waters from rocky soil formed in the region of granitite and shell formation are usually found to have contaminated excessive fluoride are fluorospar, fluroapatite, amphiboles such as hornblende, tremolite and some micas. Weathering of alkali, silicate, igneous and sedimentary rocks especially shell supply a major portion of fluoride to natural water. Apart from natural water, phosphate fertilizers which may contain fluoride as an impurity and contribute considerable amount of fluoride and these can increased level of fluoride concentration in soil which in turn may eventually result in leaching by percolating water and increase the level of fluoride in groundwater (Mahapatra *et al.*, 2004).

Contamination of groundwater by fluoride causes irreparable damage to plants, domestic animals and human beings (Mishra 2004). Fluoride also occurs in groundwater in some parts of the world, where it can give rise to disabling bone diseases in animals and man. In addition to these natural discharges these are many discharges made as a results of residual fluoride toxicity (Harvey 1952; Jolly *et al.*, 1977). In the present study we observed the amount of fluoride are following secure bore well > dug well > water bodies (Table 7.1). The fluoride average of the concerned community is 11.73 (Mahapatra *et al.*, 2004)

Mahapatra (2004) reported fluorosis in 50 locations of 19 districts of Orissa. The problem of excess fluorosis has been noticed in 15 states of India(Selvaraj, 2001; Susheela *et al.*, 1993). Global concern has been raised many times about the fluorosis as 17 countries were reported the disease since 1963 (WHO, 1963). The quantity of fluoride entitled from a given source depends upon the rate and quantity of the product produced and material compound (Samal, 1989).

The fluoride in nature, it passes to and from air, water, soil, and living organism through some definite path ways (Figure 7.1).

Fluoride in Biotic Components

Fluoride in Producer

Plants are regarded as producer in almost all arranging community.The producer get toxicated by retention and release reaction of solute with soil matrix (Brar, 2003). Plants can accumulate fluoride directly from the air, which may eventually lead to visible leaf injury, damage to fruits, productivity lessening and possible

Transpiration

Atmosphere

Volatization

Gaseous
H₂SiF₄
HF, SiF₄

Residual
Toxicity

Inhalation

Mineral

Evaporation

Artificial
sources
(Industry)

Particulates N₃F C₃F
C₃F (P₀₄)₂ N₃₃AlF₃

Natural
sources
(Soil)

Water Bodies

Decomposition

Percolation
and
Minerals

Producer
Grass, Crop
Vegetables

Root
absrbpths

Dug Well

Tube
Well

Irrigation

and
Minerals

Irrigation

Pry. consumer

Decomposer

Sec. Consumer

Iert. Consumer

Residual toxicity

Particulate Deposit by
Consuming Food

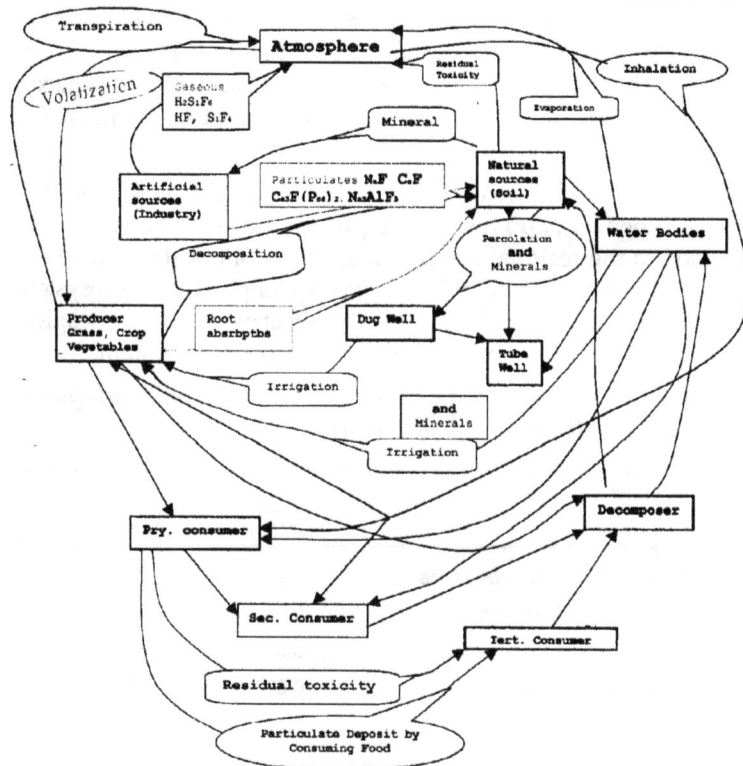

Figure 7.1: Fluoride Cycle in Community Ecosystem
Source: Mahapatra, 2004
Sketch: Er.J.N.Panda, A.M.I.E.(I)

other effects. It reduces the growth and yield of plants including change of colour (Rao and Pal, 1979). The structure and dynamics of the producer provide an indication of the nature of the concerned community.(Nivanonee and Sharma, 2003). Rice cultivation is thought to be the oldest form of intensive agriculture by man and it occupies the basil trophic level of community ecosystem, the residual fluoride content in rice need to be analysed (Shupe and Alther, 1966; Samal, 1989). Rice is the form of food, forage and straw add a significant amount of residual fluoride toxicity to cattles and livestock (Naik and Samal, 1987). The fluoride enters to the rice plant through the stomata and subsequently passes into the intercellular spaces and in eventually absorb by the mesophil cell of the leaves (Halwachs,

1963). From the mesophil cell fluoride occurs upward either by cell to cell diffusion via vascular tissue and finally it reaches to leaf margins (Jacobson *et al.*, 1966). With in chloroplast fluorine complexes with mg 2t in the chlorophyll molecule renders it unfit for photosynthesis (Lee black *et al.*, 1971).

In this study the rice plant accumulation fluoride 35.5 ppm(at 12 ppm conc. of NaF).It seems to producers has more amount of fluoride in the residual toxicity then the abiotic water and soil, it shows less productivity (Table 7.2) has great importance in ecosystem analysis as it integrates the cumulative effect of many physiological process which occur simultaneously with in the ecosystem.

Samal 1989 studied some vegetables of communities ecosystem with following fluoride concentration.

Name of the Vegetables	Fluoride in (ppm)	
	(Control)	*(Accumulated)*
Abelmoschus esculanthus Monech	0.20	24.8
Amaranthus hybridas Linn	0.35	105.8
Brassica oleralera Linn	0.56	25.0
Cucurbita maxima Linn	0.35	12.0
Dolichos labiab Linn	0.75	35.0
Momordica charntia Linn	1.20	17.3
Solanum melongena Linn	0.90	42.0
Raphanus sativa Linn	1.0	45.5

These all vegetables of the communities shows a high varieties of residual fluoride which in turn passes to the consumer level of the community ecosystem.

Fluoride in Consumer

The purpose of acute toxicity test in to assess various abnormalities caused due to administration of a chemical to animals on occasion or other and to determine the order of lethality of the chemical.In aquatic toxicology, acute lethal toxicity tests with fish or invertebrates are usually intended to assess the numerical value of toxicity to compare the potencies of toxicicants to assess the effects of environmental variables on toxicity (Harichandan *et al.*, 2003).

Fish

The present community ecosystem has a great diversity non cap fishes (Mahapatra and Mishra, 2003; Mahapatra *et al.*, 2003). The non carp fishes are of smaller size than the carp fishes. The sizes (ages) of fish also affect the fluoride toxicity level and accumulation rates and levels. Larger fishes are more tolerant of higher fluoride levels and accumulates less fluoride on a per weight basis. As the community has large no of dry (annual) pond, it has a greater diversity of fluoride toxicity. Mahapatra (2004) have prescribed 1.5 ppm as saver level for fish survivility but from different sources suggest that a uniform consensus about the maximum safe level of fluoride icon for fish in natural waters of varying hardness has not get been achieved. In fish fluoride toxicity is depends on many factors. Species is one factor. Dash (2002) suggested different accumulation levels of fluoride in different fresh water fish species. He also suggested larger fish survive larger than smaller fish, as determined by weight or length.In this community the survival rate is minimal due to high residual toxicity.

Mammals (Animal Husbandry)

Population exposure to toxic trace metals are of great concern due to their non biodegradable nature and long biological half line for elimination from the body (Palaniappan *et al.*, 2003). Excessive dietary fluoride have a serious impact on herbivores and man. Discolorations weakening and disintegration of the teeth lameness and stiffness of the joints are symptoms often ascribed to fluoride toxicity(Naik and Samal, 1986-87). The dental fluorosis and skeletal disorders are attributed to high fluoride content of water. The fluorosis in herbivores and human beings is due to consumption of highly F-contaminated water, Fluoride in (ppm F-vegetation and inhabitation of F-air discharged in the air (Singh *et al.*, 2003). The affected cattle show a loss in their body weight (Udall and Keller, 1952). Fluorosis reduce in the quality and quantity of milk of domestic animal such as cattle and goat (Hendrickson, 1961). Depending upon the levels of fluoride in the ingested food supplements the ailment can be acute or chronic. It is generally seen that acute fluorosis is rather rare under field condition because it can occur only when a very high amount of fluoride enters the animal body accidentally, in such cases of F-ingestion the animals show the symptoms of stiffness of muscles, lameness, lack of appetite, increased thirst, diarrhoea, constipation, acute abdominal pain, nausea and vomiting. Ingestion of sub-lethal concentration of fluoride by herbivores and man during the early stages of their teeth formation, produces dental changes, which are regarded to be the first detectable symptom's of fluorosis. The floristic teeth appears yellowish brown or deep dark stained, with ruptured edges and brown stretches may turn black affecting the whole teeth and chipped many get pitted, perforated and chipped off or rounded at the final stage (Lass, 1961; Rao and Pal, 1979; Samal and Naik, 1988; Naik and Samal, 1986-87). Dental fluorosis is seen in animals or human beings who are born and brought up in an endemic area. Even if they are taken in to the F-polluted area before the eruption of their permanent teeth. However if a domestic animal/man during adolescence moulds from a non-endemic area to an endemic area, he/she may not suffer from dental fluorosis, but may get affected with skeletal fluorosis and may also suffer from non skeletal manifestation (Shupe *et al.*, 1963).

Subsequent to ingestion about 96 per cent of the fluoride in the body is departed in bones and teeth and rest is excreted with urine (Susheela and Mukherjee, 1981; Sushila and Sharma, 1982). The acute lethal dose of fluoride for the average 70 kg man is 2.5–5 gm. One would need to consume 2,500 its of fluoridated water to reach this level (Leone *et al.*, 1956). In this study it is observed that the nos. of fluoridated teeth is in following sequence as group unit (Table 7.4).

Age Group	Fluoridated Teeth
3–5	Bufallow > Sheep >Cow > Goat
5–7	Goat > Bufallow > Sheep > Cow
>7	Bufallow > Goat > Cow > Sheep

The bufallow is affected most while sheep has ore affected in lower age group and goat is more affected in higher age group. This is may be due to the versatile and comparatively more feeding habit of goat than sheep. The cow is less affected then the bufallow may be due to the quality of food and forge cleared by the bufallow is male. Irrespective of maternal fluoride intake level, the maternal milk supply has not been demonstrated to exceed 0.2 ppm fluoride. This level may be beneficial to the developing embryonic teeth (Shupe *et al.*, 1963). But in the study the fluoride content of the milk is increasing particularly with the increasing age. This may be due to the higher accumulation of fluoride in the blood (Table 7.5 and 7.6). The fluoride conc. in blood of low age animal is significantly less than that of old animal. This may be due to the less accumulation of fluoride because of the less nutrient and drinking water uptake. The urine of cow, bufallow, sheep and goat shows higher rate of excretion of fluoride with increasing age. The fluoride content of urine of these animals are positively correlated with the fluoride content of flesh. The flesh of sheep content lesser residual fluoride than that of goat up to 7 years the conc. of fluoride in flesh is less than 1 ppm, although it is about 5-6 fold higher than that of the control. The animal husbandry get these fluoride content from food, forage, drinking water and from their mother milk.

Heavy metals primarily affect the kidney, which is involved in the cleaning process of the body fluid and tissues (Palaniappan *et al.*, 2003). May be due to this reason the goat of 5-7 age group excretion comparatively less amount of fluoride in urine.

Human

Man is the ultimately suffer and victim of nature due to the natural aquatic pollution of fluoride Being the tertiary consumer and omnivore in nature. It ingested higher amount of fluoride. The fluoride get deposited in its bone and teeth. It was Dr. Trendely Dean "the father of fluoridation," who first hyphothesised that fluoridation would protect teeth from cavities (Ann. 04).

Fluorides helps to present dental carries in several ways *i.e.*, strengthen teeth, remineralise teeth, kills decay causing bacteria when present is permissible limit but it excess it rupture teeth. Fluorosis is a result of abnormal deposition of fluoride in hard tissues. Mahapatra *et al.* (2004) observed the effects of fluorosis in Karlakote community ecosystem.

1. 87 per cent dental fluorosis
2. 29 per cent loss of teeth enamel
3. 100 per cent skeletal fluorosis above age 20
4. Hump on the back of 8 per cent school going children
5. 18 per cent children born physically handicapped.

The occurrences of dental fluorosis in school going children is due to the consumption of high fluoride contaminated water and F-vegetation growing in the vicinity (Samal, 1989). No significant gross or histological changes were noted in the brain, pituitary, liver, kidneys, adrenals, spleen, pancreas, thyroids, ovaries, mammary gland, uterus, rumen, reticulum omasum, abomasums (Udall and Killer, 1952).

The 87 per cent dental fluorosis and 29 per cent loss of teeth enamel may be due to the inflow of breast milk from mother which was reported to 1.67mg/l.

There is no evidence that milk production is directly affected by ingestion of low concentration fluoride. In cases of marked fluoride toxicity, many effects on milk production are probably secondary to major symptoms and lesion induced by fluoride rather than due to direct interference with lactogenesis. Milk production of cow fed 93 ppm apparently was adversely affected in the 2nd lactation and was definitely reduced in 3rd and subsequent location. Milk production of some cows on 49 ppm of NaF was adversely affected in the fourth and subsequent lactations the amount of fluoride in the milk correlated with the amount of fluoride ingested and did not exceed 0.2 ppm of fluoride when the new cows ingested 93 ppm of fluoride. Experiment shows that the mammary glad was a minor means of fluoride excretion from the body and that the concentration of fluoride in milk were low and with in save levels for human consumptions (Stoddard *et al.*, 1966).

Samal (1989) observed that 60.9 per cent of school age boys and 47.8 per cent girls had characteristic markings on their teeth various degrees of mottling,staining weakening and rupturing/hipping off edges were involved. Most of the boys and girls of age groups 5–7 years were suffering from mild to moderate types of horizontal yellow lines and yellowish brown areas on the enamel.

Siddque (1972) reported that 8–10 years boys and girls teeth were damaged by mottling with brownish stretches, worn out edges reduced thickness and nearing enamel. Samal (1989) also reported that 67.9 per cent male and 64.9 per cent female were affected in the fluorosis (Dental and skeletal).

Fluorosis in Karlakote

The author observed fluorosis in Karlakote is a note in a typical manner. Karlakote is a densely populated G.P. supported by 9 particular fluorosis village. The disease fluorosis was not observed in the till mid 1980. Then several tube wells were dug by the government. Which the author think as a turning point. In the deep soil of Karlakote fluoride compounds are found in excessive amount which comes in contact with the community. The author observed three category of fluorosis in the community ecosystem.

1. Primary fluorosis
2. Secondary fluorosis
3. Tertiary fluorosis

Primary Fluorosis

When an organism 1st ingest excess amount of fluoride water supported by F-vegetation, for a period of 10-15 years, primary fluorosis takes place. The symptoms are graying of hair, loss of visual power (decreasing), bended vertebral column, horizontal lines on teeth, curved foot, abdominal pain etc. The range of primary fluorosis is in between the age group of 40–70. It also advances the process of gerontology. Comparatively women are the most affecting in primary fluorosis.

Secondary Fluorosis

When a person ingested higher fluoride concentrated water and food since birth becomes a victim of secondary fluorosis.The symptoms are more severe in the age group of 20–40.They not only reported abdominal pain but also pain of bone. The total body pains them severely. They can not stand tough their chick, with the chest. Their fingers can not touch their toe in standing posture. There is loss of epidermal hair from most of their body surface. The most important problem for them is a socioecological one. As most of them gives birth to a handicapped child, the peoples of other community is neither interested to give their offspring to the village groom and nor to take bride from this community. They also reported unsatisfactory in copulation with their sex partner.

Tertiary Fluorosis

The new born of 1–20 age group which ingested high fluoride from surrounding water, F-group, F-vegetables, breast milk, and flesh of animal husbandry are the victim of tertiary fluorisis. 8 per cent children born physically handicapped. Most of them are not going to school and 8 per cent of school going children are bearing hump on their back. Their physical statue is bended. Dental fluorosis is common, the changes of lethality is maximum. Their poor parents are also not in a position to treat them for fluorosis.

Decomposer

Decomposer in general and particularly the mega decomposers of the community ecosystem. They also greatly affected by the higher fluoride concentration.

Samal, 1989 suggested that the population of earth worms are greatly disturbed by the fluoride concentrations. The population of earth worm is decreases significantly in the affected community than the normal one (Control pollution in 180 and affected 97–110). Their weight increases with the increasing fluoride concentration. The gut soil (Cast) shed off by them also content fluoride.

Thus it is crystal clear that the decomposer and transformers which are engaged in converting the dead and decaying material of producer and consumer again return the toxic element (residual fluoride) to the community which will be again alarming the situation by supplying fluoride nutrient to the community plant. It may be a reason due to why the vegetation in which leaf and stem are taken as food contain more of fluoride as they absorb fluoride in the form of nutrient in less time from soil then others.

Flow of Fluoride

Fluoride occurs in soil crust. It percolates the groundwater and till the water bodies through runoff. It is reported to be 11.73 ppm.which affect the community vegetation. The fluoride accumulation in rice is reported to be 35.5 ppm. It is found to be 0.42–0.83, 0.75–1.12, 0.82–1.49, 0.78–2.03 in the milk of sheep, goat, cow and bufallow respectively. The flesh of sheep and goat contain 0.56–1.13 and 0.68–1.46 respectively (Table 7.5). Ultimately in human brest milk 1.67mg/lt. Which again return to the atmosphere in a cyclic manner.

Management of Fluoride Pollution

According to Dr. Behera (2004) Fluorosis is preventable but not curable. The skeletal and dental changes may be arrested by drinking fluoridated water, but the changes are not reversible. Thus, the management include:

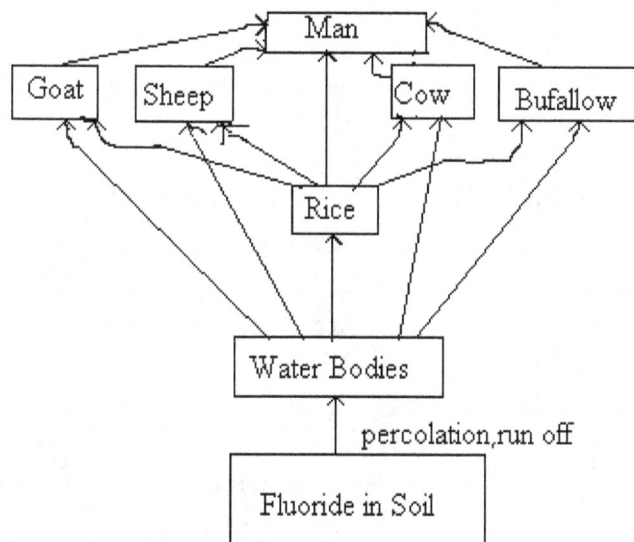

Figure 7.2: Flow of Residual Fluoride

1. Rehabitation of population.
2. Alternate water supply from lotic sources.
3. Defluoridation of water.
4. Roof top harvesting for drinking water.
5. Rain water harvesting for agriculture and animal husbandary.

Joshi *et al.* (1999) reported some modern technique for removing excess fluoride from groundwater. Such as:

Nalgonda Technique

This comprises mixing of aluminum sulfate(alum) and lime with raw water. The excess fluoride gets precipitated with the sludge precipitating at the bottom of the container.

Activated Alumina Technique

Activated alumina (Gamma aluminum oxide) has removed fluoride through adsorption.

Electrochemical Defluoridation

Electrolysis of fluoride rich water using an aluminum electrode as an anode results is the formation of aluminium hydroxide which binds the fluoride.

Ion Exchange

The method is based on the adsorption of fluoride ion on the surface of exchange materials, which is a resin.

Reverse Osmosis and Electro Dialysis

It is the desalination of water having high TDS, to bring it to potable standard.

Acknowledgement

The author is thankful to Prof..M. Shedpure, Govt. D.B. Girls P.G. College, Raipur (Chhatishgarh) Dr. B.P. Das, Govt. Autonomous College, Bhawanipatna. Prof. A. Mishra, and Mr. B.D. Patel, School of Life Sciences, Sambalpur University, Orissa for the suggestion and help tendered by them during the period of study.

References

Allen, S.E., Grim Shaw, H.M., Parkinson, T.A. and Quarmby, C., 1974. *Chemical Analysis of Ecological Materials*. Blackwell Scientific Publishers, Oxford.

Anonymous, 2004. Fluoride does not help teeth. *Internet.*

APHA, 1995. *Standard Methods for the Examination of Water and Wastewater*. American Public Health Association, Washington DC.

Athvale, R.N., 2003. *Water Harvesting and Sustainable Supply in India*. Rawat Publication, Jaipur.

Behera, M.K., 2004. Fluorosis. In: *Proc. Symposium on Flouride Pollution*. Bhawanipatna, January, pp. 11–16.

Brar, J.S., 2003. Adsorption of cadmium in selected sewer water irrigated soil of semi-arid zone. *Indian J. Env. and Ecoplan.*, 7(1): 115–118.

Cass, J.S., 1961. Fluorosis: A critical review. IV. Response of live stock and poultry to adsorption of in organic fluorides. *J. Occup. Med.*, 3: 471–477 and 527–543.

Dash, B.P., 2002. Toxicological effects of sumicidin on a fresh water fish *Sarotherodon mossambicus,* Peters and its ecological implications. *Ph.D. Thesis,* Berhampur University.

Halwachs, G., 1963. Investigation on directed active flow and material transport in leaf. *Flora,* 153: 358–372.

Hand, B.K., 1988. Fluoride occurrence in natural water in India and its significance. *Bhu-Jal News,* 3: 31–34.

Harichandan, N., Beura, B. and Panigrahi, A.K., 2003. Toxicological effects of cadmium chloride on fresh water fish *Tilapia mossambica,* peters and its ecological significance. *Indian J. Env. and Ecoplan.,* 7(1): 63–70.

Harvey, J.M., 1952. Chronic endemic fluorosis of Marino sheep in Queensland. *Queensland J. Agri. Sci.,* 9: 47–141.

Hendrickson, E.R., 1966. Dispersion and effects of air borne fluorides in Central Florida. *J. Air Poll. Control Association,* 11: 220–225.

Joshi, V.A., Nanoti, M.V. and Vaidya, M.V., 1999. *Base Paper on Control of Fluoride in Drinking Water.* Published by National Environmental Engineering Research Institute, Nagpur, India, pp. 1–81.

Jolly, S.S., Prasad, S., Sharma, R. and Raj, B., 1977. Human fluoride intoxication in Punjab. *Fluoride,* 4: 64–79.

Leone, N.C., Geever, E.I. and Moran, N.C., 1956. Acute and sub acute toxicity studies of sodium fluoride in animals. *Public Health Rep.,* 71: 459–467.

Mahapatra, M.K. and Mishra, A., 2003. Effects of limnological parameters on indigenous inland fishery diversity in western Orissa. *J. Curr. Sci.,* 3(2): 467–470.

Mahapatra, M.K., Mishra, A. and Mishra, H.S., 2003. Cooperative analysis of frequency, abundance, density and diversity of Indian major carp and available non carp fishes of western Orissa. *Indian J. Env. and Ecoplan.,* 7(3): 621–624.

Mahapatra, M.K., Mishra, A. and Dash, B.P(2004. Fluorosis first reported in Nuapada district of Orissa. *Ecology, Environment and Conservation,* No. 2 (In Press).

Mahapatra, M.K., 2004. Three dimensional analysis of lentic fish biology in ecobiology and economic dimension of fishes, (Ed.) Arvind Kumar. Daya Publishing House, Delhi (In Press).

Megregion, S., 1958. *Fluorine in Chemical Analysis,* 8th edn. Biotic D.F. Inter Science Publishers, New York, p. 231–259.

Mishra, B., 2004. Some facts about occurrence and contamination of fluoride. In: *Proc. Symposium on Fluoride Pollution,* Bhawanipatna, January, pp. 33–36.

Mohapatra, L.N., 2004. Groundwater pollution: Basic conceptual perspective. In: *Proc. Symposium on Fluoride Pollution,* Bhawanipatna, January, pp. 51–61.

Nagarajan, P., Ramachandramoorty, T., Raja, E., Kabita, B., Lakhmi, R. and Selvaraj, C., 2003. Analysis of fluoride level in drinking water and an attempt for defluoridation using lignite, wood carbon *cocosnucifera* (fiber) and *Vetiveria zizanoidoe. Indian J. Env. and Ecoplan.,* 7(2): 311–314.

Naik, B.N. and Samal, U.N., 1996. Dental fluorosis is domestic animal ground an aluminium factory. In: *Env. and Ecotoxicology Proc. 7th Annual Session of AEB and National Symposium on Man, Development, Bioresources and Environment,* H.G. Gaur University, Sagar, pp. 353–360.

Naik, B.N and Samal, U.N., 1987. Effects of fluoride pollution on cattle teeth at Hirakud, Orissa. *Environment and Ecology,* 5(1): 114–118.

Nivononee, C. and Sharma, B.M., 2003. Ecological study of the macrophytes of Ikop lake, Manipur: Morphometry and qualitative analysis. *Indian J. Env. and Ecoplan.,* 7(2): 243–250.

Palaniappan, P.R., Jagadeesan, G., Venkatachalan, P., Krushna Kumar, N. and Kartikeyan, S., 2003. Bioaccumulation and elimination of cadmium in freshwater finger lings, *Labeo rohita* (Ham). *J. Curr. Sci.,* 3(1): 191–195.

Patel, B.D., Mahapatra, M.K., Shedpure, M. and Mishra, H.S., 2004. Impact of fluoride on chloroplast content and growth of rice (*Oryza sativa* Linn) plant. *J. of Toxicology* (Accepted).

Pedini, B., 1967. Clinical observation of fluorosis in cattle. *Vet. Italiana,* 18: 23–26.

Rao, D.M. and Pal, D., 1979. The effects of fluoride pollution in cattle environmental pollution and toxicology (Symposium volume), p. 281–290.

Reddy, K.S., Prasad, K.S.S. and Raju, A.N., 1999. Physio-chemical property and fluoride content of termite moulds and their adjacent grand water of podili area, Prakasam District, A.P., India. *Pollution Research*, 18: 129–132.

Rosenberger, 1964. Investigation of fluorosis in cattle cause by emission of a hydrofluorosis acid plant. *Symposia on the Toxicology of Fluoride*, Bern October, 1962, pp. 144–146.

Samal, U.N. and Naik, R.N., 1988. Dental fluorosis in school children in the vicinity of an aluminium factory in India. *Fluoride*, 21(3): 142–148.

Samal, 1989. A study of fluoride pollution around an aluminium factory of Orissa, *Ph.D. Thesis*, Sambalpur University.

Sahu, S.K., 2004. Fluoride pollution: Its source, path, Toxicity and control. *Proc. Symposium on Fluoride Pollution*, pp. 5–10.

Selvaraj, 2001. Importance of environmental science. *Medical Geology: Current Science*, 79: 8–9.

Short, H.E., McRobert, J.R., Barnard, T.W., Manadi and Nagar, A.S., 1937. Endemic fluorosis in Madras presidency. *Indian J. Medical Research*, 25: 553–568.

Shupe, J.L. and Alther, E.W., 1996. The effects of fluorides on live stock with particular reference to cattle. In: *Handbook of Experimental Pharmacology*, (Eds.) Eicher, A. Faran, H. Herkeen, A.D. Weich and F.A. Smith. Springer-Verlag, New York, 20(1.0): 307–359.

Shupe, J.E. and Nielson, H.M., 1963. The effects of fluoride on dairy cattle v. fluoride is the urine as an estimator of fluoride intake. *American J. Vert. Res.*, 24: 300–306.

Singh, A., Jolly, S.S., Devi, P., Bansal, B.C. and Singh, S.S., 1962. Endemic fluorosis: An epidemiological, biochemical and clinical study in the Bhatinda district of Punjab. *Indian J. Med. Res.*, 50: 387–389.

Susheela, A.K. and Mukherjee, D., 1981. Fluoride poisoning and the effects on collagen biosynthesis of osseous and non-osseous tissues of rabbits. *Toxicological European Research*, 3(2):99–104.

Susheela, A.K. and Sharma, Y.D., 1982. Certain factors of F-action on collagen protein in osseous and non-osseous tissues. *Fluoride*, 15: 177–190.

Udall, D.M and Keller, K.P., 1952. A report on fluorosis in cattle in Colombia in river valley. *Cornell Vet.*, 42: 159–184.

World Health Organization, 1966. *Study of Crippling Fluorosis in the Development and Developed Countries.*

Chapter 8

Impact of Mercury Contamination on the Hydrobiology of Rushikulya Estuary of Bay of Bengal at Ganjam, Orissa

Alaka Sahu, S.K. Sahu & A.K. Panigrahi

Environmental Science Research Center, Department of Botany,
Berhampur University, Berhampur – 760 007, Orissa

ABSTRACT

Caustic soda industries occupy an important place so far as pollution of the aquatic and terrestrial environment with mercury is concerned. Keeping in view the role of these industries in polluting the surrounding environment, M/s Jayashree Chemical Ltd., Ganjam producing caustic-soda was selected as a study object. The industry has been releasing significant amount of mercury along with other pollutants as effluents. While the solid waste is dumped near by the factory as huge deposits, liquid effluent is discharged into the Rushikulya river estuary. Thus both the aquatic and terrestrial environments around the factory are being polluted by the industry. Hence, it was felt necessary to survey these areas to understand the mercury pollution problem, alteration of the natural estuarine environment.

Significant fluctuation in temperature was recorded at different stations studied in the Rushikulya estuary. Temperature of water is strongly dependent on the season. Higher temperature in summer months and lowest temperature in winter months was recorded in all the stations studied. The pH also varied at different stations studied. Highest pH was recorded in the effluent channel. Out of the rest four stations, Station-II showed the highest pH, when compared to the rest of the stations. The pH declined, at the downstream due to dilution by the tidal water. Highest suspended solid load was marked at the junction point, Station-II and IV. Heavy load of suspended solids at Station-II can be attributed to the effluent mixing at this point and high amount of suspended solids at station-IV can be attributed to the tidal wave influx and mixing at the estuary. The dissolved solids content was highest in Station-IV, when compared to other stations. The dissolved solids content was much higher in summer months, when compared to rainy months, due rainwater and floodwater. Lowest dissolved solids were recorded in the Station-I, which is located in the upstream. Variable BOD was noticed at all stations studied. BOD level was low in the upstream and high in the downstream. The junction point recorded the highest BOD level. Dissolved oxygen content was high at Station-I and Station-IV, when compared to other stations. The dissolved oxygen content indicated the productive nature of the ecosystem. Station-I and Station-IV is highest productive, when compared to other stations. The dissolved oxygen of water depends on temperature of water, hence, low dissolved oxygen was recorded in summer months and high dissolved oxygen was recorded in winter months at all stations studied. Highly variable values were recorded in all the stations studied. The junction point showed the highest sechi depth in some periods and station-I and station-IV showed the highest sechi depth in some months. The transparency varied with season. Variable salinity level was recorded in all the five stations studied. The station-IV showed the highest salinity due to tidal influx. During rainy months, the salinity decreased due to floodwater and in summer the salinity level was record high. Station-I showed the lowest salinity level in all seasons, except in summer, when the tidal water reaches up to station-I. Highest level of mercury was recorded in the effluent channel. A maximum of 0.3891 mg l⁻¹ was recorded, which is much more than the stipulated limit of 0.01 mg l⁻¹, set by Pollution Control Board. Except the effluent channel, Station-II showed highest mercury concentration. The upstream station-I showed mercury concentration, which is much less than the stipulated standard. Otherwise, the rest all stations showed higher mercury level much more than the stipulated level prescribed by the Pollution Control Board. Sediment samples showed variable levels at different stations. The value ranged from minimum 0.016 mg to maximum 489.62 mg Kg^{-1} dry weight. Except, station-E, station-II showed higher load of mercury in the sediment samples. Lowest load of mercury was noticed at station-I, which is located in the upstream. At different stations and sites in estuarine water and sediments the level of mercury concentration showed great variation. The concentration of mercury varied seasonally and also with the high tide times. However, very low values were obtained in rainy season, due to over dilution by rainwater and floodwater. Highest level of mercury in both water and sediment was obtained at the junction zone where effluent channel meets the river water. A considerable stretch of the estuary was contaminated as revealed from bed sediment studies. The present study indicated the poor status of hydrobiological condition of the Rushikulya River and Rushikulya estuary, which also indicated that in future an incident similar to Minamata Bay incidence is not far away.

Keywords: Hydrobiology, Physico-chemical parameters, River, Estuary, and Mercury.

Introduction

Contamination of the environment is considered a sequel to ecosystem development, and to a great extent, it is purified by the nature itself. It is the excessive contamination that creates environmental hazards. The secondary manifestations of human activity, commonly referred to as contamination or pollution, are the basis for the emergence of a new discipline of science, 'eco-toxicology', which combines a whole group of scientific field relating to the study of environment. The pollutants act in many ways on the living systems. The effect of sub-lethal concentrations of some pollutants is by far the most frequent eco-toxicological problem and it can be far more noxious in the long run for the exposed species resulting in chronic deterioration of the ecosystem. The persistent non-biodegradable pollutants are absorbed by living organisms and are concentrated many times more than the concentration of the surrounding medium. Further these bio-accumulated pollutants are passed from one trophic level to another and exhibit increasing concentration in organisms related to their trophic status resulting in bio-magnification. Industry is responsible for creating a fantastic array of new chemicals every year, all of which eventually find their way into the environment. For most of these chemicals, not even the chemical formulae are known and much less are known about their acute, chronic or genetic effect on the living organisms. At present, the industry is the focus of attention, the world-over, as the most important source of pollution of the environment. Chemical industry in India has grown up phenomonally since independence. There are to-day about 4000 chemical factories in India. They release large quantities of chemicals in the form of gas, liquid and solid wastes, into the environment. Many of these chemicals are toxic and create pollutional problem. The problem of toxic hazard has already reached alarming proportions in this country and is bound to grow with increasing industrialization. The wastes from the industry are generally disposed by land filling or released into water bodies. Generally the toxic effluents from industry are neutralized before their discharge, but still they contain substantial amount of toxic substances that can cause pollution. At present it is believed that rivers are most severely polluted by industries followed by estuaries, lakes and ocean in declining order. The chlor-alkali industry in India has grown ten fold since 1947. Large quantities of caustic soda and soda ash produced are used in Paper and Rayon industries. Majority of the chlor-alkali factories adopt old process of using mercury as cathode. This has long since been discarded in Japan and in many other countries after the Minamata Bay mercury-poisoning episode. The country imports around 200 tons of mercury every year of which 180 tons are used by the chlor-alkali industry alone and most of it finally goes out along with wastewater and sludge, and ultimately enters into the environment (CSE report, 1985).

Mercury is a unique pollutant because of its apparent indestructibility and ubiquitous ness. It attracted worldwide attention as a potential pollutant because of a tragic episode, which occurred in the Japanese villages surrounding the Minamata Bay. The victims were struck with a mysterious nervous illness commonly called "Minamata disease" which was later shown to be caused by consuming fish and shell fish that has lethal accumulation of methyl mercury (Takeuchi, 1961). It resulted in 121 cases of poisoning and 41 death. A search ultimately traced the source to be the waste discharge containing high amounts of methyl mercury and other organic mercurials from industries using mercury as catalyst in the manufacture of vinyl chloride and acetyladehyde. Another similar incident of mercury poisoning was reported again in Japanese villages near Niigata prefecture during 1969, which left 6 people dead with 47 poisoned cases. The catastrophic mercury poisoning in Iraq, where a record number of 6530 persons were hospitalized, of whom 459 persons died. This was caused by exposure to methyl mercury, when people ate home made bread prepared from wheat seed, which had been treated with methylmercurial fungicide. Mercury appears in industrial discharges in five principal forms, divalent mercury (Hg^{2+}), metallic mercury (Hg^{0}), phenyl-mercury (C_6H_5Hg), alkoxyalkyl-mercury ($CH_3O–CH_2CH_2Hg^+$) and methyl-mercury (CH_3Hg^+) of which methyl-mercury is the most toxic form. Amalgamation of mercury with electrolytically formed sodium is the key factor in Chlor-alkali industry. The chlor-alkali industry uses 25 per cent of the total mercury consumption of our country. Out of 38 chlor-alkali plants in India 23 units are based on mercury cell electrolysis, in which a flowing mercury cathode is used. The chlor-alkali industry uses the continuous flow mercury cathode cell to produce chlorine and caustic soda. The 1968 figures indicate that over 1.3 million pounds of mercury or 23.1 per cent of the total consumption in the U.S. was used to produce over 8.4 million tons of chlorine. The percentage of chlorine and caustic soda produced by the continuous mercury cathode cell has steadily increased from 4 per cent in 1946 to over 12 per cent in 1956, to 21 per cent in 1963 and swelled to more than 28 per cent by 1968. Production was switched from diaphragm cells to mercury cells because of the demand for a purer caustic soda–purer in the sense of having less chloride ions. Mercury losses from the mercury cell process have been estimated on the order of 0.4 lb of mercury for each ton of chlorine so produced. In the chlor-alkali industry, mercury can be lost during several phases of the process through the caustic soda product, through the overall cleaning of the mercury cell room, through the brine solution, through the hydrogen gas by-product, through the brine saturation and through the air by volatilization. The quantity of mercury that ultimately reaches the environment depends largely upon the overall operating efficiency of the particular mercury cathode cell being used.

Chlor-alkali industries are considered as the most significant source of mercury in the environment (Hortung and Dinman, 1974; Suckcharoen and Nourteva, 1982). India imports around 200 tonnes of mercury every year of which 180 tonnes are used by the Chlor-alkali industry alone. Larsson (1970) has reported that in 1967, 15-19 metric tonnes of mercury were discharged into the air from Swedish Chlor-alkali factories. Chlorine, caustic soda and potash containing mercury are used in other industrial processes and products, thus the chlor-alkali industry gained importance as a potential source of mercury in the environment. The hydrogen gas produced during preparation of chlorine by Hg-cell process also contains mercury. Without proper monitoring, as much as 50 lb of Hg may be emitted per 100tonnes of chlorine produced. Van Wamboke (1976) gave a detail report of chlorine produced and mercury released for some of the countries. Though a detailed report is not available with regard to the loss of mercury from chlor-alkali plants, an official report states that when 180tonnes of mercury are released into the environment, out of which, as high as 166tonnes are contributed exclusively from chlor-alkali industries. In India, mercury contamination due to chlor-alkali industry is local and also alarming. Although incident of mercury poisoning similar to Minamata disease has not been reported in India, this may be only because nobody has yet looked for it. Indian scientists are well aware that the mercury load in estuaries and other water sources near chlor-alkali, paper and Rayon plants, on an increase, giving sufficient indication of the hazardous consequences (CSE report, 1985).

The chlor-alkali industry (M/s Jayshree Chemicals Ltd.) presently under study is situated on the bank of Rushikulya river estuary in Ganjam, 30 Km away from Berhampur city. The river originates from the Rushimal hills situated on the north of Ganjam. The factory started production during August 1967 and since then it is continuously discharging its effluent (Wastewater) into the estuary at the rate of 50,000litres per day. The rated capacity of the factory is 50tonnes per day. Initially, the electrolytic cells of the industry were charged with 54tonnes of mercury and there after it consumes approximately 3 tonnes of mercury per year. The solid waste (Brine mud) from the factory is periodically removed and dumped in pits adjacent to the effluent channel of the industry or dumped nearby as huge deposits. The liquid effluent (Wastewater) of the industry is continuously discharged through a channel that opens into the river estuary. Generally the effluent is colourless with slight turbidity, but periodically it looks milky white in colour due to the discharge of mercury cell washing. The physico-chemical characteristics of the effluent have been reported at different times by Shaw (1987) which show much variation in pH, alkalinity, chlorinity, dissolved oxygen, total nitrogen, total phosphorus, biological oxygen demand (BOD), chemical oxygen demand (COD) and the mercury content. Shaw (1987) has reported that physico-chemical characteristics of the effluent after storing for three months have not changed significantly. In the present piece of study the physico-chemical properties of the effluent was characterized by mixing the effluents of three consecutive mercury cell washings collected at different points of the discharge channel and analysed in the laboratory. The greyish white solid waste (brine mud), which is dumped in pits in the nearby field contain large quantity of mercury. Shaw (1987) reported mercury in the sediment of effluent channel to be fluctuating between 456.67 to 2053.3 mg/Kg^{-1} dry wt. The liquid effluent and the brine mud once measured by the "Central Board for Prevention and Control of Water Pollution" contained 0.5 mg 1^{-1} and 2.4 g/Kg^{-1} dry weight of mercury respectively. The minimum national standard (MINAS) for effluents of chemical industry, fixed for the discharge of Hg from a chlor-alkali industry is 0.01 mg 1^{-1} in the effluent and the total mercury discharged from a 100tonnes/day capacity caustic-chlorine factory should not exceed 1.5 gram per day. A review of work done on this chlor-alkali industry indicates that it is seriously polluting the surrounding environment, both terrestrial and aquatic. Shaw (1987) carried out a critical study pertaining to mercury pollution by this chlor-alkali plant. Both terrestrial (Shaw *et al.*, 1986 and Shaw and Panigrahi, 1986 a, b) and aquatic (Shaw *et al.*, 1989) habitats were heavily contaminated by mercury. The authors have expressed their concern to a potentially dangerous situation. The solid and liquid wastes of the industry have also been critically studied for their supposed toxicity. Thus, the over all review produces a gloomy picture regarding the habitats and the inhabitants of the adjoining areas of the industry. Studies on loss of mercury into the environment from the chlor-alkali industry were carried out by Flewelling (1971), Bouveng (1968), and Shaw *et al.* (1986).

Rushikulya Estuary–Type Study

The Indian sub-continent is bordered by the Arabian Sea on the West Coast and the northern Indian Ocean at the South coast, and the Bay of Bengal on the East Coast. The Bay of Bengal is situated between 0-23°N latitude and 80-100°E longitude. It receives about 2329 Km^3 run off, which carries about 14 x 10^8 tonnes of silt from different rivers flowing through different parts of India, Bangladesh, Burma and Sri Lanka. The exclusive economic zone on the East Coast of India covers an area of about 2 million km^2. Majority of the 220 different major, medium and minor rivers flowing through India, carry enormous amount of freshwater inundation and introduction of silts along with all types waste materials that severely alters the natural water quality and profoundly affects the biological diversity of marine organisms. The seawater mass of the coastal region is characterized by the presence of warm and low saline water at surface layers (40-60 m depth) and cold and high saline water at the sub-surface layers. The tides are of semidiurnal type with two floods and two ebbs of varying amplitude. The east coast of India has a coastline stretching about 3010 km and comprises of sandy, muddy and rocky shores. However, sandy beaches are more frequent than the other two beaches/shores. The entire coast line between Kalingpatnam (18° 21′ N Latitude and 84° 08′ E Longitude) and Chandrabhaga (19° 52′ N Latitude and 86° 08′ E Longitude) is represented by a stretch of sandy beach without being interrupted by either muddy or rocky shores. This stretch of the east coast beach provides an excellent surf-zone, which facilitates for ideal ecological research and a best beach for the tourists. There are many small and big river inlets into the sea creating good and biologically active estuaries and this stretch also contains small coastal lagoons. The Rushikulya river estuary under the present investigation comes in between the stretch of east-coast surf-zone beaches. Mankind knows estuaries since the dawn of civilization. The scientific study in these environments began only around 1900 AD onwards with the survey of British estuaries. Extensive studies dealing with different physical, chemical, biological, geological, productivity and pollution studies of different estuaries of the world have been made during the past five decades by different scientists belonging to different laboratories, institutes and Universities of different countries. More attention was focused on the estuaries for the peculiar habitat of the estuary, biodiversity pattern, periodic availability of few species during a particular time period. Ecologically, the estuary became more important and significant due to the reason that the estuary enjoys three different distinct zones, namely–a fresh water zone, a mixture or buffer zone and a saline zone or marine water zone. The salinity gradient is probably more important in estuary than any other parameter because this salinity plays a crucial role in biodiversity.

Much work has been done on some aspects of the estuary. A wealth of information is available on the physico-chemical nature of the estuary, phytoplankton productivity studies. The status of estuarine research has been reviewed periodically through several national and international bodies, associations and in conferences. Many textbooks, review articles, research proceedings are now available on estuaries and estuarine ecosystems of the world. Very little information is available on pollution studies of the estuaries. Most of the industrial wastes, domestic wastes, city sewages, effluents etc. are discharged into fresh water bodies like canals, nallahs, small, medium and big rivers. The fresh water of the aquatic bodies were drawn and used by the industries and the effluents and wastes are thrown into fresh water bodies either after treatment or before treatment. These freshwater aquatic systems suffer because of these liquid or solid wastes. In the down stream, these contaminated waters flows down to estuary and ultimately enter into the sea or ocean.

Hydrobiological Studies of the Rushikulya River Estuary

Hydrographical properties of estuaries considerably affect the biotic communities and govern various chemical and geological processes. Much work has been done on the hydrography of different estuaries of the world. In India Annandale and Kemp (1915) and Sewell (1929 and 1932) initiated the hydrographical observations in Chilika Lake, a brackish water lagoon situated on its east coast. A lot of information on the spatio-temporal variations of various hydrographical parameters like tides, transparency, temperature, salinity, dissolved oxygen and pH of different estuaries were available. Informations are plenty in this line. We hand picked few to explain and discuss with our findings.

Tide is one of the most important requisite that governs the spatio–temporal distribution of various physico-chemical and biological characteristics in the estuarine environment. The estuaries in India are affected by semidiurnal tides comprising of two flood and two ebbs of unequal amplitudes in a tidal day. Marked variations, as regards to the height of tides at different places and in the time of occurrence of high and low waters, are also seen along its coasts. Some of the important observations with regard to the tidal fluctuations in estuaries are given here. Qasim and Gopinathan (1969) and later on Joseph and Kurup (1990) have studied the tidal variations in Cochin Backwaters. Their observations indicated considerable variations relating to tidal heights at different locations and different seasons. Zingde and Desai (1987) have summarized tidal fluctuations in a 9 different estuaries of Gujarat Coast. In all these estuaries a progressive decrease was discernible from the mouth towards the head region. The tidal levels in Godavari, Cauvery and Narmada river estuaries varied between 0.53–1.34m, 0.27-0.62m and 4.02-6.43m respectively. On the east coast, tidal variations in Vellar estuary have been well documented (Vijayalakshmi and Venugopalan, 1973; Chandran and Ramamoorthi, 1984a,b). A variation of 0.15-1.45m was reported in the gradient zone of this estuary. Keeping in view of the above observations and findings, survey of literature, the present piece of work was designed to study the hydrographical parameters of Rushikulya river, the junction point where the effluent of the chlor-alkali industry joins the river, two more locations where the leachates of the solid waste deposit of the chlor-alkali industry enter into the river and a site near the estuary.

Materials and Methods

Measurement of Hydrological Parameters

The water transparency characteristics of each station was measured by the Secchi disc method. A minimum-maximum centigrade thermometer with 0.1°C accuracy was used for measuring both the air and water temperature. For salinity measurements, the samples were brought to the laboratory and analysed colorimetrically following the procedure of American Public Health Association (APHA, 1984). Dissolved oxygen concentration was estimated by modified Winkler's method as described in APHA (1984). The obtained values of oxygen were presented as mg of Oxygen l^{-1}. The pH of water samples were measured with the help of a portable field pH meter (Systronics, India). Suspended solids content of the water samples was determined by filtering a known volume of the water sample through a Whatman filter paper No.40, previously weighed in an electric single pan balance (Dhona make), and then the filter paper was oven dried and weighed. The difference in weight was due to the suspended solids present in the known volume of water. Dissolved solids content of the water sample was estimated after filtration of the samples through a whatman filter paper. A known volume of the filtrate was taken in a pre-weighed beaker and oven dried. After complete drying, the weight of the beaker was taken. The difference in weight gives the amount of dissolved solids present in the known volume of sample taken. The data was calculated with proper corrections. The level of BOD was estimated following the methods described in Analytical Methods Manual (APHA, 1984). In the present study, two micronutrients, *viz.*, nitrate and phosphate were estimated. For all nutrients, the methods described by Strickland and Parsons (1972) were adopted.

Methods of Studying Aquatic Contamination

Pollution of the estuary all over the world has been a subject of greater concern. Taking into account the large water bodies and natural phenomenon such as temperature, precipitation, flow of river etc. a year long study was carried out to know the degree of contamination. Four sites along the estuary were earmarked as study stations. One at the junction (Station II, place of union of the effluent channel with the estuary), one at upstream (Station I) and two at down stream (Station III and IV). All the four stations were at a distance of about 0.75 to 1 km from each other. The effluent channel was taken as fifth station of study (Station E). Water and sediment samples were collected at those sites monthly for the analysis of mercury levels. Samples were collected in triplicate. To study the concentration of mercury in bed sediment of the estuary, sediment samples were collected from 10 points, covering all the study station. Samples were collected only once during low tide in every month to avoid mixing of saline waters from sea. Eight types frequently available fishes of different age groups were collected during catch and analysed. Water samples from different stations were collected in pre-washed 500 ml plastic bottles. Soil samples from different stations of study were collected in plastic packets. These were brought to the laboratory, dried in air and then in oven at 35°C, ground to powder with the help of a mortar and pestle and kept separately for further analysis. Measurement of mercury in the samples followed the basic principle of Wanntorp and Dyfverman (1955), which has undergone substantial modification in the light of recent development. The process is described in "Analytical Methods for determination of Hg with Mercury Analyser MA 5800 A" issued by ECIL, 1981.

Results

Figures 8.1 shows the monthly variation of temperature from January 2002 to December 2002 in different stations studied. The temperature showed wide variation in different months, which is strongly seasonal and also variation was recorded in different stations of the same month. The temperature varied between 19.4°C to 31.4°C in station-I, which is located at upstream. Highest temperature was recorded in June (Summer months, pre monsoon period) and the lowest was recorded in January (late winter). The temperature recorded at station-II was higher when compared to Station-I, III and IV. Because station-II was the junction zone, where the effluent from the chlor-alkali industry joins river Rushikulya. The temperature of the effluent channel was highest, when compared to all other stations in every month assessment. The temperature varied between 22.6°C to 32.6°C in station-II, which is located at the junction point. Highest temperature was recorded in June (Summer months, pre monsoon period) and the lowest was recorded in December (winter). The station-II showed higher temperature when compared to stations-I. Station-III was located at the down stream after station-II. The temperature recorded at this station was higher than Station-I and lower than station-II. The temperature varied between 20.2°C to 30.5°C in station-III, which is located at the down stream. Highest temperature was recorded in July (late summer months, pre monsoon period) and the lowest was recorded in January (winter). Station-IV was located nearer to the estuary and after station-III, where a mixture value of temperature was recorded. The temperature varied between 20.6°C to 27.2°C in station-IV, which is located at the estuarine point. Highest temperature was recorded in August (pre monsoon period) and the lowest was recorded in December. Station-E indicated a station in the effluent channel. The temperature varied between 27.4°C to 38.8°C in station-E, which is located at the effluent channel. Highest temperature was recorded in May and the lowest was recorded in January (Figure 8.1).

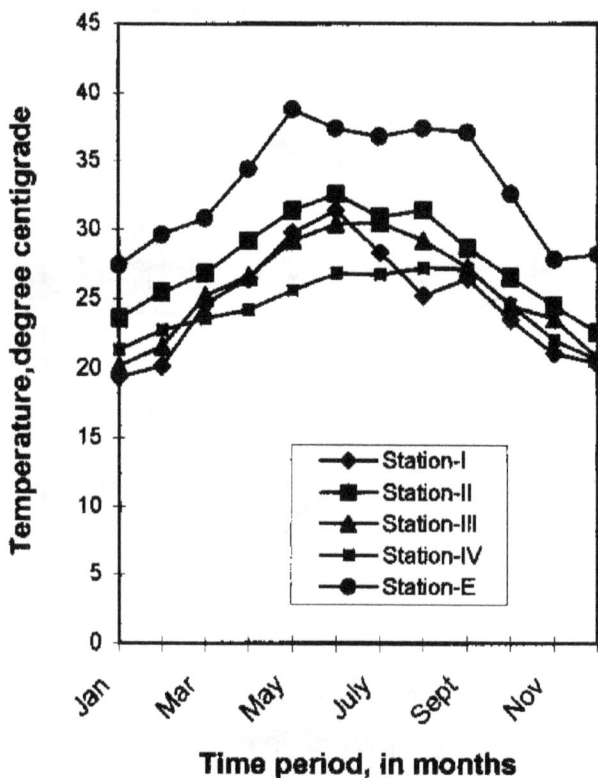

Figure 8.1: Monthly Variation of Temperature of Water Samples at Different Stations Near Rushikulya Estuary

Figure 8.2 indicated the variation in pH in different months in the 5 stations studied. Significant variation in pH level was marked at all the five stations studied. Strong monthly variation in the parameter was observed. The pH of water at the upstream, at Station-I varied within 7.1 to 7.8. No regular trend was marked in pH change. Though variation in the parameter was noted but the change cannot attributable to monthly variation. High pH (7.8) in January and low pH (7.1) in March, July and September cannot be explained. The high pH might be due to the tidal influx at the estuary, which might have carried the effluent water towards the upstream. Otherwise the pH was stable at all other months observed. Station-II showed the

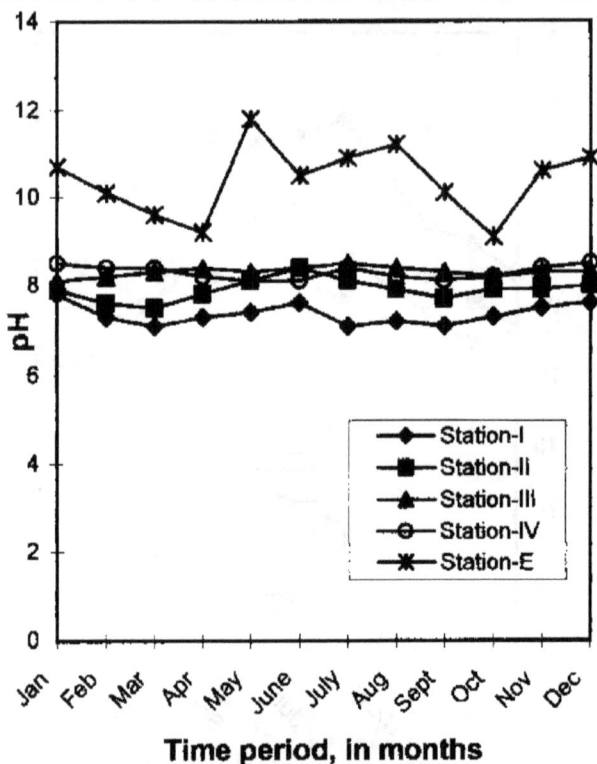

Time period, in months

Figure 8.2: Monthly Variation of pH at Different Stations Located Near the Rushikulya Estuary from January 2002 to December 2002

highest pH in June (8.4) and lowest pH (7.7) in September. Station-III showed higher pH values in all months studied, when compared to Station-I and Station-II. Station-III showed the highest pH in July (8.5) and lowest pH (8.1) in January. Station-IV showed lower pH values in all months studied, when compared to Station-I, Station-II and Station-III. Station-III showed the highest pH in December and January (8.5) and lowest pH (8.1) in May and August. Station-E showed the highest pH 11.9 in July (late summer) and lowest pH in 9.1 in October (Rainy month). The pH in this station was highest in all months, when compared to the rest of 4 stations (Figure 8.2).

Figure 8.3 showed the monthly variation of suspended solids in different stations studied. Lowest suspended load was marked in

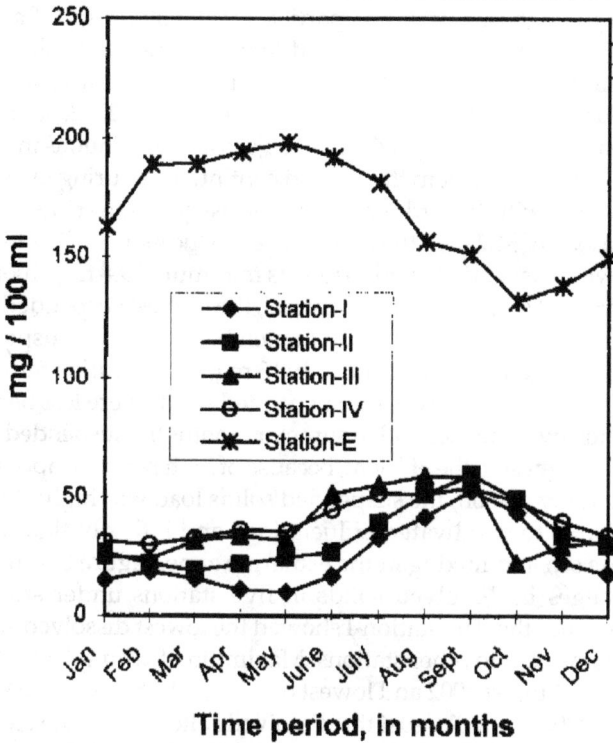

Figure 8.3: Monthly Variation of Suspended Solids in Different Stations at Rushikulya Estuary from January 2002 to December 2002

Station-I. The suspended solids (mg/100ml) varied between 9.6 mg/100ml in the month of May (summer month) to 52.9 mg/100ml in the month of September (rainy month). Heavy loading of suspended particles at Station-I at the upstream in the month of September was due to floodwater coming in the river along with suspended particles. In summer months, the suspended particle load was low due to slow moving water in the river. At station-II, moderate levels of suspended solid load was noticed. Station-II was the junction zone, where we should expect still higher load of suspended particles, but lower value was noted. Highest suspended particle load of 59.8 mg/100ml was noted in the month of September and lowest suspended solid load was minimum 18.6 mg/100ml in March. In station-III, highest suspended particle load of 59.2 mg/100ml was noted in the month of August and lowest suspended solids load was minimum 23.4 mg/100ml in February. Highest suspended solids load was marked between June to September 2002. Station-IV was nearer to the estuary, which is subjected to continuous influx of periodic tidal wave from the sea and continuous pouring in of river water along with the effluent of the industry. Highest suspended particle load of 54.2 mg/100ml was noted in the month of September and lowest suspended solid load was minimum 29.8 mg/100ml in February 2002. In the effluent channel, the highest suspended load was marked in between July to September 2002. The suspended solid load varied between 131.6 to 198.3 mg/100ml during the entire year. During rainy season, the suspended solids were less because of dilution by rainwater and in summer months the suspended solid load was highest in the effluent, because of high rate of evaporation. During rainy season, the suspended solids load was highest in the estuary due to floodwater, effluent load and influx of tidal water from the sea and mixing in the estuarine zone. Figure 8.4 showed the changes in dissolved solids in five stations under study in different months. The station-I showed the lowest dissolved solids, when compared to other stations. Maximum of 1.9 mg/100ml was marked in August 2002 and lowest of 0.6 mg/100ml was marked in January 2002. From August the dissolved solids decreased steadily up to January and March then slowly the dissolved solids increased gradually till August. This showed a trend with the season. In case of station-II, highest dissolved solids was noted to be 2.9 mg/100ml in June and lowest of 1.2 mg/100ml October to January. At station-III, the dissolved solids varied from 1.0 to 4.2 mg/100ml, which is located at the downstream after the junction point. In station-IV, the dissolved solids varied from 0.9 mg/100ml in September 2002 to 4.8 mg/100ml in May 2002 (Figure 8.4). Figure 8.5 shows the changes in biological oxygen demand at different stations in different months from January 2002 to December 2002. The BOD value ranged from 10.1 mg l^{-1} in the month of June to 14.5 mg l^{-1} in the month of September and December in Station-I, which is located in the upstream. In Station-II, the BOD value ranged from 18.4 mg l^{-1} in the month of August to 41.1 mg l^{-1} in the month November, which is located at the junction point. The BOD value ranged from 15.2 mg l^{-1} in the month of February to 20.2 mg l^{-1} in the month of July in Station-III, which is located in the downstream. The BOD value

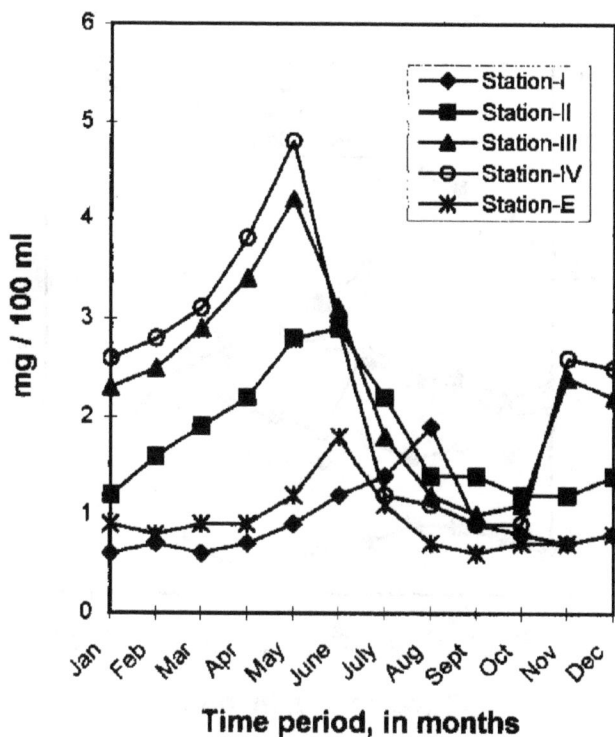

Time period, in months

**Figure 8.4: Monthly Variation of Dissolved Solids in Samples
Collected from Different Stations of the Rushikulya Estaury from
January 2002 to December 2002**

ranged from 13.2 mg l^{-1} in the month of November to 19.4 mg l^{-1} in
the month of May in Station-IV, which is located in the downstream
but nearer to the estuary. The BOD value ranged from 2.14 mg l^{-1} in
the month of August to 6.24 mg l^{-1} in the month of November in
Station-E, which is located in the effluent channel. The BOD value
was lowest in the effluent channel, when compared to the rest four
stations studied. Highest BOD value was recorded at Station-II and
III. Figure 8.6 indicate the changes in dissolved oxygen content of
water at different stations in different months. The dissolved oxygen
content varied from 2.1mg l^{-1} to 5.4mg l^{-1} between January to
December 2002, in Station-I. Highest dissolved oxygen was recorded
in the month of October and the lowest dissolved oxygen was

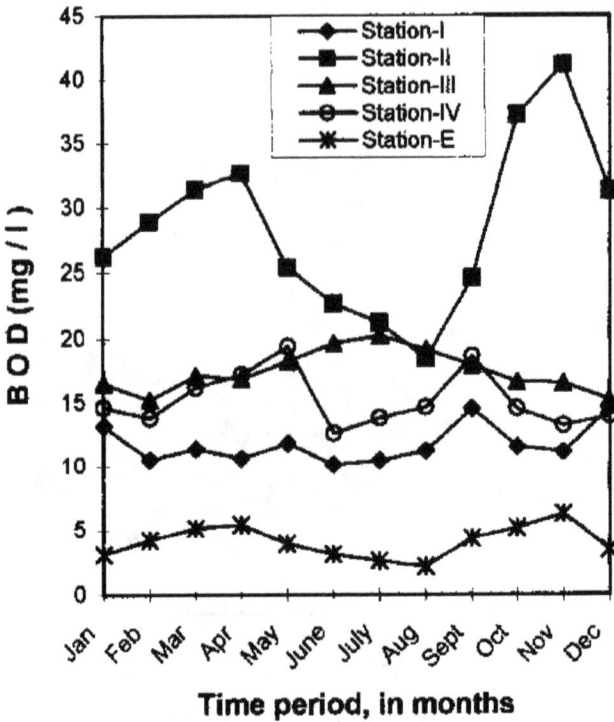

Figure 8.5: Monthly Variation of Biological Oxygen Demand (BOD) from Samples Collected from Different Stations of Rushikulya Estaury from January 2002 to December 2002

recorded in the month of April. The dissolved oxygen varied with the change in water temperature. In winter months the dissolved oxygen content was high and in summer months the dissolved oxygen was low. The dissolved oxygen content varied from 1.2 mg l⁻¹ to 4.4mg l⁻¹ between January to December 2002, in Station-II. Highest dissolved oxygen was recorded in the month of November and the lowest dissolved oxygen was recorded in the month of April. The dissolved oxygen varied with the change in water temperature. In winter months the dissolved oxygen content was high and in summer months the dissolved oxygen was low. The dissolved oxygen content varied from 2.2 mg l⁻¹ to 5.2 mg l⁻¹ between January to December 2002, in Station-III. Highest dissolved oxygen was recorded in the

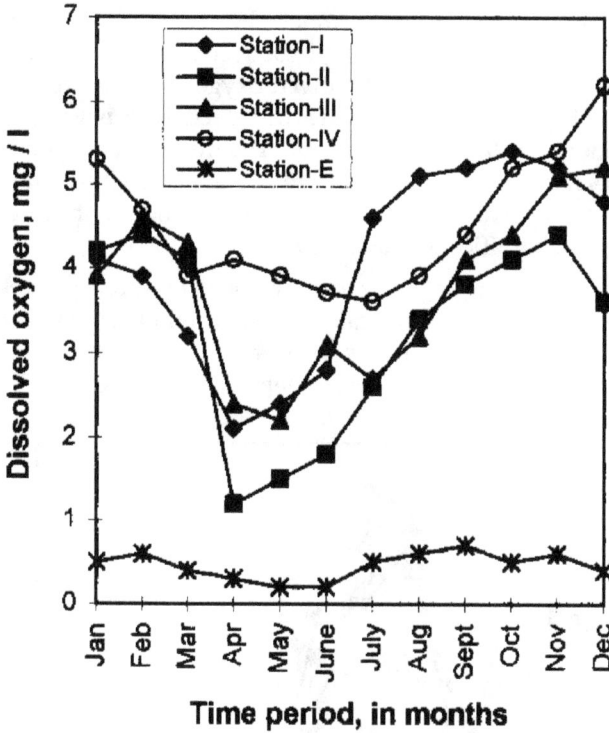

Figure 8.6: Monthly Variation of Dissolved Oxygen in Samples Collected from Different Stations of Rushikulya Estuary from January 2002 to December 2002

month of December and the lowest dissolved oxygen was recorded in the month of May. The dissolved oxygen varied with the change in water temperature. In winter months the dissolved oxygen content was high and in summer months the dissolved oxygen was low. The dissolved oxygen content varied from 3.6 mg l^{-1} to 6.2 mg l^{-1} between January to December 2002, in Station-IV. Highest dissolved oxygen was recorded in the month of December and the lowest dissolved oxygen was recorded in the month of July. The dissolved oxygen varied with the change in water temperature. In winter months the dissolved oxygen content was high and in summer months the dissolved oxygen was low. Among all stations studied, the Station-IV was the most productive and active station, which is located nearer to the estuary. Lowest dissolved oxygen was recorded in the effluent of the effluent channel. The dissolved oxygen content varied from 0.2 mg l^{-1} to 0.7 mg l^{-1} between January to December 2002, in Station-I. Highest dissolved oxygen was recorded in the month of September and the lowest dissolved oxygen was recorded in the month of May and June 2002. The dissolved oxygen varied with the change in water temperature. In winter months the dissolved oxygen content was little high and in summer months the dissolved oxygen was significantly low.

Figure 8.7 indicated the measurement of transparency by Sechhi disk depth at four stations located in the Rushikulya river estuarine zone. The water depth was very less at many times during measurement; hence Sechhi disk depth was not measured in the

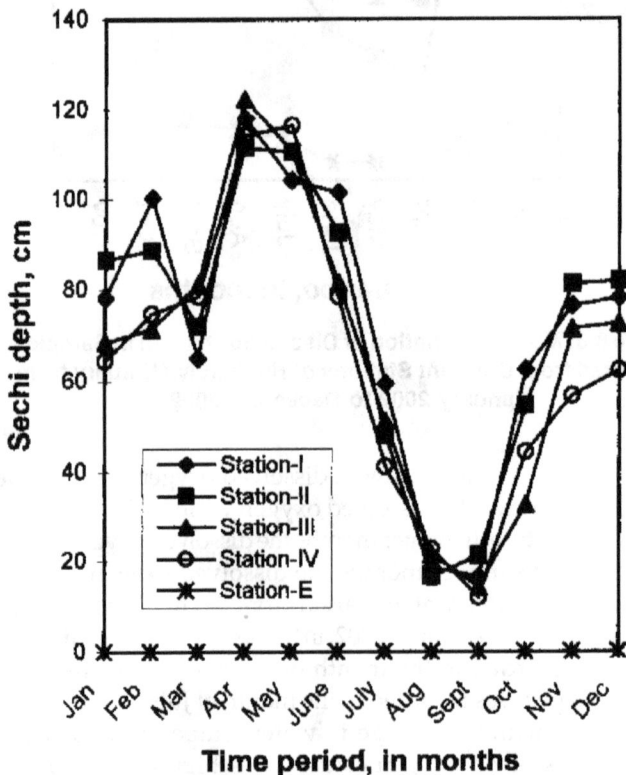

Time period, in months

Figure 8.7: Monthly Variation of Transparency by Sechhi Disk Method in Samples Collected from Different Stations from January 2002 to December 2002

effluent channel. In case of Station-I, the sechhi depth varied from 16.5 to 118.0cm in the entire year. Highest depth was recorded in the month of April and lowest depth was recorded in the month of September 2002. In case of Station-II, the Sechhi depth varied from 16.5 to 111.4 cm in the entire year. Highest depth was recorded in the month of April and lowest depth was recorded in the month of August 2002. In case of Station-III, the Sechhi disk depth varied from 14.2 to 122.4 cm in the entire year. Highest depth was recorded in the month of April and lowest depth was recorded in the month of September 2002. In case of Station-IV, the sechhi depth varied from 11.8 to 116.5 cm in the entire year. Highest depth was recorded in the month of May and lowest depth was recorded in the month of September 2002. The analysis of variance ratio test indicated the existence of significant difference between stations and significant difference between months studied from January 2002 to December 2002. Figure 8.8 indicate the changes in salinity ($^0/_{00}$) level in four stations studied. The salinity level of the effluent channel was not estimated. Significant variations were marked in the salinity content of different stations in different months. Higher salinity levels were recorded in the summer months and lowest salinity levels were marked in the rainy months at all the stations studied. At station-I, the salinity level ranged from 1.8 to 46.8 $^0/_{00}$ in the entire period of study. Lowest salinity level was in the month of August 2002, where 1.8 $^0/_{00}$ was recorded and highest salinity was in the month of May 2002 and the salinity level was 46.8 $^0/_{00.}$ The salinity level declined with rainfall and flow of floodwater and the slowly the salinity level increased with evaporation and non-flow of rainwater. At station-II, the salinity level ranged from 32.4 to 288.2 $^0/_{00}$ in the entire period of study. Lowest salinity level was in the month of October 2002, where 32.4 $^0/_{00}$ was recorded and highest salinity was in the month of May 2002 and the salinity level was 288.2 $^0/_{00.}$ The salinity level declined with rainfall and flow of floodwater and the slowly the salinity level increased with evaporation and non-flow of rainwater. This station was the junction point where the effluent of the industry joins the river. Hence, higher salinity level was noted. At station-III, the salinity level ranged from 6.5 to 385.1$^0/_{00}$ in the entire period of study. Lowest salinity level was in the month of September 2002, where 6.5$^0/_{00}$ was recorded and highest salinity was in the month of May 2002 and the salinity level was 385.1$^0/_{00}.$ The salinity level declined with rainfall and flow of floodwater and the slowly the

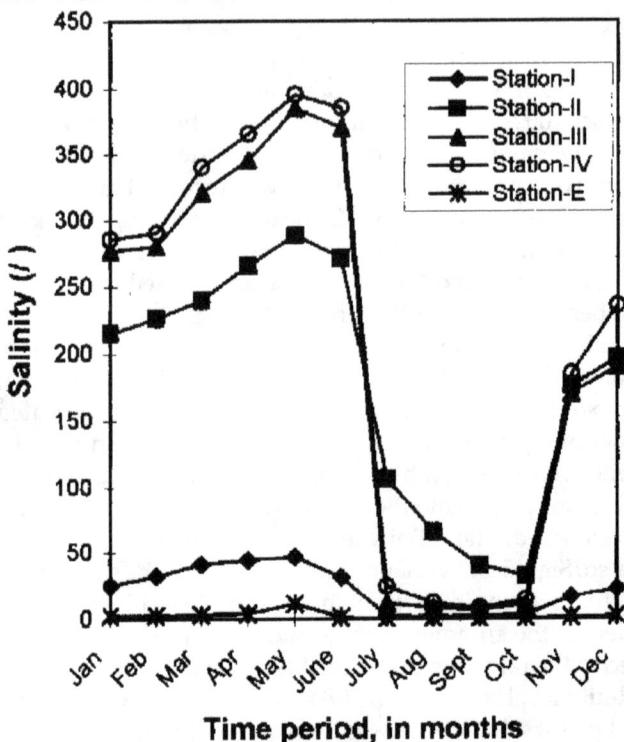

Figure 8.8: Monthly Variation of Salinity at Different Stations of the Rushikulya Estuary from January 2002 to December 2002

salinity level increased with evaporation and non-flow of rainwater. At station-IV, the salinity level ranged from 8.2 to $395.2^0/_{00}$ in the entire period of study. Lowest salinity level was in the month of September 2002, where $8.2^0/_{00}$ was recorded and highest salinity was in the month of May 2002 and the salinity level was $395.2^0/_{00}$. The salinity level declined with rainfall and flow of floodwater and the slowly the salinity level increased with evaporation and non-flow of rainwater. Station-IV was nearer to the estuary, where seawater influx helped to raise the salinity level.

Figure 8.9 indicate the monthly changes in mercury concentration in surface water samples at different stations of the estuary. In case of Stations-I, mercury was not detectable from the month of July to October 2002. From November onwards, mercury

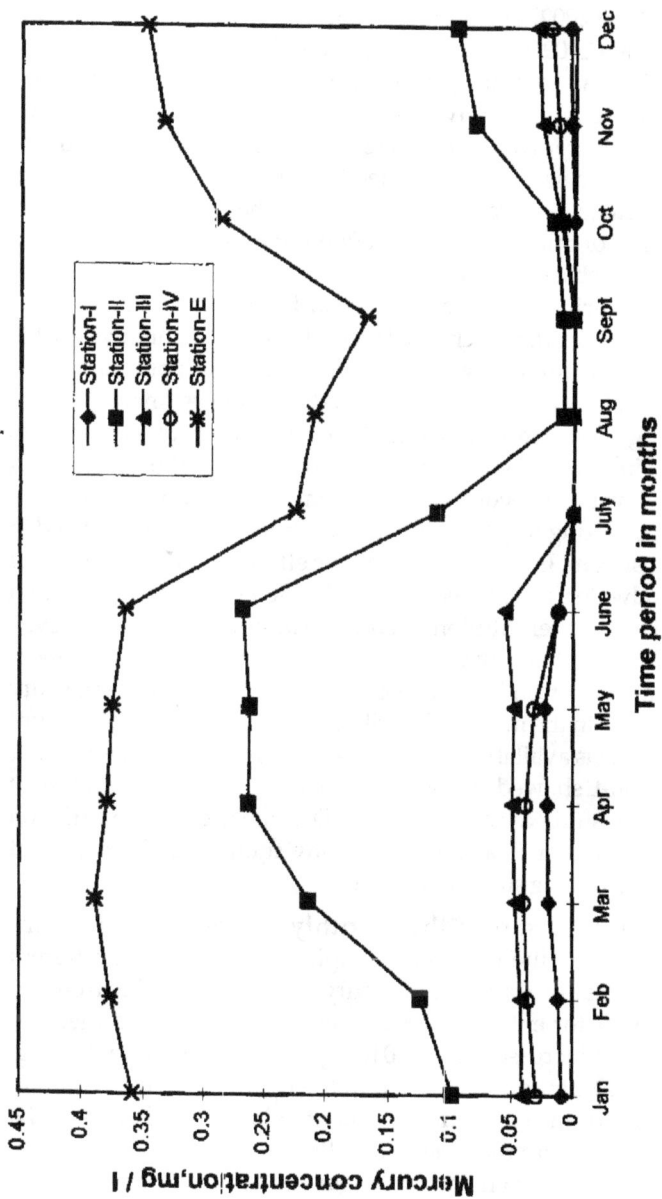

Figure 8.9: Monthly Variation of Mercury Concentration in Samples Collected from Different Stations from January 2002 to December 2002

concentration increased from 0.0034 mg l⁻¹ in November to 0.0226 mg l⁻¹ in May 2002. The mercury concentration depleted to 0.0121 mg l⁻¹ in June 2002. Station-I is located in the upstream of the river and much away from the junction point, where the effluent of the industry mixes with the river water. But due to periodic high tidal influx of seawater from Bay of Bengal, the effluent along with the river water and seawater is pushed to station-I. Hence, a very low level of mercury was recorded from December to June. After June, due to onset of rainy season, fresh floodwater washes away all the mercury available in the area into the sea. Hence, virtually no mercury was found during rainy season. Station-II is located in the junction point, where the effluent channel meets the river Rushikulya. In this station highest amount was recorded in the month of June, where 0.2684 mg of mercury l⁻¹ was recorded. The lowest mercury level (0.0085 mg l⁻¹) was recorded in August. Availability of mercury was recorded in all the months in this station only. In station-III, highest mercury level was recorded in the month of June and no mercury was found during July to September. Station-III has been located in the downstream. In rainy season, all the effluent water moves along with floodwater into the sea and the dilution is highest during that period. Due to over dilution, mercury was not at record able level. Station-IV is located in the downstream nearer to the estuary, where maximum dilution and mixing occurs. Maximum amount of mercury was recorded in the month of April 2002,and no record able amount of mercury was available during the period from July to September 2002. Station-E showed the highest amount of mercury in the effluent at all months studied from Jan.2002 to December 2002. The variation was due to either heavy rain in rainy months or high rate of evaporation during summer months.

Figure 8.10 shows the monthly variation of mercury concentration in the sediment samples collected from different stations. In case of Stations-I, mercury was not detectable from the month of June to September 2002. From October onwards, mercury concentration increased from 0.016mg g⁻¹ in October to 0.031 mg g⁻¹ in May 2002. Station-I is located in the upstream of the river and much away from the confluence point, where the effluent of the industry mixes with the river water. But due to periodic high tidal influx of seawater from Bay of Bengal, the effluent along with the river water and seawater is pushed to station-I. Hence, a very low level of mercury was recorded from October to May. After June, due

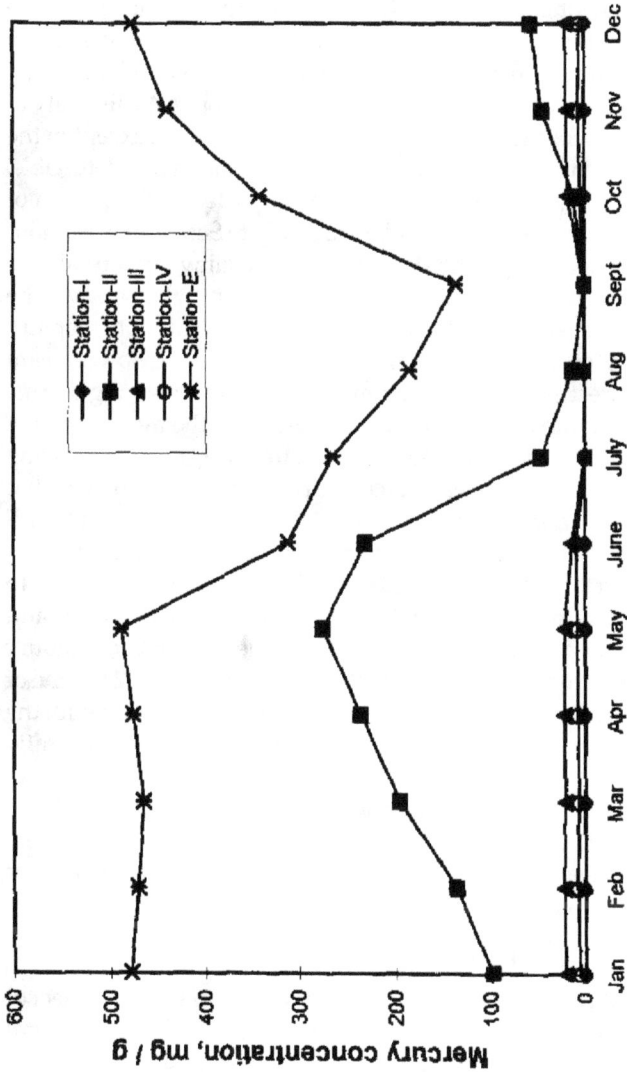

Figure 8.10: Monthly Variation of Mercury Concentration in Sediment Samples Collected from Different Stations from January 2002 to December 2002

to onset of rainy season, fresh floodwater washes away all the mercury available in the area into the sea. Hence, virtually no mercury was found during rainy season from June to September. Station-II is located in the junction point, where the effluent channel meets the river Rushikulya. In this station highest amount was recorded in the month of June, where 231.2 mg of Hg g^{-1} was recorded. The lowest mercury level (11.65 mg g^{-1}) was recorded in October. Availability of mercury was recorded in all the months in this station except in the month of September, where no mercury concentration was detectable. In station-III, higher mercury level was recorded in the month of April and no mercury was found during July to September. Station-III has been located in the downstream. In rainy season, all the effluent water moves along with floodwater into the sea and the dilution is highest during that period. Due to over dilution, mercury was not at recordable level. Station-IV is located in the downstream nearer to the estuary, where maximum dilution and mixing occurs. Maximum amount of mercury was recorded in the month of June 2002, and no recordable amount of mercury was available during the period from July to August 2002. During September and October 2002, insignificant amount of mercury was recorded. Station-E showed the highest amount of mercury in the effluent at all months studied from Jan.2002 to December 2002. The variation was due to either heavy rain in rainy months or high rate of evaporation and sedimentation during summer months. Highest significant amount of sediment mercury was noted in the month of May 2002 to tune of 489.62 mg of Hg. g^{-1}. The mercury concentration decreased with the increase in rainwater or floodwater. The analysis of variance ratio test indicated the existence of significant difference between rows and significant difference between columns. No specific trend was marked in all the parameters studied at all the stations. The differences observed were either due to dilution by rainwater or floodwater or evaporation due to summer.

Aquatic Contamination by Mercury

Pollution of the estuary all over the world has been a subject of great concern. Taking into account the large water bodies and natural phenomenon such as temperature, precipitation, flow of river etc. A yearlong study was carried out to know the degree of contamination. Four sites along the estuary were earmarked as study stations. All the four stations were located within a distance of about 0.75 to 1.0 km from each other. The effluent channel was taken as the fifth station of study (Station E–0.25 km away from the industry and 0.25 km away from the confluence point.

Water and sediment samples were collected at those sites monthly for the analysis of mercury levels. Samples were collected in triplicate for estimation and analysis. Water samples from different stations were collected in pre-washed 500 ml plastic bottles. Soils samples from different stations of study were collected in plastic packets. These were brought to the laboratory, dried in air and then in oven at 35°C, grounded to powder with the help of a mortar and pestle and kept separately for further analysis.

The effluent of the chlor-alkali industry is highly toxic in nature. The average temperature recorded was 31.3 ± 1.4°C. The effluent was alkaline and the pH of the effluent recorded was 9.8 ± 0.5. The dissolved oxygen content was much less, when compared to any other aquatic body. The recorded dissolved oxygen in the effluent was 1.84 ± 0.4 mg of oxygen l^{-1}. The alkalinity (as $CaCO_3$ in mg l^{-1}) of the effluent recorded was 232.4±12.6 as $CaCO_3$ in mg l^{-1}. The hardness of the effluent was 468.4 ± 25.8 as $CaCO_3$ in mg l^{-1}. The BOD and COD of the effluent were 12.4± 2.2 and 321.5± 4.8 mg l^{-1}, respectively. The salinity in terms of Chlorinity of the effluent as recorded was to the tune of 1954.2±32.6 in mg l^{-1}. The total suspended solids present in the effluent was 104.6±6.5 mg l^{-1}. The total Nitrogen content and total phosphorus content in the effluent were 2.2 ± 0.8 mg l^{-1} and 0.22 ± 0.08 mg l^{-1}, respectively (Table 8.1). The effluent which when released from the factory finds its way in to the Rushikulya river estuary, was found to contain very high amount of mercury. Out of twelve analyses carried out in twelve months, only once, in the month of March, a lower concentration of mercury was observed Though the concentration in March was low, the value was in itself much more higher than the permissible limit of 0.01 mg l^{-1}. Maximum concentration of mercury, as recorded in the month of April, was to the tune of 0.3798 mg l^{-1}. Concentration of mercury in the effluent was found to be much fluctuating having a mean value of 0.3758 ± 0.0108 mg l^{-1}. In 1996, the value increased to 0.3894 ± 0.0258 mg l^{-1}. Mercury levels at this station in the effluent channel were found to be dependent on the levels of mercury in the effluent. When, we compare the data obtained in 1986, 1996, 1999 and the present data, it seems no correlation exists regarding the retention or residual mercury concentration. Neither a decreasing trend nor an increasing trend was marked particularly linked to residual mercury level in the effluent channel.

Table 8.1: Physico-chemical Properties of the Effluent

Sl.No.	Parameters	Present Data of June, 2002
1.	Temperature (°C)	31.3±1.4
2.	pH	9.8±0.5
3.	Alkalinity (as $CaCO_3$ in mg l^{-1})	232.4±12.6
4.	Hardness (as $CaCO_3$ in mg l^{-1})	468.4±25.8
5.	Chlorinity (in µg l^{-1})	1954.2±32.6
6.	Dissolved Oxygen (in mg l^{-1})	1.84±0.4
7.	BOD (in mg l^{-1})	12.4±2.2
8.	COD (in mg l^{-1})	321.5±4.8
9.	Suspended solids (in mg l^{-1})	104.6±6.5
10.	Total N (in mg l^{-1})	2.2±0.8
11.	Total P (in mg l^{-1})	0.22±0.08
12	Total Hg (in mg l^{-1})	0.3891± 0.0132

Table 8.2: Shows Mercury Concentration in Water and Sediment at Different Stations and in the Effluent Channel. Data represented were the concentrated analysis over a week time to find out the status of the estuary.

Stations	Sample Size	Water (mg of Hg l^{-1})	Sample Size	Sediment (mg of Hg.Kg^{-1} dry weight)
I	10	0.0226±0.0086	5	0.031±0.012
II	10	0.2622±0.0092	5	276.11±19.35
III	10	0.0485±0.0112	5	21.78±3.45
IV	10	0.0316±0.0174	5	8.65±2.16
E	10	0.3758±0.0108	5	489.62±36.54

The Table 8.2 indicates the residual mercury levels in the water and sediment collected from different stations and the effluent channel. A significant variation was recorded in all the stations studied. Out of the four station selected for studying mercury dynamics in the estuary, station II, which is the junction point was found to be containing the highest amount of mercury in comparison to other station. The range was 0.0226±0.0086 to 0.2622±0.0092 mg of Hg.l^{-1} in all the stations studied. The mercury level in station-I was less, when compared to other stations, as Station-I was located in the upstream of the river, where no contamination is expected. However, a significant amount of mercury was recorded at station-I. The reason for accumulation/availability of mercury in water sample at station-I was because for back water flush during high tide time, when the entire estuarine water along with the effluent gets circulated in the back water area. During low tide, again this mixed water passes through the estuarine mouth into Bay of Bengal carrying the effluents and backwater. High level of mercury was recorded at Station-II. (Table 8.2). Mercury levels at this station were found to be dependent on the levels of mercury in the effluent to some extent except in the month of July and August. High mercury level at station-II was because of the fact that Station-II was at the junction point, where the effluent channel meets the river and the effluent channel carries a huge load of mercury along with the effluent. Station-III is located in the downstream, little away from the junction point. Dilution of mercury level was noted at station-III, when compared to Station-II. But the mercury concentration at station-III was significantly higher, when compared to Station-I (Table 8.2). In Station-IV, still less amount of mercury was reported, when compared to Station-III and Station-II. This was probably due to highest dilution of mercury level by the river water and tidal water during high tides. The effluent channel (Station-E) enjoyed the highest amount of mercury in the effluent, when compared to water samples at the rest 4 stations (Table 8.2). Within 10 years time, significant increase in mercury level was recorded in station II and III and sediment mercury level in station I, III and IV. The decrease in mercury concentration in the effluent channel was due to periodic removal of sediments from the channel and dumping the wastes in a nearby site. Levels of mercury at other stations were low in comparison to station II. Out of the three station, station IV was found to contain the lowest amounts of mercury. Levels of mercury at station I and III were nearly identical with a tendency of little higher levels at station III than at station I, except in the month of April when the value was less than at station III. No particular trend of increase or decrease in the levels of mercury was noted at any of the stations. However, all the three stations showed their minimum and maximum levels in the monsoon and pre-monsoon seasons, respectively. The levels of mercury during monsoon season at these stations were much less and very similar to each other due to high rainwater diluting the level of mercury. The trend of decrease or increase in the levels of mercury was also similar in all the stations studied. A strong seasonal variation was recorded pertaining to residual mercury level.

Sediment analysis showed the presence of a remarkable quantity of mercury. Maximum amount of mercury was found in the sediment collected from the effluent channel. The levels of mercury fluctuated much in different study stations, the maximum being 276.11 ± 19.35 mg kg^{-1} dry weight in Station-II and the minimum being 0.031 ± 0.012 mg kg^{-1} dry weight in Station-I. The residual mercury level decreased with the increase in distance from station-II to station-IV. Obviously, station-I showed the lowest amount of mercury in the sediment sample (Table 8.2). Table 8.2 indicates the residual mercury levels in the sediment collected from different stations and the effluent channel. A significant variation was recorded in all the stations studied. Out of the four station selected for studying mercury dynamics in the estuary, station II, which is the confluence point was found to be containing the highest amount of mercury in comparison to other station. The range was 276.11 ± 19.35 to 0.031 ± 0.012 mg of Hg. kg^{-1} dry weight of the sample in all the stations studied. The mercury level in station-I was less, when compared to other stations, as Station-I was located in the upstream of the river, where no contamination is expected. However, a significant amount of mercury was recorded at station-I. The reason for retention/availability of mercury in sediment sample at station-I was because of back water flush during high tide time, when the entire estuarine water along with the effluent gets circulated in the back water area. During low tide, again this mixed water passes through the estuarine mouth into Bay of Bengal carrying the effluents and backwater. High level of mercury was recorded at Station-II (Table 8.2). Mercury levels at this station were found to be dependent on the levels of mercury in the effluent to some extent. High mercury level at station-II was because of the fact that Station-II was at the confluence point, where the effluent channel meets the river and the effluent channel carries a huge load of mercury along with the effluent. Station-III is located in the downstream, little away from the confluence point. Dilution of mercury level was noted at station-III, when compared to Station-II. But the mercury concentration at station-III was significantly higher, when compared to Station-I (Table 8.2). In Station-IV, still less amount of mercury was reported, when compared to Station-III and Station-II. This was probably due to highest dilution of mercury level by the river water and tidal water during high tides and continuous flush of sediments into the sea. The effluent channel (Station-E) enjoyed the highest amount of mercury in the sediment, when compared to sediment samples at the rest 4 stations (Table 8.2). Within 10 years time, significant increase in mercury level was recorded in station II and III and sediment mercury level in station I, III and IV. The decrease in mercury concentration in the effluent channel was due to periodic removal of sediments from the channel and dumping the wastes in a nearby site. Levels of mercury at other stations were low in comparison to station II. Out of the three station, station IV was found to contain the lowest amounts of mercury. Levels of mercury at station I and III were nearly identical with a tendency of little higher levels at station III than at station I, except in the month of April when the value was less than at station III. No particular trend of increase or decrease in the levels of mercury was noted at any of the stations. However, all the three stations showed their minimum and maximum levels in the monsoon and pre-monsoon seasons, respectively. The levels of mercury during monsoon season at these stations were much less and very similar to each other due to high rainwater diluting the level of mercury. The trend of decrease or increase in the levels of mercury was also similar in all the stations studied. A strong seasonal variation was recorded pertaining to residual mercury level. No particular trend of increase or decrease was marked. However, lower concentration of mercury was observed in the monsoon and the post monsoon season, *i.e.*, from June to December. No relationship could be noticed between the concentration of mercury in sediment of this site and that of the effluent channel. However, it can be inferred that mercury was available in the contaminated area both in water and sediment. No particular trend of increase or decrease in the mercury concentration was noticed at any station during the entire period. However, the levels of mercury were lower in the monsoon and the post-monsoon seasons at all the three stations.

Discussion

Industry is responsible for creating a fantastic array of new chemicals every year, all of which eventually find their way into the environment. For most of these chemicals, not even the chemical formulae are known and much less are known about their acute, chronic or genetic effect on the living organisms. At present, the industry is the focus of attention, the world over, as the most important source of pollution of the environment. Sundaresan *et al.* (1983) have given the growth of industries dealing with toxic chemicals and generating toxic and hazardous wastes during 1950's '1970's in India. Generally the toxic effluents from industry are neutralized before their discharge, but still they contain substantial amount of toxic substances that can cause pollution. At present it is believed that rivers are most severely polluted by industries followed by estuaries, lakes, and ocean in declining order. Variations in temperature of the effluent throughout the period of study have also been presented. The effluent showed the highest temperature followed by the junction zone, *i.e.* station II. Temperature of the effluent varied from 27° C to 39.4° C, and that act station II from 24.5° C to 31° C. There was no particular trend of increase or decrease; the variation followed simply a zigzag pattern (Shaw, 1987). Temperature at station II was under the influence of the effluent temperature as indicated by the significant correlation obtained for the variation in temperature between the effluent and station II. Monthly variations in temperature at station I, III and IV followed nearly a similar pattern except in the month of November, when station I showed the lowest temperature of 21°C. This was also the lowest temperature recorded those three stations, the maximum being 28.2°C in the month of April '84 at station I and III. One interesting point to note is that the variations in temperature at station III were more closely related to the variations at station I up to the month of October, whereas after October the variations at station III were more closely related to other variations at station IV (Shaw, 1987). This was probably because of the fact that till October station III was under the influence of floodwater, whereas after October the same was under the complete influence of seawater. It is also evident that except in the month of November temperature at station IV was the lowest when compared to the other stations, certainly due to influence of the seawater. The lowest temperature at station III and IV was recorded in the month of January, *i.e.* the winter season. Present study confirms the report that the temperature at the mouth of the estuary to be lower than the mid estuary. Difference in temperature between stations was also found to be significant. However, a least significant difference of 0.85 for between stations analysis indicated that the temperature at stations I, III and IV did not differ significantly. Had there been no mixing of the effluent water at station II, there would not have been any difference in temperature recorded at station I, II, III and IV. Nevertheless, from the results it appears that with respect to temperature the estuarine water was not heavily polluted by the effluent. Correlation analysis for the temperature Vs different parameters done separately for each station, revealed a significant correlation of temperature with suspended solids, dissolved oxygen, and NO_2-N at station II; DO, BOD and gross primary productivity (GPP) at station III; BOD and GPP at station IV. No significant correlation was obtained for station I and the effluent water. Positive correlation of the temperature with the GPP might be because of the fact that up to certain degree productivity increases with temperature. Correlation's obtained with other parameters are difficult to explain, rather appear to be simply a matter of coincidence, Presence of significant positive correlation between temperature and DO, which should rather be negative, confirms the same (Shaw, 1987).

In comparison to other stations pH of the effluent showed maximum variations. The variation did not fallow any particular trend. Minimum and maximum pH values recorded were 6.93 and 11.12, respectively. Though the minimum pH value recorded was well above the lower pH tolerance limit of 5.5. fixed by the Indian Standard Institute, the maximum, pH value recorded was not below the upper pH tolerance limit of 9.0 fixed by the same institute. Like temperature, variations in pH at station II followed the pattern of the effluent's pH variation. This is also well evident from the highly significant co-efficient of correlation obtained for pH variations between the two stations. The minimum and maximum pH recorded were 7.32 and 9.96, respectively (Shaw, 1987). Seasonal variations in pH at station III and IV were found to be quite related, where as the variations at station I showed a little deviation. The maximum pH recorded at station I was in the month of December (8.35) and the minimum in the month of July (7.35). In general the pH values were lower in the monsoon season than in the pre and post-monsoon seasons. This might be due to the fact that in monsoon season station I generally remained under the influence of fresh river water in which the pH generally has been ground to vary between 7.45 to 8.20. On the other hand, in pre and post-monsoon seasons station-I remains under the influence of sea water (Shaw, 1987) in which the pH has been reported to vary from 7.5 to 8.0 (Verlencer 1984,87), 7.9 to 8.2 (Sankaranarayana and Reddy, 1968); 5.5–8.0. pH at station III and IV showed more or less similar variations. Reverse to the case as obtained with station I, the pH was the maximum in the monsoon season at both the stations. Influence of the effluent, carried to those stations along with the floodwater or during the low tide, might be the probable reasons. Maintenance of pH to a particular range is important since all the biochemical activities depend on pH of the environment. pH lower than 6.7 has been found to increase the ventilator frequency of blue gill, whereas with an increase in pH toxicity of Ni to microbes has been reported to be reduced. Similarly, the acute lethality of dissolved Zn to rainbow trout increased 2 to 5 times with an increase in pH from 5.5 to 7.0. Variance ratio test analysis revealed that the pH at different stations differed significantly (Shaw, 1987). However, there occurred no significant difference in the levels of pH between different seasons. Least significant difference of 0.36 revealed that station III and IV did not differ significantly with respect to the variation in pH. Correlation coefficient analysis for the pH Vs different parameters revealed a significant correlation of pH with SS, DS, COD, salinity, TN, SiO_3-Si and Respiration at station I, with none at station II, with NO_2–N and total Hg at station III, with none at station IV and E. Significant correlation's obtained are not easily explainable because of the lack of any consistency in the correlation values obtained at different stations and hence can be regarded as a coincidence (Shaw, 1987).

The variations in suspended solids content of the water at different stations during the period April' 84 to March' 85. Variations in the SS content followed a Zigzag pattern at all the stations. Maximum suspended solids were recorded at station II, followed by the effluent channel. The variations in SS content of the water at station I, III and IV were closely related showing peak during the monsoon season. High SS levels at station II might be because of the stirring of the bed sediment at the point of mixing of the effluent with the estuarine water (Shaw, 1987). Besides, high amount of suspended particles present in the effluent water might also be responsible for increasing the levels of the same at station II. This is well evident from the highly significant correlation value obtained between the suspended solids content of the water at station II and the effluent. Higher levels of suspended solids at station I, III and IV during monsoon indicate the influence of high river discharge. Suspended materials play the role of scavenger for the particulars metals in an aquatic systems. They cause ecological imbalance by mechanical abrasive actions (clogging and irritation of tissues), blanketing action and sedimentation, reduction of light penetration, availability as a surface for growth of bacteria and fungi, reduction of temperature fluctuation and absorption of various chemicals, specifically the heavy metals (Shaw, 1987). High loads of suspended solids cause a significant deterioration in the survival conditions for aquatic organisms. In freshwater, at 100 mg 1^{-1} of slit, the pumping rate of the oyster was reduced to 57 per cent. Suspended solids of more than 350 mg 1^{-1} affect the fish adversely by causing mechanical injuries to the gill. These reports suggest that the load of SS in the effluent and the estuarine water was well the limit of pollutional hazards. Relationships of SS with DS, salinity and SiO_3-Si are well explainable but not the relationships with other parameters. Since a major part of DS is because of salinity, both salinity and DS can be taken as one. Negative correlations of SS with salinity indicates that suspended load of the estuarine water was largely because of the river discharge as increase in its level was accompanied by decrease in the level of salinity. Positive correlation with SiO_3-Si also favour the same since a larger amount of SiO_3-Si is present in the river water than the estuarine water (Shaw, 1987). Also an increase in the levels of suspended solids with decrease in the ionic strength of the water has been reported. Since station I was under more influence of the effluent than the river discharge, no significant correlation of SS with DS, salinity and SiO_3-Si could be obtained for this station. Highly significant correlation of SS with Hg at station E might be due to the fact that most of the mercury particles remain adhered to the suspended particles. These three parameters (DS, salinity and hardness) are closely related. Minimum quantity of DS was recorded in the effluent followed by station II and then station I, III and IV. Salinity and hardness variations also showed a similar trend. All the three parameters at station I, III and IV showed their peak in the pre-monsoon season, the levels declined in the monsoon followed by again an increase in the post monsoon till they reached their peak in the pre-monsoon season. Decline in the levels in the monsoon season was certainly because of the influx of fresh water into the estuary. Quantitatively station IV showed the highest values for all the parameters followed by station III and then station I. It appears from the figure that station II was largely under the influence of the effluent water (Shaw, 1987). Trend of increase in the quantitative values from station I to III to IV is well understood as we move towards the mouth the estuarine water becomes more and more oceanic. Decrease in the levels at all the stations in the month of January might be due to a winter rain. Salinity is widely used as an indicator of the estuarine conditions. Distribution of salinity varied from time to time depending upon the influence of fresh water influx and penetration of seawater into the estuary. In the present study values of salinity ranging from purely marine to the almost freshwater were recorded (Shaw, 1987). More or less similar variations in salinity in estuarine water have been reported. Salinity and hardness, which may together be represented as dissolved solids, play an important role in distribution and maintenance of many organisms in the estuary. A decrease in hardness from 386 to 31 mg $CaCO_3$ 1^{-1} increased Zn toxicity to the rainbow trout several times. Role of water hardness in increasing the ventilator frequencies in fishes has also been reported. All the three parameters showed significant difference in their levels between different stations as well as between seasons at all the stations. Salinity variations showed significant correlation with DS, SS, pH, hardness, SiO_3–Si, total Hg, GPP and R at station I, with DS, hardness and respiration at station II, with DS, SS, hardness, NO_2-N, SiO_3–Si, total Hg and GPP at station III, with COD, DS, SS, hardness, NO_2-N, SiO_3–Si, total Hg, GPP and respiration at station IV and with DS, NO_2-N at station E. Highly significantly correlation between salinity, DS and hardness are due to the fact that these parameters are inter related. Salinity represents the halogenated anions, and hardness Mg and Ca ions whether halogenated or not. Both in turns represents dissolved solids content of the water together with other ions. Decrease in the value of any of them will results in decrease in the value of others. Highly significant negative correlation of these parameters with SiO_3-Si suggests quantitative decreases of the same due to influx of the river water, which contains large amount of SiO_3-Si. Influx of the river water, causing dilution in the levels of DS, salinity and hardness along with dilution in the levels of mercury must also be responsible for significant positive correlation positive correlation of these parameters with mercury at station I, III and IV. Significant positive correlation of DS, salinity and hardness with GPP and respiration suggests that metabolic activities in the estuary were higher under the saline environment. Other significant correlations are not readily explainable.

The variations in dissolved oxygen levels of the water at different stations in different seasons. The figure clearly indicates that station E (effluent channel) and station II had a lower DO concentrations than the other stations through out the period of investigation (Shaw, 1987). Minimum DO level recorded for the effluent was 2.28 mg l^{-1}, the maximum being 5.25 mg l^{-1}. However, the values generally ranged between 3 to 4 mg l^{-1}. The levels of DO at station II also ranged generally between 3 to 4 mg l^{-1} like the effluent water with a minimum value of 3.08 mg l^{-1} and a maximum value of 5.48 mg l^{-1}. The dissolved oxygen levels at station II were certainly under the influence of effluents DO levels as it appears from the significant correlation between the levels of DO at station E and station II. With respect to DO levels station III and IV were closely related to each other. Though station I was related to them to some extent, it did show variation. No parameters trend of increase or decrease could be noticed at any of the stations. Levels of oxygen in any aquatic system are of great importance to the inhabiting organisms. One of the most important chemical analysis in determining the water quality is the determination of dissolved oxygen levels. It has been recommended that a minimum of 4 mg DO per liter should be maintained in estuarine and coastal water for healthy growth of fish population. In estuarine environment DO concentrations varying from 3 to 6.9 mg l^{-1} have been reported. These reports suggest that the estuarine system under present investigation was not under stress with respect to the dissolved oxygen level. However, a concentration as low as 1.20 mg l^{-1} does recall attention. Depletion of oxygen to such an extent suggests that there was an influx of heavy organic load in recent past. Since the depletion was maximum during the post monsoon season, this may be correlated with decrease in the volume of water and degradation of the organic materials brought along with the flood, or released along with the effluent. Though several significant correlations were obtained, important to note–are correlation of DO with BOD, COD and R at station I, with temp. at station II, with temp., BOD, R and P/R at station III and with BOD, R and P/R at station IV. Negative correlation of DO with BOD and COD was expected since the litters are mainly responsible for the depletion in DO levels. Since the presence of organic load increase respiration and since BOD and R are positively correlated, a negative correlation of DO with R and P/R is obvious. Significant positive correlation of DO with temperature at station II and III, as well as a tendency of positive correlation of DO with temperature at other stations indicates a highly unusual condition. Theoretically, the correlation should be negative. Distortion of the relationship might be because of the presence of high organic load. Biological oxygen demand (BOD) and chemical oxygen demand (COD) represent the biologically and chemically oxidisable load respectively present in water. The two parameters are more or less related with the each other. The variations in BOD and COD levels of the effluent and estuarine water did not follow any particular trend. However, a lower level of BOD and COD during the monsoon season at station I, III and IV were noticed, probably because of the dilution effect. With respect to the BOD and COD levels also station II was largely under the influence of the effluent water as it is evident from the significant correlation obtained between station II and E for those parameters. The levels of BOD and COD were mostly higher at station E and II than at station I, III and IV. This strongly suggests input of organic and chemical loads along with the effluent. Further, higher levels of COD and BOD mostly at station III than at station I and IV indicate that former was in more easy reach of the effluent water than the later. COD levels of the effluent were much higher and significant than the BOD levels, the maximum and minimum values being 620.28 and values being 620.28 and 211.39 mg l^{-1}, respectively. Effluent from a Paper and pulp mill had COD level of 315 mg l^{-1}. BOD levels of the effluent were though lower but never insignificant. The values obtained were mostly higher than the tolerance value of 30 mg l^{-1} fixed by Indian standard Institute. The same for COD is 250 mg l^{-1}. BOD level as high as 51.3 mg l^{-1} has been reported in the effluent from a chemical industry. BOD and COD levels of the estuarine water, except at station II were though lower in comparison to the effluent, the values were still significant. It was reported that COD values as high as 3797 mg l^{-1} for a polluted estuarine water. Prati *et al.* (1971) classified water bodies into five classes depending upon the BOD, COD and other values. He put water body with BOD and COD respectively of above 12 and 80 mg l^{-1} into class V, of 12 and 80 mg l^{-1} into class IV, of 6 and 40 mg l^{-1} into class III, of 3 and 20 mg l^{-1} class II and of 1-5 and 10 mg l^{-1} into class I. Referring to this classification the present estuarine system mostly falls under class IV and V. Similar to the other parameters the levels of alkalinity in the effluent were the highest, followed by station II. Station IV showed the lowest levels preceded by station I and III. No particular trend of increase or decrease in the alkalinity was marked at any of the stations. The levels of alkalinity in the effluent fluctuated very much. The minimum value recorded was 5.79meq l^{-1}, whereas the maximum value was 406.04meq l^{-1}. Fluctuation in the levels of alkalinity at station II followed a similar trend. This was because the levels of alkalinity levels of the effluent. This dependency is evident from the highly significant correlation obtained between station E and II with respect to the alkalinity level. Influence of the effluent discharge on the alkalinity levels of the water was equal at station I and III since they were at equal distance from the discharge point. Station IV being at a larger distance from the point of discharge, showed the least concentrations of alkalinity. Variation in the levels of alkalinity at station IV was from 1.38 to 2.42 m eq l^{-1}. The upper range was more or less similar to the range 2.14 to 2.48 m eq l^{-1} reported for seawater. However, the lower range was much less than the reported range for seawater (Shaw, 1987). This must be because of the influx of fresh river water in which a value as low as 76.9 mg l^{-1} has been reported. Elevated levels of alkalinity at station I and III might be because of the effluent discharge. The effluent showed much elevated levels than the reported value of 1.98 m eq l^{-1} for the effluent from a chemical industry. The alkalinity of seawater is a measure of quantity of weak acid present in it and of the cations balanced against them. In most sea water cations of weak bases are present in negligible concentration and the only anions that need be considered are those of carbonic and boric acids. However, in polluted system OH$^-$ions released from chemical factories play on important role in increasing measured is not the true alkalinity. Alkalinity and acidity play an important role in controlling different enzyme activities. This was probably because of the large residual variation obtained. Any significant correlation of alkalinity with different physico-chemical parameters is unlikely to be reported. Concentrations of the reactive silicate were found to be the highest in the effluent, the minimum and maximum values being 274.96 and 546.21 mg l^{-1}, respectively. Levels of the reactive silicate at station II was largely governed by the effluent silicate concentrations, similar to some of the other parameters described earlier, as is evident from the correlation analysis between station E and II with respect to the SiO_3–Si levels. Concentrations of the reactive silicate at station II were only next to the effluent silicate concentrations. Concentrations at station I, III and IV were more or less similar with station I and III showing little higher levels than station IV. In the monsoon season the reactive silicate concentrations showed an increase up to 170 µg l^{-1}. Purushothaman and Venugopalan (1972) also reported similar values in their study with Vellar Estuary. Present investigation shows reactive silicate concentrations a little higher than the reported values, probably because of the effluent discharge containing high amount of the same. Chlor-alkali plants are widely accepted as a potential industrial emission of mercury (Flewelling, 1971). This dragged our attention to the environmental problem caused due to build up of mercury at various zones of the Rushikulya river mouth to which the effluents of a chlor-alkali factory are released. A high level of mercury concentration was noticed in water and sediment.

The environmental pollution is growing at a tremendous speed and it seems ultimately there must be recognition that the environmental crisis is a confrontation between Man and Nature, human system, where influence is now global and the natural ecosystem that has built and maintained the biosphere. With the wide use of killer chemicals for higher production, no doubt, the ecosystem gets polluted, but in addition to that the non-target organisms suffer from toxic stress. The microorganisms (economically important) suffer the most. The fresh water fish could not survive in this toxic environment. A survey indicated disappearance of these fishes from the ponds gradually in the contaminated areas. Two factors such as concentration and duration of exposure have been identified as having paramount roles in heavy metal accumulation process. Accumulation, distribution and loss of metals from tissues are generally influenced by intrinsic factors including age, size, weight, reproductive condition and heavy metal body burden attained, and by extrinsic environmental factors including the hydro-climatic conditions, particularly temperature and salinity concentration, duration of exposure and the chemical form of the metal in aquatic medium. Effects of mercury on different fishes have intensively been studied as a result of the global nature of the pollution problem and mercury contaminated fish have been implicated as the causative agent in different diseases. Even with the environmental awareness of the 1940's it is still difficult to attribute any reason for such enormous increase in the use of mercury. Despite warnings of hazards during the early 1950's, production of organo-mercurical fungicides increased many time. After repeated warnings even the chlor-alkali industries dump their solid wastes at nearby water bodies. It is difficult to predict which toxicant will likely endanger aquatic ecosystem in the future. Contamination of aquatic environment may be a manifest in a number of ways, none of them being pleasant. The organic and inorganic mercurials have been investigated more extensively than many other single groups of toxicants mercury based pesticides are used both as disinfectants and as preservatives. But there has always been an objection against the use of mercury derivatives due to their acute toxic properties. Now a day, this is very common in developing countries. Because, the funds necessary for careful disposal and preservation of the waste becomes a limiting factor for industries and in addition, a space problem, for damping the wastes in a safer place. A considerable knowledge of the toxicology of the mercury has accumulated which, makes it possible to avoid at least the occupational hazards. The effects of mercury have gained an increasing interest during later years. Much attention has been focused on the possible poisoning of aquatic life specifically fish with mercury based/contained wastes. In countries like Sweden and Japan extensive investigations on the mercury pollution problem have been carried out during the last decade. Some species of freshwater fishes in Sweden have been found to contain high concentrations of mercury as a result of industrial pollution and possible also as a result of atmospheric precipitation. The reports of Shaw (1987) and the present data clearly indicated the Rushikulya estuary is totally contaminated with mercury contained effluent of the Chlor-alkali industry situated at Ganjam.

Acknowledgements

Authors sincerely acknowledge the financial assistance given by DOD (OSTC), New Delhi in the form of a project.

References

APHA, 1984. *Standard Methods for the Examination of Water and Wastewater*. Environmental Protection Agency, USA.

Annandale, N. and Kemp, S., 1915. Introduction to the fauna of Chilka lake. *Mem. Indian Mus.*, 5: 1–20.

Bouveng, H.O., 1968. The chlorine industry and mercury problem. *Modern. Kemi.*, 3: 45.

Butler, G.C. (Ed.), 1978. *Principles of Ecotoxicology.* Scope–12, John Wiley and Sons, Chisester, Brisbane, New York, p. 350.

Chandran, R. and Ramamoorthi, K., 1984a. Hydrobiological studies in the gradient zone of the Vellar estuary. I. Physico-chemical parameters. *Mahasagar Bull. Natl. Inst. Oceanogr.*, 17: 69–77.

Chandran, R. and Ramamoorthi, K., 1984b. Hydrobiological studies in the gradient-zone of Vellar estuary. II. Nutrients. *Mahasagar Bull. Natl. Inst. Oceanogr.*, 17: 133–140.

CSE (Center for Science and Environment, New Delhi), 1985. *The State of India's Environment*, 1984–85. The Second Citizen's Report.

ECIL (Electronic Corporation of India Limited), 1981. Analytical methods for determination of mercury with mercury analyser, MA 5800 A.

EPA (Environmental Protection Agency), 1979. Mercury in sediment. In: *Methods for Chemical Analysis of Water and Wastes*, p. 134–138.

Flewelling, F.J., 1971. Loss of mercury to the environment from chlor-alkali plant. In: *Mercury in Man's Environment.* Proc. Special Symp. Soc. Canada, p. 34.

Holdgate, M.W., 1979. *A Perspective of Environmental Pollution.* Cambridge University Press, Cambridge, London, New York, Melbourne, p. 278.

Hortung, R. and Dinman, B.D. (Eds), 1974. *Environmental Mercury Contamination.* Ann. Arbor. Sci. Publishers Inc., 349 p.

Joseph, J. and Kurup, P.G., 1990. Stratification and salinity distribution in Cochin estuary, South-West coast of India. *Indian J. Mar. Sci.*, 19: 27–31.

Larsson, J.E., 1970. Environmental mercury research in Sweden, Stockholm. *National Swedish Environmental Protection Board.*

Miller, D.R., 1984. Chemical in the environment. In: *Effects of Pollutants at the Ecosystem Level*. John Wiley and Sons, Chichester, p. 7.

Purushotaman, A. and Venugoapalan, V.K., 1972. Distribution of dissolved silicon in the Vellar estuary. *Indian J. Mar. Sci.*, 1: 103–105.

Qasim, S.Z. and Gopinathan, C.K., 1969. Tidal cycle and environmental features of Cochin Backwater: A tropical estuary. *Proc. Indian Acad. Sci.*, 3: 146–149.

Sankaranarayanan, V.N. and Reddy, C.V.G., 1968. Nutrients of the north-western Bay of Bengal. *Bull. Natl. Inst. Sci., India*, 38: 148–163.

Sewell, R.B.S., 1929. Geographic and oceanographic research in Indian waters, Part V: Temperature and salinity of the surface waters of the Bay of Bengal and Andaman Sea with reference to Laccadive Sea. *Mem. Asiatic Soc., Bengal*, 9: 207–356.

Sewell, R.B.S., 1932. Geographic and Oceanographic research in Indian waters, Part VI: The temperature and salinity of the deeper water of the Bay of Bengal and Andaman Sea. *Ibid.*, 9: 357–424.

Shaw, B. P., 1987. Eco-physiological studies of the waste of a chlor-alkali factory on biosystems. *Ph.D. Thesis*, Berhampur University, Orissa, India.

Shaw, B.P. and Panigrahi, A.K., 1986a. Uptake and tissue distribution of mercury in some plant species collected from a contaminated area in India: Its ecological implications. *Arch. Environ. Contam. Toxicol.*, 15: 439–446.

Shaw, B.P., Sahu, A. and Panigrahi, A.K., 1986b. Mercury in plants, soil and water from a caustic-chlorine industry. *Bull. Environ. Contam. Toxicol.*, 36: 229–305.

Shaw, B.P., Sahu, A., Choudhury, S.B. and Panigrahi, A.K., 1988. Mercury in the Rushikulya river estuary. *Mar. Pollut. Bull.*, 19: 233–234.

Strickland, J.D.H. and Parsons, T.R., 1972. A practical handbook of sea water analysis. *Bull. Fish. Res. Bd. Can.*, 167: 311.

Suckcharoen, S. and Nourteva, P., 1882. In: *Bio-Accumulation of mercury*. Department of Environment Conservation, University of Helsinki, Finland, p. 4.

Sunderesan, B.B., Subrahmanyam, P.V.R. and Bhinde, A.D., 1983. An overview of toxic and hazardous waste in Indian Industry and Environment. Special issue (1983): In: *The State of Indias Environment*, 1984–85. The Second Citizen's Report, p. 196.

Takeuchi, T., 1961. A pathological study of Minamata disease in Japan. In: *Symposium on Geographic Neurology with Special Reference to Geographic Isolated*. World Federation on Neurology in conjuction with the 8th International Congress of Neurology in Rome.

Van Wambeke, L., 1976. In: *Environmental pollution and Control*. Can. Polyt. Est. Science. Ole The Rekelsen. Vatten.

Veriencar, X.N., 1984. Dissolved organic nutrients and phytoplankton in the Mandovi estuary and coastal waters of Goa. *Mahasagar, Bull. Natn. Inst. Oceanogr.*, 17: 141–149.

Veriencar, X.N., 1987. Distribution of nutrients in the Coastal and estuarine waters of Goa. *Mahasagar, Bull. Natn. Inst. Oceanogr.*, 20: 205–215.

Vijayalakshmi, G.S. and Venugopalan, V.K., 1973. Diurnal variation in the physico-chemical and biological properties in Vellar estuary. *Indian J. Mar. Sci.*, 1: 125–127.

Wanntorp, H. and Dyfverman, A., 1955. Identification and determination of mercury in samples. *Arkiv. fur Kemi*, 9(2): 7.

Zingde, M.D. and Desai, B.N., 1987. Pollution status of estuaries of Gujarat: An overview. *Contribution in Marine Sciences*, Dr. S.Z. Qasim, Sastyabdapurti Felicitation Volume, p. 245–267.

Chapter 9

Mercury Pollution in Aquatic Ecosystem

N. Varadarajan, Veerabasawant Reddy & S. Chandra Kumar

National Institute of Hydrology, Hard Rock Regional Centre
Hanuman Nagar, Belgaum – 590 001

ABSTRACT

A wide variety of pollutants–Physical, Chemical, Biological and Radiological have been identified in the environment consequent to urbanization, industrialization and new technological developments. Heavy metals through naturally occurring can be present in some areas in sufficient concentrations and physico-chemical forms that might create pollution problems. Sources of pollution of trace metals Hg, Pb, Cd, As, Cr, Zn, Cu, Mn and Fe are mainly aquatic releases from industrial operations, atmospheric releases from fossil fuel burning, domestic sewage discharges and land run-off. These elements exhibit varying environmental behaviour and toxicity to aquatic organisms and man. The background levels of some of the trace metals vary widely depending upon the location. Among the heavy metals, Mercury is commonly regarded as a great threat to man's welfare. It is steadily increasing in amounts as a result of its use in many products and subsequent release in the environment. So, therefore in this context, the problems associated with mercury, it sources and environmental behaviour with level have been clearly discussed with some specific cases. Also, some of the removal processes of mercury from contaminated water bodies have been briefly discussed.

Keywords: *Heavy metals, Aquatic releases, Mercury, Environmental behaviour and level, Contaminated water bodies.*

Introduction

Mercury, a liquid volatile metal found in rocks and soils is present in air as a result of human activities as the use of mercury compounds in production of fungicides, paints, cosmetics, paper pulp etc. Mercury enters water naturally as well as through industrial effluents. It is a potent hazardous substance. Both inorganic and organic forms are highly poisonous. Mercury occurs in the environment as metallic Hg and HgS. Annual production in the world is estimated to be about 9000 tons. About 50 per cent of it is estimated to be lost to the environment. Mercury is discharged into streams from crop lands treated with mercurial fungicides into the air from incineration of electrical equipment's, from exhausts, from metal smelters and from the combustion of coal and petroleum.

The mercury present into waste products get converted into two forms by anaerobic bacteria. The dimethyl form is volatile and escapes into the water from sediment and is converted into the methyl form at low pH. The ion is soluble, and it is in this form that it is absorbed by fishes. Methyl mercury is highly persistent and thus accumulates in food chain. Methyl mercury is soluble in lipids and thus after being taken by animals it accumulates in fatty tissues. Fish may accumulate the methyl mercury ions directly. There may be nearly 3000 times more mercury in fish than in water. Methyl mercury is the dominant toxic mercury species in the environment. Consumption of fish and crustacea being the main hazard to humans and higher animals. Methyl mercury appears to show strong teratogenic effects and carcinogenic and mutagenic activity has also been implied.

Table 9.1: Properties of Mercury

Sl.No.	Properties	Levels
1.	Atomic Weight	200.60
2.	Atomic Number	80
3.	Boiling point	356.60°C
4.	Melting point	-38.9°C
5.	Density	13.6
6.	Abundance in earth's crust	0.5 mg/kg
7.	Colour	Silver White
8.	Structure	Liquid, Metal

Specific Cases of Mercury Pollution

There are some specific cases of mercury contamination in the environment have been reported. Mercury enters the body through the food chain principally in the methyl mercury form; it affects the central nervous system and the brain. In Japan, methyl mercury poisoning through consumption of methyl mercury contaminated fish and shell fish caused disease known as "Minamata disease". This was the major incident involving mercury poisoning reported from Minamata bay, Japan (Figure 9.1). The Minamata bay incident was the first known modern case of mercury pollution in the aquatic environment. The "Minamata" disease characterized by madness, paralysis, loss of speech, vision and emotional control. Crippling of limbs followed by wasting muscle, coma and finally death. This disease had precipitated among 111 fishermen and their families living along the shore of Minamata bay, who consumed methyl mercury contaminated fish and 44 people died during the years 1953-1960.

A second outbreak of Minamata disease occurred in Japan during 1964-1965 among inhabitants again mainly fishermen and their families living near Niigata, who regularly eat seafish and shell fish from this area and also river fish from the inflowing Agano river carrying effluents from the Showa electrical industrial plant. In case of contamination of Agano river, shell fish levels of 40 ppm were reported. A third case of mass poisoning by mercurial effluents in Japan was reported during May 1973 by a research team of the Kumaoto University. They established that 59 inhabitants from the

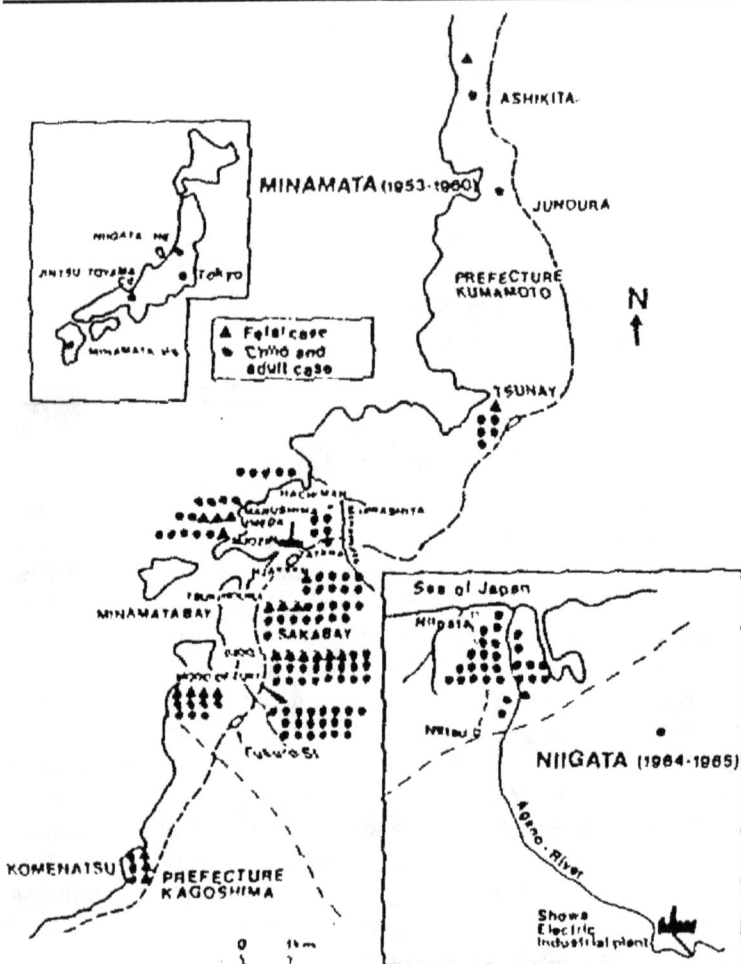

Figure 9.1: Mercury Poisoning in the Minamata Region

city of Goshonoura on the island Amakusu, opposite of Minamata, had become afflicted with Minamata disease.

Several other alarming incidents of mercurial poisoning have been reported from other continents. In Sweden many rivers and lakes are already polluted due to wide spread use of mercury compounds as fungicides and algicides in paper and pulp industries and in agriculture. Investigations of the aquatic environment in

Sweden revealed abnormally high concentration of mercury compounds in fresh and salt water fish and other aquatic organisms, which led to a ban in 1967 by the Swedish Medical Board on the sale of fish from approximately 40 lakes and rivers. In one particular case the source of mercury pollution was traced to discharge of mercurial fungicides by a paper mill. However, even after the use of the fungicide had been discontinued, high mercury concentrations prevailed in fish from the vicinity of the plant. By 1964 Swedish Ornithologists had noted that a significant decrease in wild bird population had occurred.

The most severe incident of man made mercury poisoning occurred during 1972 in Iraq. The first outbreaks were reported from northern Iraq, where farmers had received Wheat seed treated with mercurial fungicides (ethylmercury P-toluene Sulfonanilide) from Mexico, and ate the seed instead of planting it. When local government authorities began to note this self-inflicted plaque it was announced that any farmer possessing treated seeds was liable to prosecution involving the death sentence. Subsequently, the peasants disposed of these hazardous seeds into nearby rivers and lakes. The combined effects of this blunder have caused the loss of an estimated 5000 to 50,000 lives and the permanent disability of more than 100,000 people (possibly more than 500,000). The WHO has proposed a permissible level of 0.5 ppm for fish. In the Minamata incident Crabs contained as much as 24 ppm, while Kidney's from human victims contained 144 ppm.

Sources of Mercury Pollution

Mercury from Industry

Chlor alkali plants seem to be the chief source of mercury containing effluents. Maximum consumption of mercury is in the Chlor alkali plants. Mercury cells are used in the manufacture of chlorine and caustic soda. Losses estimated are 0.3 to 0.4 lb of mercury per ton of Cl_2 produced. With better designs losses are about 0.25 lb per ton of Cl_2 produced. A plant producing 100 ton Cl_2 per day may release 4000-8000 kg of mercury per year in waste effluents. The products of chlor alkali industry can also be source of long term mercury pollution. For example, it has been reported that household bleaching solutions contain 17-24 ppb of mercury and sodium hydroxide 53-1290 ppb. $HgCl_2$ is used as a catalyst in the manufacture of vinyl chloride plastics and acetaldehyde. Effluents from such plants contribute mercury into the aquatic environment. Mercury is used in production of Batteries, Street lamps, fluorescent tubes, Circuit breakers, Thermometers etc. all of which are finally discarded as wastes. Mercury compounds are used in antifouling paints. Mercury is used to prevent formation of slime and is lost to the wastewater effluents. Paper and Pulp industries of Japan and Canada also cause mercury pollution. Sediments from lakes in Southern Sweden were examined for mercury pollution originating from Pulp and Paper mills and from chlor alkali plants. Phenyl mercuric acetate and ethyl mercuric chloride have been used as fungicides. Other sources of mercury are in research, jewellery, moulding processes, in Pharmaceuticals and felt hat production, atmospheric fall out from thermal power plants, influences from mining and smelting processes of ores in particular Copper and Zinc smelting.

Mercury from Fossil Fuel Burning

Fossil fuel burning and cement manufacture cause emission of mercury into the atmosphere. Average value of mercury reported in coal is 0.012 ppm and in oil 10 ppm. Enhanced levels of mercury have been reported around fossil fuel burning plants.

Mercury from Domestic Sewage

Stronger increases in mercury concentrations from municipal sewage effluents often originate from plants accepting industrial wastes. Mercury content in surface sediments of Madison lakes, Wisconsin, USA were reported and it was due to disposal of sewage. Sludge from several sewage treatment plants in the area were reported to contain upto 25 ppm mercury and in treated effluents 0.002 ppm. Typical concentrations of mercury levels in sewage sludge are in the range of 4–320 ppm. Therefore, Mercury may get remobilize by burning of these sludge's or when used as fertilizers. One of the most serious forms of environmental pollution is the contamination of lakes by mercury wastes. This is due to the specific toxic behaviour of mercury arising from the biosynthesis of mercury-alkyls and to a distinct residue forming tendency of these substance in aquatic ecosystem. A comprehensive survey of the effects of mercury pollution in lakes, rivers and coastal regions was first undertaken in Sweden. After severe cases of bird poisoning resulting from seed dressings containing methylmercury dicyanidiamide in the early

1960's, the ditsribution of mercury was investigated in other parts of the environment. These studies led to the detection of industrial effluents as the major source of pollution. Table 9.2 shows the concentration and sources of mercury in lake sediments.

Table 9.2: Concentrations and Sources of Mercury in Lake Sediments (Examples)

Lake/Dam	Max. Conc. ppm (Background)	Source
Orly (Paris)	0.18 (0.15)	Municipal–background
Lake Winnipeg	0.23 (0.07)	Industrial–Winnipeg R
Lake Lucerne (Switzerland)	0.49 (0.10)	Municipal–Steinibach
Rietvlei Dam (Rep.S.Africa)	0.52 (0.12)	Agricultural
Lake Vattern (Sweden)	0.52 (0.10)	Municipal–Jonkoping
Lake Sangehris (Illnois)	0.57 (0.28)	Coal-fired power plant
Lake Oahe (South Dakota)	0.63 (0.03)	Gold-mining (Historical)
Lake Beldany (Poland)	1.77 (0.15)	Agricultural, Industrial
Lake Eklon (Sweden)	2.10 (0.10)	Industrial–Uppsala
Lake Saoseo (Switzerland)	2.20	Atmospheric fallout
Lake Trummen (Sweden)	3.30 (0.68)	Municipal–Vaxjo
Lake Coeur (d' Alene)	6.10	Mining–d'Alene River
Moste Dam (Yugoslavia)	17.70 (0.50)	Smelter–Jesenice
Lago Maggiore (Italy)	20.10 (0.24)	Industrial–Toce River
Pickwick Res. (tenn./U.S.A)	32.6 (1.30)	Industrial–Tennessee R.

Environmental Behaviour and Levels of Mercury

Mercury occurs in the environment as metallic Hg and HgS. Both elemental Hg and HgS are volatile. Mercury may be present in industrial wastes in the form of elemental mercury, inorganic or organic mercury compounds. Methylation of mercury takes place in the aquatic environment as a result of activities of bacteria, fungi or by enzyme systems. Bacteria also converts:

1. Phenyl ethyl and methyl mercury to elemental mercury.
2. Phenyl mercuric acetate to elemental mercury and diphenyl mercury.
3. Hg^{2+} ion to elemental mercury.

Other possible reactions in the environments are conversion of methyl mercury into dimethyl mercury in alkaline condition, production of insoluble HgS in anaerobic conditions and oxidation of HgS to sulphate in aerobic conditions and conversion to methyl mercury.

In fresh water, Pike accumulate 3000 times as much mercury as in the water. Tuna and Swordfish also accumulate high levels of mercury from water. Mercury in sea water probably exists as $HgCl_4$ or $HgCl_2$. In marine plankton, mercury levels reported are 2-10 ppb most of which in the inorganic form, where as at the higher trophic levels as in fish they are in the form of methyl mercury. Some mammals in the sea like seals accumulate mercury in the liver in the inorganic form. Mercury in fish being bound to the protein, its mercury concentration is related to the size, age and its trophic level.

The worldwide mercury production and release to the environment was estimated by Anderson and Nriagu. According to their estimate, the annual anthropogenic release of mercury on a global basis was about 3×10^6 kg around the year 1900, and had increased to about three times during the 1970's. Around 45 per cent was emitted to air, 7 per cent to water and 48 per cent to land. The continental degassing of mercury was estimated at 1.8×10^7 kg annually. As per WHO (1971) prescribed limits for mercury levels in soils have been reported ranging from 0.01 to 2 ppm (*Higgins and Burns, 1975). Higher values of mercury in soils/sediments have been reported near outfalls around industrial units (*Laghate et al.1972). Table 9.3 shows the concentration of mercury near outfalls of five chlor alkali industries.

Table 9.3: Typical Concentration of Mercury in Chlor-alkali Industrial Outfall Locations

Location	Hg ng/g of Soil Sediment (dry)
Anik Chempur	9
Shahabad	180
Shahabad	224
Kalyan	258
Thana, Belapur	806
Udyogamandal, Kerala	110

*Source: Laghate *et al.* (1972), * Paul and Pillai (1976).

In preliminary survey of mercury levels in fish in Mumbai and Thana region, fish muscles showed concentration varies from 35 to 566 ng/g wet tissue (*Tejam and Haldar, 1975). Table 9.4 shows the concentration of mercury in muscle and bone tissues in one of the species Tilapia Mozambique obtained from various locations.

Table 9.4: Mercury Levels in Tilapia Mozambique from Different Areas Around Mumbai

Location	Hg in ng/g (Wet)	
	Muscle	Bone
Colaba Market	243 ± 6	297 ± 8
Thana Market	417 ± 15	444 ± 17
Mardeswar lake, Thana	160 ± 5	189 ± 9
Siddeswar lake, Thana	215 ± 7	292 ± 8
Mumbai Creek	566 ± 22	621 ± 31
Thana Creek	512 ± 20	603 ± 29

Source: *Tejam and Haldar (1975).

The specimens from Creek areas indicates maximum accumulation possibly due to mercury discharges into the aquatic environment. A study was conducted in Mumbai (*Anand, 1976) about mercury concentrations in drinking water were reported to be in the range of 0.72–3.12 ng/ml with an average value of 2.30 ng/ml. Higher concentrations due to the leaching of mercury from large catchment areas of the lake from which drinking water is obtained. The daily intake of mercury through drinking water for an industrial was found to be 8 µg/day. The studies were conducted (*Anand, 1978) on 22 fish specimens obtained from Mumbai markets have shown mercury concentrations ranging from 4.6 to 283 ng/g. According to the studies conducted by the IIT Mumbai on Powai and Vihar lakes mercury concentration in various parts of fish species (Table 9.5) very well supports the proposition of biomethylation.

Removal of Mercury

Heavy metals in the environment are detrimental to all living species. The heavy metals are characterized by their non-biodegradability and tendency to accumulate in living material. Activated carbon and some other low cost adsorbents like coconut shell, rice husk, chitin, waste tea leaves and coffee powder etc. have already been utilized to remove heavy metals. Activated carbon adsorption appears to be effective process for the removal of heavy metals at trace level, but due to high cost, its practical applications are limited. The paper mill sludge is carbonaceous in nature which can be processed effectively for production of active carbon. The waste Bamboo dust and its carbon from pulp and paper mill can be used as effective adsorbents for removing mercury from wastewater. A very small amount of this can remove mercury effectively from contaminated water provided the pH is low. The experiments were conducted by using Bamboo dust and Bamboo dust carbon for removing mercury from wastewater. The result has shown that, the rate of adsorption of mercury on bamboo dust is almost twice as fast as that on bamboo dust carbon. Bamboo dust carbon is a much better adsorbent compared to Bamboo dust itself and much effective at low pH values.

Table 9.5: Mercury Content of Various Parts of Fish from Powai and Vihar Lakes (in ppb)

Parts	Powai	Vihar
Flesh	360	230
Gut	300	225
Scales	440	225
Gills	500	250

Restoration of Mercury Contaminated Lakes

There are many methods for restoration of mercury contaminated water bodies (*Jernelov and Lann, 1973).

1. Dredging of polluted sediments.
2. Converting mercury to mercury sulfide with a low methylation rate (anaerobic)
3. Binding of mercury to silica or coprecipitation with hydrous Fe and Mn oxides (aerobic)
4. Increasing the pH to give volatile dimethyl mercury rather than monomethyl mercury

5. Isolating the polluted sediments from the water body by means of physical barriers, such as polymer film overlays (*Widman and Epstein, 1972), blanket plugs of waste wool (*Tratnyek, 1972), sand and gravel overlays (*Bangers and Khattak, 1972, *Fujino, 1977).

One of the main disadvantage was due to dredging processes will disperse unrecovered sediment over large areas and would be likely to result in an increase in the mercury content over a period of time in the flora and fauna of the water course. On the otherhand, the major advantage by dredging, the possibility of mercury recovery at higher concentrations. Chemical leaching was found to be the most effective method in recovering mercury treatment with 1 per cent hypochlorite at pH 6.0–6.5 proved to be rapid and effective (upto 98 per cent–99 per cent) and no reactions occurred with carbonate. Mercury solubilized by this method could be recovered through the leachate method used to treat chlor alkali plant effluents (*e.g.* metal reduction by iron, adsorption on carbon, precipitation with aluminium, sulfide precipitation etc.) and subsequently the spent leachate could be returned to the water course.

The alternatives to the dredging process, the covering of mercury contaminated sediments with material that decreases the rate of release of methylmercury to the water (*Jernelov. *et al.*, 1975), polyethylene-nylon films, sands and gravels are mostly used for covering contaminated sediments indicate only a low sorption capacity for inorganic mercury and almost none for methylmercury; the inhibition mechanism must be due to stabilization of the colloidal sediment by reduction of sediment agitation (*Reimers *et al.*, 1975) An iron overlay using crushed automobiles has been proposed as an another approach for the treatment of mercury contaminated sediments. Iron will convert soluble mercury in water to elemental mercury; during this process hydrated ferric oxide is formed, which is an effective coprecipitator of mercury ions. In addition iron reduces toxic methylmercury elemental mercury. It was suggested by *Smith (1972), that iron scrap in combination with sand clay or other environmentally acceptable material should function as an effective barrier preventing any methylmercury found in the sediment from entering the overlaying water and contaminating the flora and fauna of the water body.

Conclusion

Heavy metals in the aquatic environment poses serious problems to aquatic ecosystem. Mercury is one among them, more toxic to human and aquatic ecosystem. The effects of mercury in aquatic environment and its removal have been discussed earlier. There are several methods of removal of mercury identified by experimental basis. When these methods are suggested for large water bodies like river, lake etc. the process become more complicated and practically limited due to high cost of implementation

Suggestions

1. There is an urgent need for protection of aquatic environment from pollution.

2. Avoid granting permission to new industries which discharges more effluents to the stream or water course without any survey and careful investigations.

3. Selection site for dumping solid wastes should be done after careful investigations and maximum possible avoid sites situated near the stream or water course.

4. Laws and Legislation's should be strictly adopted to protect the aquatic environments from pollution.

5. Educating people on environmental pollution, prevention and control measures.

6. Reuse and Recycling of wastewater according to the level of acceptance for various uses.

7. Removal of pollutants by chemical treatment has to be studied for its wide spread applications to large water bodies.

8. In any aquatic environment, sediment-water interactions has to be studied to understand the chemical behaviour and dispersal pattern.

References

Alloway, B.J., 1990. *Heavy Metals in Soils.* John Wiley and Sons, Inc. New York.

Bais, V.S. and Gupta, U.S., 1991. *Environment and Pollution.* Northern Book Centre, New Delhi.

Bhattacharya, K.G. and Sarma, Nikhileswar, 1995. Removal of Hg (II) from water by Adsorption on Bamboo Dust and Bamboo Dust Carbon. In: *3rd Intl. Conf. on Appropriate Waste Management Technologies for Developing Countries*, NEERI, Nagpur, Feb. 25–26.

Sarma, P.D., 1997. *Ecology and Environment.* Rastogi Publications, Meerut.

Sahu, K.C. and Prusty, B.G., 1984. Certain trace metal concentrations in recent sediments of Powai lake in Mumbai. In: *India's Environment Problems and Perspectives, Proceedings of the Seminar* at Trivandrum, 26–28, November.

Sinha, A.K., Ram Boojh and Vishwanathan, P.N., 1989. *Water Pollution.* Gyanodaya Prakashan, Nainital, U.P.

Varshney, C.K., 1991. *Water Pollution and Management.* Wiley Esatern Limited.

---

* Cross references are not cited.

Chapter 10

Effects of Different Environmental Factors on the Survival Rate of *Moina micrura* (Crustacea : Cladocera : Moinidae)

R. Ramanibai* & A.T. Sumitha

Unit of Biomonitoring and Management Lab,
Department of Zoology, University of Madras,
Guindy Campus, Chennai – 600 025

ABSTRACT

Survival rate and neonate production experiments were performed to study the influences of few experimental factors on *Moina micrura* from a tropical pond (Chetpet pond, Chennai). Different experimental designs were adopted for each environmental parameter. Our results show that the environmental factors have a significant role in the neonate production and survival of *Moina micrura*. We did find, however, the mixed algal diet can also took part in the process of neonate production and its longevity.

Keywords: Factors, Moina micrura, survival rate.

* Corresponding Author: E-mail: r_ramanibai@yahoo.com.

Aquatic Ecosystem and its Management

Introduction

Zooplankton plays an important role in the trophic dynamic of aquatic ecosystems. Among zooplanktons, cladocerans constitute an important component of most freshwater zooplankton communities (Dodson and Frey, 1991) and plays an important role in the field of freshwater aquaculture.many available reports mostly focusing either on the culture of cladocerans particularly the genus moina and daphnia or the role of algal feed, which promote the growth of organisms. The group cladocerans are known for their quick responses towards chemical compounds and by the predators (Rose *et al.*, 2001). Among zooplanktons, cladocerans play an important role in the field of freshwater aquaculture. Cladoceran cultures offer the possibility of obtaining large number of individuals with shorter time duration with appropriate conditions of temperature, food and water quality. Increase knowledge in the field of biology and the influences of abiotic factors on the population growth of plankton will ensure culture activities in great success (Tavares and Bachion., 2002).

Moina micrura Kurz is a quite large and filter feeding cladoceran, which lives in a variety of freshwater habitats. According to Rojas *et al.* (2001), *Moina micrura* used as live feed for fish larvae for aquaculture purposes.

The rate of growth and fecundity of cladocerans influenced by various physical factors were already studied elsewhere (Kibby *et al.*, 1971, Jeronimo and Valdivia., 1991, Shrivastava *et al.*, 1999, Aladin., 1991, Berner *et al.*, 1991). The aim of this study is to understand the influences of few selected abiotic factors (pH, salinity, temperature and photoperiod) on the survival and fecundity of *Moina micrura* maintained under laboratory conditions.

Material and Methods

Zooplankton samples were collected from a nearby natural pond situated at the back of Kilpauk Medical College, Chennai (Figure 10.1) by using a plankton net (250 μm). From the zooplankton samples, *Moina micrura* were isolated with the help of a micropipette and released into aquaria tank (5 l) filled with filtered pond water and was well aerated.The experimental set up was maintained at a temperature of $25 \pm 2°C$ with a photoperiod of 12 hours light at an intensity of 650 lux. The culture water was renewed every week. The

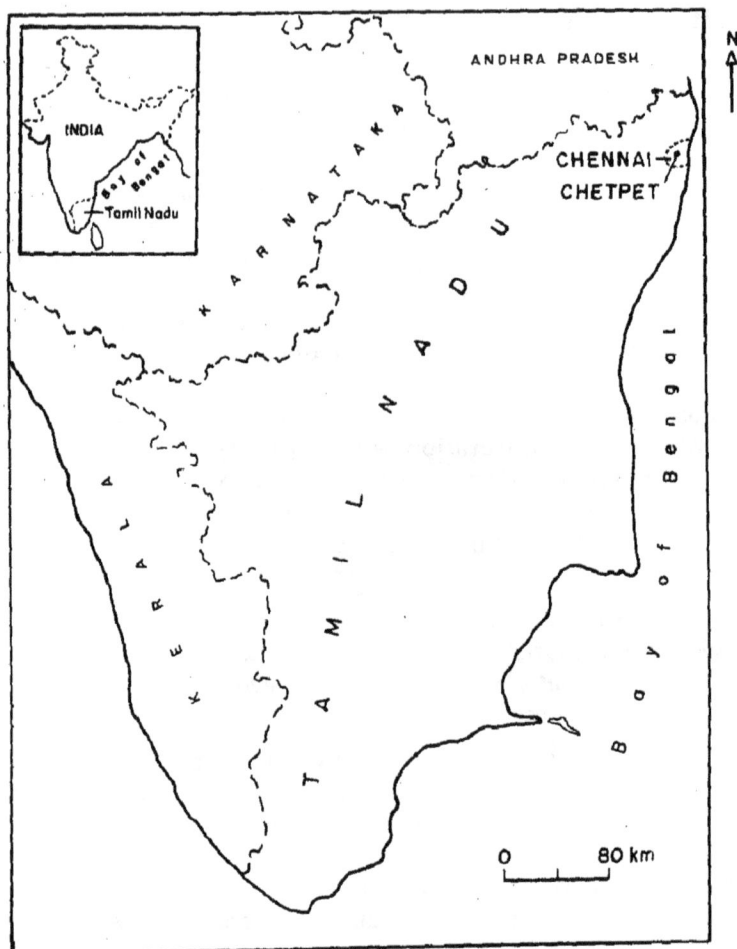

Figure 10.1: Location Map of Chetpet Pond in the State of Tamil Nadu, India

Moina micrura were fed with mixed algae (*Scenedesmus bijugatus, Chlorella vulgaris* and *Chlamydomonas globosa*) in the concentration of 3×10^5 cells/ml.

Four experimental setups were maintained in the laboratory to study the influence of each entity with varying ranges. 10 individuals were introduced along with food suspension of mixed algae feed.

The medium was renewed everyday with fresh food suspension and these experiments were lasted for 7 days. The factors selected for the study are pH (5.0–9.0), salinity (0–8 ppt), temperature (5–30°C) and photoperiod (0, 8, 12, 16 and 24 hours light). The results obtained through these experiments were subjected to student't' test analysis (Sokal and Rolf., 1981).

Results

The results showed that there was no significant change noticed in the survival of *Moina micrura* exposed from the pH range 5.0 to 9.0 ($t=1.605$; $p<0.05$). The rate of survival was directly proportional to the increase in the pH ranges. The highest rate of survival was noticed at pH 8.0 and 9.0 (Figure 10.2a). From the salinity experiments, the survival was high at 0 ppt (75 per cent) and 60 per cent survival was noticed at 2 and 4 ppt whereas nil survival was showed at 8 ppt. The survival rate was decreased with an increase in the concentration of salinity (Figure 10.2b).

As was to be expected, there was no significant changes in the temperature effects. Figure 10.2c demonstrates that the survival rate was very high at 25°C than the other ranges. Survival rate increased with increasing temperature upto 25°C and was declined at 30°C. The figure shows typical functional responses at all temperature, but the maximum growth rate differed with temperature. From the results, it showed a fitness landscape with a pronounced maximum at 25°C.

In the photoperiod experiments, our results showed that there was no significant difference in the survival rate of *M.micrura*, which exposed to different photoperiods varying from 0 to 24 hrs light (Figure 10.2d). Among different photoperiod ranges, the survival was very high at 12 hrs duration than the others.

Discussion

The results of the present study agrees with the results of Rojas *et al.* (2001) who monitored the optimum hatching conditions for *M.micrura* between pH 5.0–9.0.The temperature effect may be explained by the correlation of body size with clutch size of *Daphnia magna* (Giebelhausen and Lampert, 2001). Davison (1969) obtained 100 per cent hatching of decapsulated eggs of *Daphnia pulex* at temperatures of 24°C or lower, whereas at 30°C, only 20 per cent of

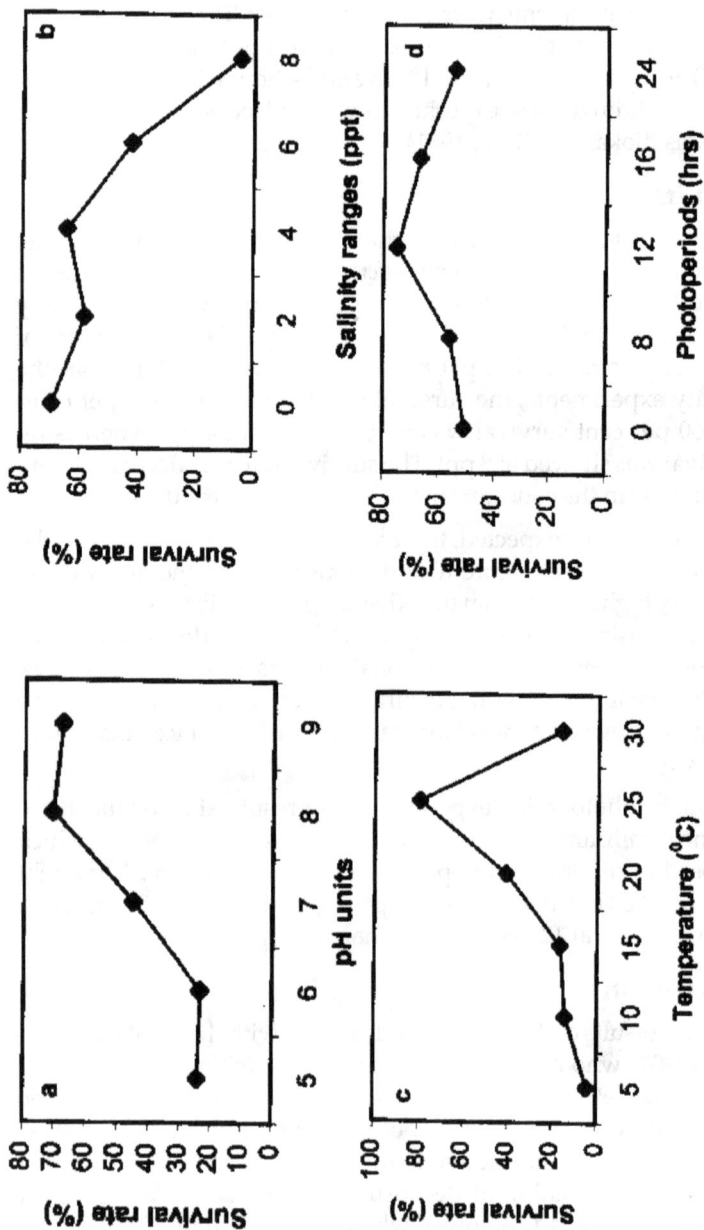

Figure 10.2: Survival Rate (Mean±SD) of *Moina micrura* at Different Environmental Factors (a) pH; (b) Salinity; (c) Temperature and (d) Photoperiod

the embryos completed their development. In the present study, survival rate was sensitive to temperature variation, especially at the highest values (Figure 10.2c). The survival rate occurred within a wide range of temperature (5 to 30°C).

Stross and Hill (1968) found similar enhancement of photoperiod in *Daphnia pulex*, and they demonstrated that the per cent diapausing broods in their cultures increased with decreasing photoperiods of particular interest is that our culture animals did not produce ephippial females under laboratory conditions simulating the temperature and photoperiod. Our results showed that there was no significant difference in the survival rate of *Moina micrura*, which exposed to different photoperiods varying from 0 to 24 hours light (Figure 10.2d). This is consistent with the results of Rojas *et al.* (2001) who found significant results on hatching rates of *Moina micrura*.

In conclusion, optimum conditions for the survival of *Moina micrura* were found to involve a water pH of around (7.0–9.0), and a temperature of 25°C, salinity at 2–4 ppt and photoperiod with minimum of 8 hours exposure.

Acknowledgement

We wish to thank UGC for financial support in the form of Post Doctoral Research Award (PDR) to one of the author (RR).

References

Aladin, N.V., 1991. Salinity tolerance and morphology of the osmoregulation organs in cladocera with special reference to cladocera from the Aral Sea. *Hydrobiologia*, 225: 291–299.

Berner, D.B., Nguyen, L., Nguy, S. and Burton, S., 1991. Photoperiod and temperature as inducers of gamogenesis in a dicyclic population of *Scapholeberis armata* Herrick (Crustacea : Cladocera : Daphniidae). *Hydrobiologia*, 225: 269–280.

Davison, J., 1969. Activation of the ephippial egg of *Daphnia pulex*. *J. Gen. Physio.*, 53: 562–575.

Giebelhausen, B. and Lampert, W., 2001. Temperature reaction norms of *Daphnia magna*: the effect of food concentration. *Freshwater Biol.*, 46: 281–289.

Jeronimo, N.M. and Valdivia, A.G., 1991. Fecundity, reproduction and growth of *Moina macrocopa* fed different algae. *Hydrobiologia*, 222: 49–55.

Kibby, H.V., 1971. Effect of temperature on the feeding behaviour of *Daphnia rosea*. *Limnol. Oceanogr.*, 16: 580–581.

Rojas, N.E.T., Marins, M.A. and Rocha, O., 2001. The effect of abiotic factors on the hatching of *Moina micrura* Kurz, 1874 (Crustacea : Cladocera) ephippial eggs. *Braz. J. Biol.*, 61(3): 371–376.

Rose, R.M., Warne, M.St.J. and Lim, R.P., 2002. Some life history responses of the cladoceran *Ceriodaphnia cf. dubia* to variations in population density at two different food concentrations. *Hydrobiologia*, 481: 157–164.

Shrivastava, M., Mahambre, G.G., Achuthankutty, C.T., Fernandes, B., Goswami, S.C. and Madhupratap, M., 1999. Parthenogenetic reproduction of *Diaphanosoma celebensis* (Crustacea : Cladocera): Effect of algae and algal density on survival, growth, life span and neonate production. *Marine Biol.*, 135: 663–670.

Sokal, R.R. and Rolf, F.J., 1981. Single classification analysis of variance. In: *Biometry: The Principles and Practice of Statistics in Biological Research*, 2nd edn, (Ed.) Wilson J. Cotters. W.H. Freeman and Company, New York, pp. 209–262.

Stross, R.G. and Hill, J.C., 1968. Photoperiod control of winter diapause in the freshwater crustaceans, Daphnia. *Biol. Bull.*, 134: 176–198.

Sumitra, V., 1970. Seasonal events in a natural population of *Daphnia carinata* King. *Proc. Indian. Acd. Sci.*, Vol.LXXI, No.5, Sec.B, 193–203.

Tavares, L.H.S. and Bachion, M.A., 2000. Population growth and development of two species of Cladocera, *Moina micrura* and *Diaphanosoma birgei*, in laboratory. *Braz. J. Biol.*, 62(4A): 701–711.

Chapter 11

Description of Freshwater Copepods of Chennai and Kanchipuram Districts

K. Renuga* & R. Ramanibai

Unit of Biomonitoring and Management Lab,
Department of Zoology, University of Madras,
Guindy Campus, Chennai – 600 025

ABSTRACT

The ultimate purpose of this review is to enhance the current state of information on the occurrence of freshwater copepods especially the cyclopoids from few freshwater habitats of Chennai city (Tamil Nadu, India) and few near by districts.

Keywords: Copepoda, Cyclopoids, Freshwater, Taxonomy, Zooplankton community.

Introduction

Copepods serve an important ecological part of both marine and fresh water habitats. Copepod constitute biggest source of

---

* Corresponding Author: E-mail: renu_2311@yahoo.co.in.

protein in the ocean, they are the predominant source of food for other organism (Davis, 1955). *Halicyclops hurlberi*, a new species of cyclopoid copepod, was described from San Diego, California. (Da Rocha, 1991).

More than sixty different species have been described by Dussart and Defays, (1985). *Mesocyclops leuckarti* from lake Biwa (Japan) was re-described as *Mesocyclops-thermocyclopoidis* Harada, 1931 (Kawabata, 1989). Cyclopoid copepods, *Diacyclops biceri* n.sp., were identified from lake Baikal. It was found in the interstitial water of a sandy beach at buchta peschanaye on the western shore of the central basin of Baikal. Stella (1988) reported that taxonomy and distribution of *Cyclops abyssorum* Sars in several deep and shallow lake and in long lasting ponds of Central Italy. Interspecific competition among predatory and herbivorous cyclopoids results in different vertical distributions from lakes.

The aim of our study is to identify and describe few copepod species from freshwater habitats in Chennai and Kanchipuram district of Tamil Nadu, India.

Materials and Methods

Surveys were made from few fresh water bodies including natural lakes, man made lakes, and pools located in the Southern region of Chennai Tamil Nadu. Four sampling locations were selected in Chennai which lies between lat. 13.00°–13.20°N; long. 80.20°–80.35°E and six sites from Kanchipuram district which lies between lat. 12.30°–13.00°N; long. 79.60°–80.30°E. (Figure 11.1).

Plankton samples from each water body was collected (200 μm mesh size) and preserved with 4 per cent formalin.One time sampling was done during the summer months of the year 2003(March-April 2003). Copepods were isolated, identified, and dissected under dissection microscope. Identification was done by standard keys Edmondson (1959) and Battish (1992). Drawings were made with the help of camera lucida attached to light microscope (Carl Zeiss).

Results

Five copepods belong to the order cyclopoida were recorded.

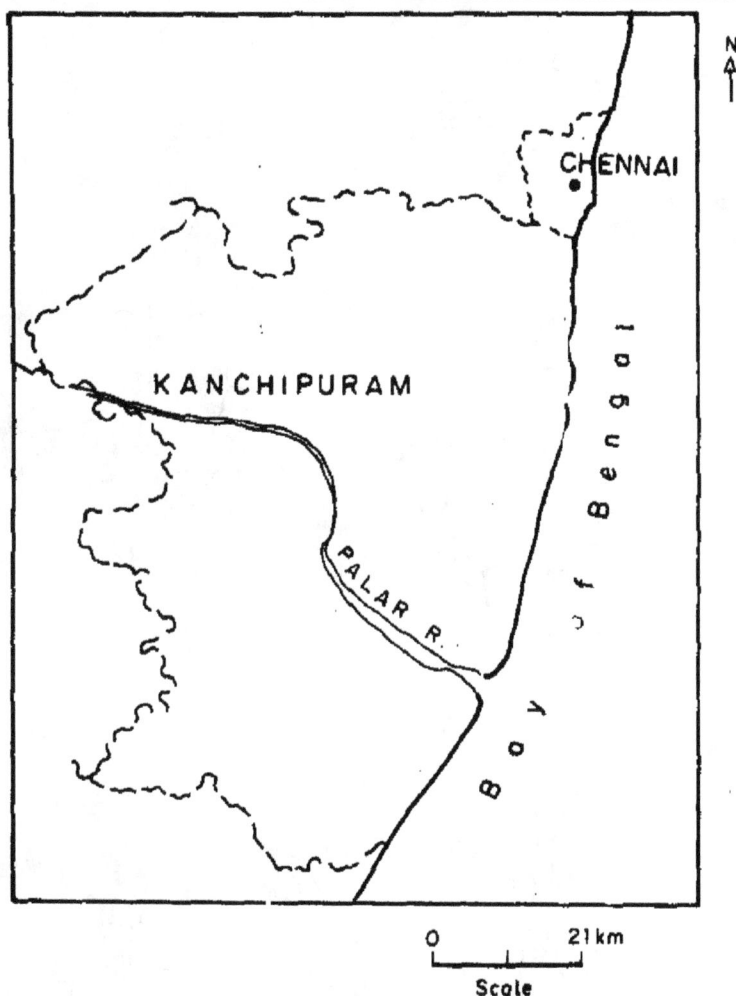

Figure 11.1

1. *Ectocyclops phaleratus* (Koch, 1838) (Figure 11.2)

Occurrence

1. Kiliyar River (Kanchipuram District)

Description

The shape of the body was long and dorsoventrally flattened and the total length was calculated as 1.12 mm. Cephalothorax was

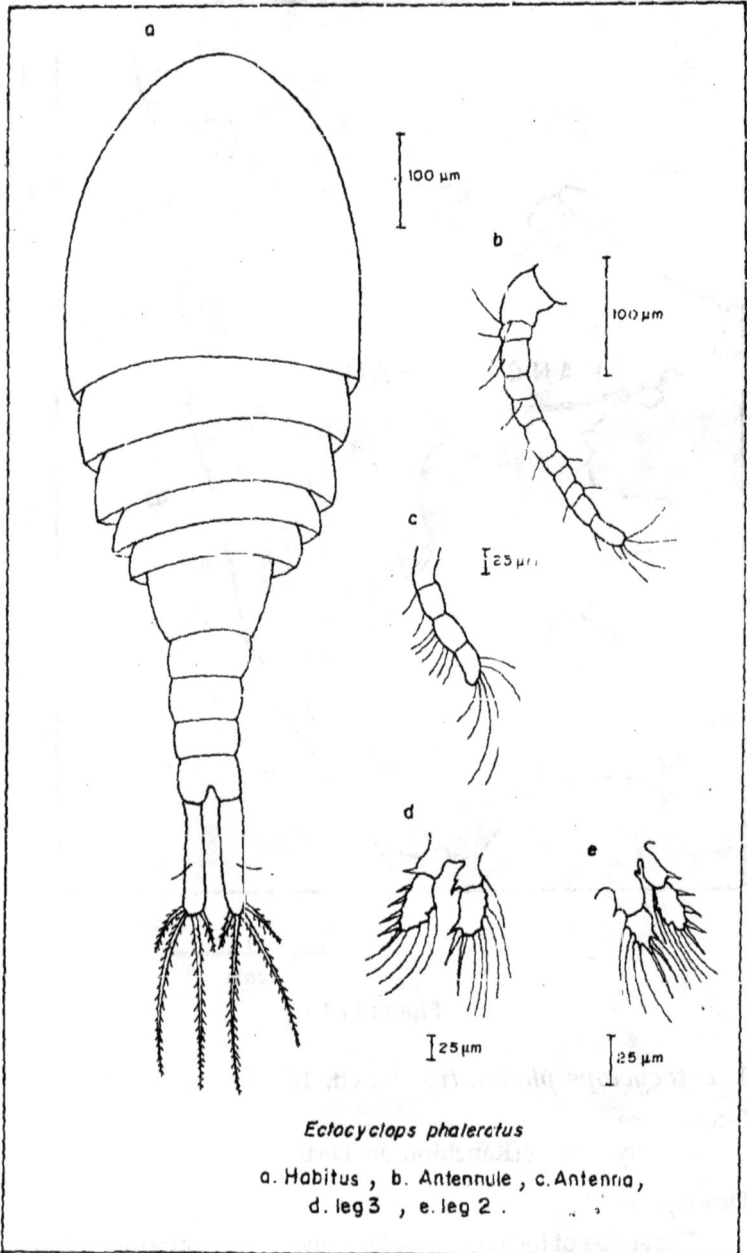

Ectocyclops phaleratus

a. Habitus , b. Antennule , c. Antenna,
d. leg 3 , e. leg 2 .

Figure 11.2

oval in shape and covered by carapace, consists of 5 segments. The antennules were long with 11 segments each. The antenna was shorter than the antennule and it was 4 segmented. The terminal segments of the antenna possess unequal setae. Swimming legs were biramous with 2 segments. The urosome consists of 5 segments, the genital segment slightly broader than the other 4-urosomal segments. The urosome ends with short caudal rami. The morphometric feature of this organism was agreed with the details of the same species given by Koch (1838).

2. *Metacyclops campestris* n.sp (Reid, 1987) (Figure 11.3)

Occurrence

1. Chetpet pond (Chennai) 2. Vaiyavoor lake and 3. Padappai lake (Kanchipuram district)

Description

The total length of the body was measured as 1.06 mm. The shape of the organism was cylindrical. Cephalothorax consists of 4 segments. The antennules were uniramous and 11 segmented. The lengths of the antennules were shorter than the cephalosome. The antenna was shorter than the antennules and consists of 4 segments. The terminal segment ends with 4 setae. Swimming legs were biramous. In leg 4 both endopod and exopod consist of single spine. The urosome comprised of 4 segments, the genital segments were broad and longer than the other 3 segments. The first urosome segment possess spine on each side in the posterior margin. The urosome ends with caudal rami.

3. *Microcyclops varicans* (Sars, 1863) (Figure 11.4)

Occurrence

1. Kolavai lake and 2. Padappai lake (Kanchipuram district)

Description

The cephalothorax was oval in shape with 4 segments the total length was measured as 1.23 mm. The antennules were uniramous and divided into 11 segments. The antenna was short and 4 segmented. Swimming legs were biramous with 3 segments. The urosome consists of 4 segments. The genital segment was broader than the other segments. The urosome ends with caudal rami.

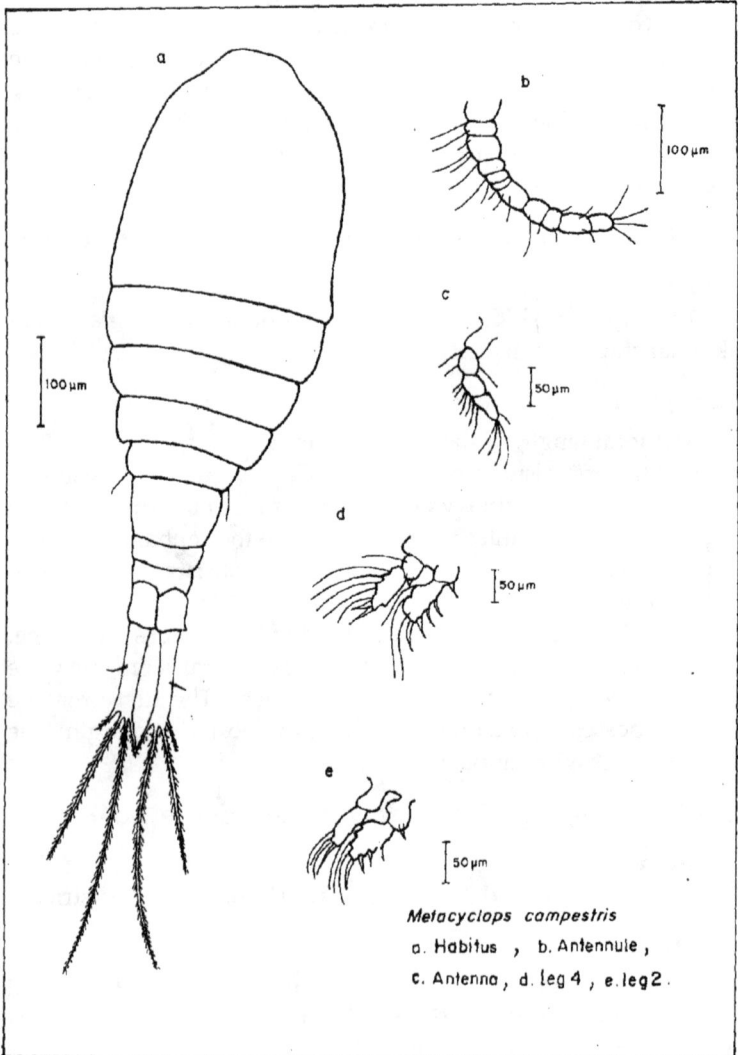

Metacyclops campestris
a. Habitus , b. Antennule ,
c. Antenna , d. leg 4 , e. leg 2.

Figure 11.3

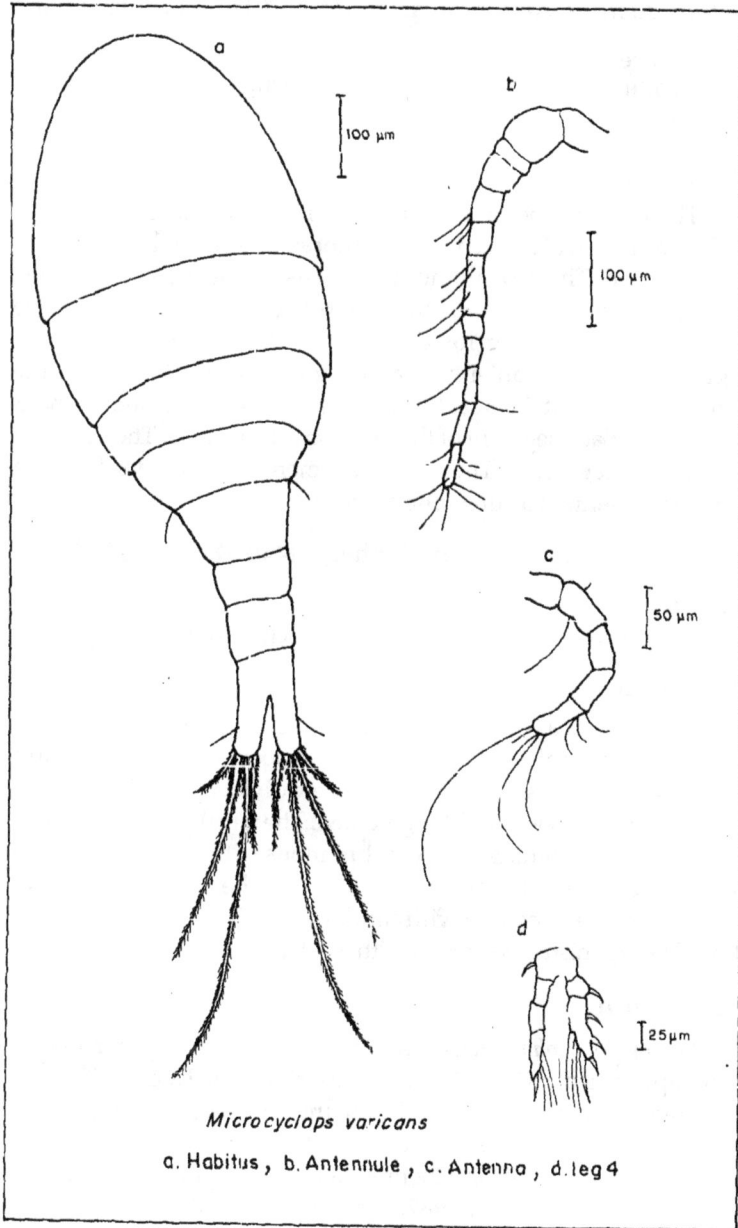

Microcyclops varicans

a. Habitus , b. Antennule , c. Antenna , d. leg 4

Figure 11.4

4. *Paracyclops carectum* n.sp (Reid, 1987) (Figure 11.5)

Occurrence

1. Chitlapakkam lake, 2. Perumbakam Chinna yeri, and 3. Retteri (Chennai).

Description

The body of the organism was cylindrical and calculated as 1.03 mm. The oval shaped cephalothorax was much broader than the abdomen. The cephalothorax consists of 4 segments. Prosome were dorsoventrally depressed. Posterolateral margins of last three prosomites ends with long spinules. The antennules were 8 segmented and the antenna was 4 segmented the basipod of the antenna consist of 2 setae. Swimming legs were biramous and 3 segmented. Each segment of the legs consists of spine. The urosome consist of 4 segments. The genital segment was broader than the other 3 segments. The urosome ends with caudal rami.

5. *Tropocyclops prasinus* (Fisher, 1860) (Figure 11.6)

Occurrence

1. Palar river, 2. Parundhur lake (Kanchipuram district).

Description

The body was short and slender and measured as 0.91 mm inclusive of caudal rami. The cephalothorax was oval shaped and consist of 4 segments. The antennules were divided into 12 segments. The antenna was short and 4 segmented. The terminal segment ends with 4 setae. Swimming legs were biramous with 3 segments. Setae were present in both endopod and exopod. The urosome was long and 5 segmented and end with caudal rami. The caudal rami were short. The organism was agreed with the detail given by Fisher (1860).

Discussion

Ectocyclops phaleratus was present in the kiliyar river in Kanchipuram district. This was described by Koch, 1838. He considered this species differed from the other species because the presence of fine bristles on the row of caudal rami. *Metacyclops campestris* was present in Chennai and Kanchipuram districts; it was identified by Reid, 1987. He considered that the seminal receptacle is less broadened and divided anteriorly.

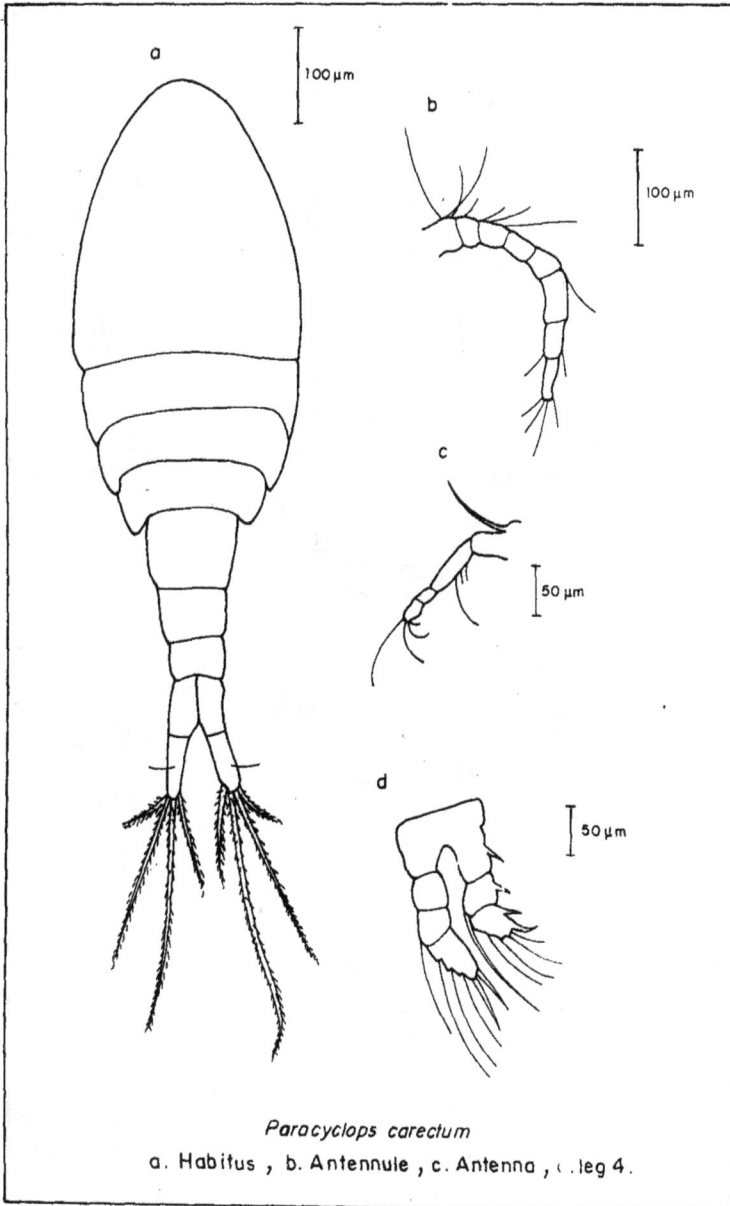

Paracyclops carectum

a. Habitus , b. Antennule , c. Antenna , c.leg 4.

Figure 11.5

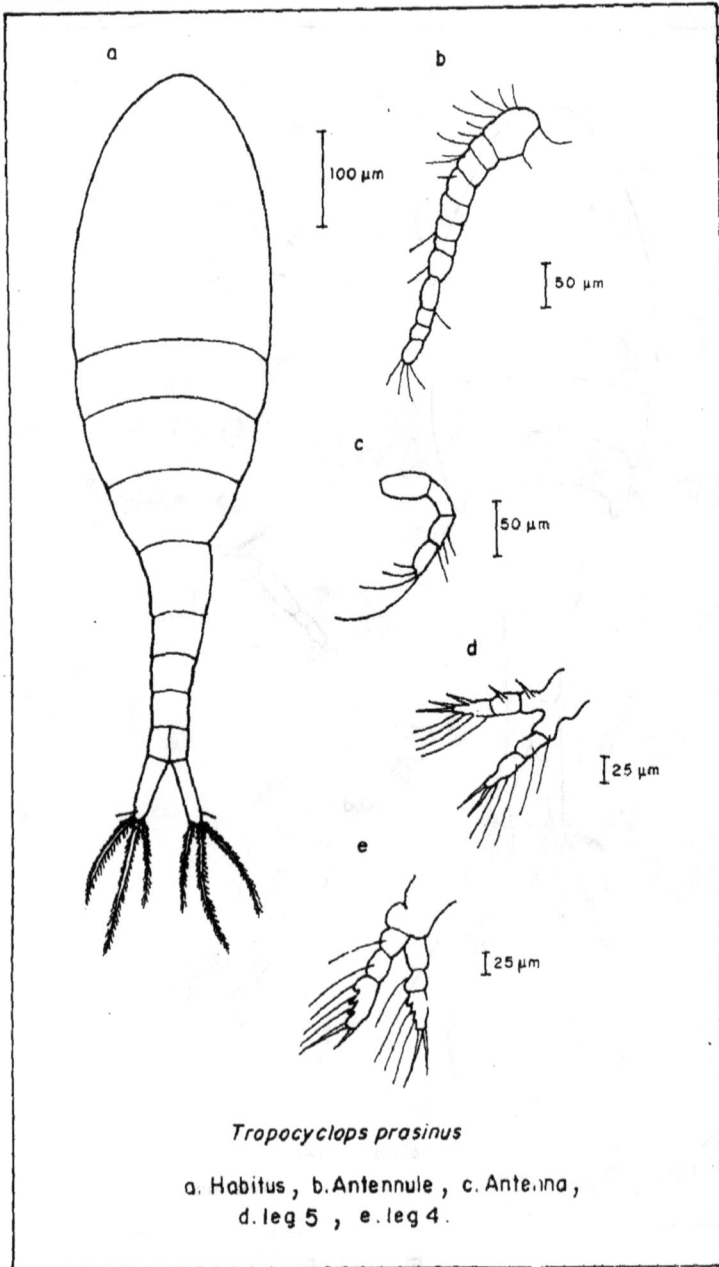

Tropocyclops prasinus

a. Habitus , b. Antennule , c. Antenna,
d. leg 5 , e. leg 4 .

Figure 11.6

Microcyclops varicans was present in the Kolavai lake in Kanchipuram district. This was described by Sars, 1863. This species is distinguished from the other species by the presence of caudal rami 3 to 4 times as long as wide. *Paracyclops carectum* was present in Chennai district. This was identified by Reid in 1987. *Tropocyclops prasinus* were present in Kanchipuram district and it was described by Fisher, 1860. The key to easily identify this species by its short caudal rami.

In the study areas of Chennai and Kanchipuram districts copepods were abundant in the temporary freshwater lakes and ponds.

Conclusion

We assume that a number of subspecies occur for each genus where it requires reconfirmation of species based on detailed analysis of large gametogenetic population of copepods. Where it requires for their contributions towards copepod taxonomy.

Acknowledgement

One of the author (RR) thanks UGC for financial support as postdoctoral research Award (2002-2005).

References

Battish, S.K., 1992. *Freshwater Zooplankton of India*. Oxford and IBH Publishing Co. Pvt., New Delhi, India.

Davis, C., 1955. *The Marine and Freshwater Plankton*. Michigan State University Press.

Da Rocha, C.E.F., 1991. A new species of halicyclops (Copepoda : Cyclopidae) from California, and a revision of some halicyclops material in the collection of the US museum of natural history. *Hydrobiologia*, 226: 29–37.

Dussart, B. and Defaye, D., 1985.Repertoire mondial des crustaces copepods cyclopoides. Editions CNRS, Bordeaux/Paris, 236 pp.

Edmondson, W.T., 1959. *Freshwater Biology*, 2nd edn. John Wiley and Sons Inc., New York.

Harada, I., 1931. Studien uber die susswasserfauna formosas. IV. Susswasser-cyclopioden aus Formosa. Annol. Zool. Japan, 13: 149–168.

Kawabata, K., 1989. Seasonal changes in abundance and vertical distribution of *Mesocyclops thermocyclopoides, Cyclops vicinus* and *Daphnia longispina* in lake Biwa. *Jpn. J. Limnol.,* 50: 9–13.

Stella, E., 1988. Contribution to the taxonomy and distribution of *Cyclops abbyssorum* Sars (Crustacean : Copepoda) in several lakes and ponds of Central Italy. *Hydrobiologia,* 167/168: 381–385.

Chapter 12

Studies on Pollution Sources and the Need for Early Steps for Abatement Along the Coastal Karnataka

T. Ananda Rao[1] *& Doda Aswathanarayana Swamy*[2]

[1]*Karnataka Association for the Advancement of Science,
Mathematics Building Annexe, Central College, Bangalore – 560 001*
[2]*Department of Forest, Ecology and Environment,
Government of Karnataka, M.S. Building, Bangalore – 560 001*

ABSTRACT

Environmental pollution has become a potential problem in respect of water, air and soil for the welfare of human community. The coastal environment in Karnataka needs immediate attention so as to apply measures to mitigate the severe adverse consequences of pollution. The hazards involved due to the water bodies, industries, ballast water from the anchored ships, ash pond, coir-ratings, caustic soda plant, waste products from urban areas, and pollutants through stack from industrial complex are reviewed. The lack of sufficient data especially on Environmental Impact Assessment and Environmental Management Plants so as to monitor their toxic effects on the Coastal environment systematically in a coordinated manner along the coastal Karnataka have been emphasized.

Keywords: Pollution sources, Coastal Karnataka, Issues for abatement.

Introduction

Environment pollution is a world wide phenomenon of severe adverse consequences. It is infact a side effect in the process of development in respect to the human society. Of late, it has become a potential problem in respect of water, air and soil for the welfare of human community. The water bodies in the coastal zone are polluted mainly due to effluents from industries, urban sewage, insanitary flow of used water, night soil, garbage, detergents of varied kinds and coir rettings. It appears from the information available to us presently that there has been not sufficient concerted effort to monitor the marine ecosystems systematically in a coordinated manner along the Karnataka coast.

An attempt is made, here to visualise the likely pollution impact on the obtaining natural and sociosphere of the coastal area with the hope that it would help to mitigate the harmful effects in the coastal zone, by adopting remedial measures.

The wastewater effluents of Mangalore Chemicals and Fertilizers Ltd. (MCF) is conveyed through a pipe into the sea after spray-stripping, diffusion and cooling. The wastewater as such is rich in ammonia/urea concentration which would certainly be harmful to biolife near the Mangalore shore. This industry manufactures urea, and DAP fertilizer at an installed capacity of 30.900 MT/M and 19.800 MT/M and generates about 11,800 M^3/ day of effluent and this, effluent after treatment is discharged into Arabian Sea during monsoon and during summer the effluent is partially discharged and partially used for green belt. The effluent discharge point is 100 mts. from low tide line. However, ammonia nitrogen levels should be constantly moniotored along the coast particularly as this nutrient is a dominant chemical component of the effluent from the Mangalore Chemicals and Fertilizers.

The Kudremukh Iron Ore Ltd. (KIOCL) discharges coloured wastewater into the sea near Thannirbhavi shore. The iron concentration in the wastewaters would certainly affect the quality of micronutrients available to marine fauna. This industry is engaged in the production of 3 million tons per annum of Iron ore pellets and is having licence capacity of 6 million tons per annum and discharges 14 thousand cubic meter/day effluents into Arabian Sea at a distance of 50 mts. From the shore line.

Investigations of the ecological conditions prior to the discharge of effluents from MCF into the Arabian Sea off Mangalore shore was attempted by Menon *et al.* (1977). Since 1976, the monitory programme has been continued, on a regular basis (Gupta *et al.*, 1988). In their account an assessment of the impact of the effluents from the MCF on the zooplanktons is recorded. The collections were made in the vicinity of discharge point covering an area of 10 kms over a period of one year and the data thus obtained was compared with those of corresponding periods in the latter years. Dissolved oxygen content has shown to marked variations while the salinity values of post-monsoon seasons have been lower, but pre-monsoon salinity values were similar. The plankton distribution had exhibited marked qualitative, spatial and temporal variations. It has been observed that copepods, copepodities, cladocerans, rotifers and decapod larvae have also showed temporal shifts in their abundance. Similarly hydrographic parameters like water temperature, pH, transparency, salinity, dissolved oxygen, phosphate-phosphorus, ammonical nitrogen and nitrate-nitrogen have also been monitored.

From the data available from the researches of the College of Fisheries on the quality of coastal water of Dakshina Kannada it can be concluded that it created certain impacts on the biota, so far. However, there is a scope to intensify the research to find out the accumulation of trace elements in the benthic flora and fauna and also in sediments to evaluate their impact on the marine life (consultancy in put to DANIDR-EMPS-Report pp. 1–33) (Setty *et al.*, 1995).

In correction with the establishment of a three million tonne Grass Root Refinery at Mangalore, a multi-purpose vessel with buoy laying capability is in stream which will also function as an anti-pollution craft.

This industry has laid a submarine pipeline to a distance of 900 mts. from the shore. They propose to discharge 6360 KLD of effluent into sea. This industry was commissioned during March/April 1996 and presently the effluent is stored in the reservoir and is using the part of the treated effluent for processing fire extinguisher, gardening etc. the submarine pipeline is commissioned. This expansion programme has increased the quantity of effluent and same is proposed to be discharged into the sea. Also, the Government of India in its environmental Clearance and the Government of

Karnataka has stipulated a condition to the industry to recycle the effluent before discharging into the sea.

The old Mangalore Port has a maritime history of its own. It has an important gateway to the illustrious Hindu Empire of Vijayanagar. It is situated at the confluence of the Gurpur and Nethravathi rivers with a bar at the mouth of the port. This minor port served as a road stead, and also used to be closed during monsoon months.

The Mangalore Port Trust (MMPT) is located 9 kms north of the minor port and it is a all-weather port. Now, it is the deepest inner harbour catering to 66,000 DWT vessels drawing a draft of 12.5 meteres. The bay of the port covers the w following facilities: berth, cargo handling, dredging, port crafts, bunkering, and storage investigations. There is a discharge of oil ballast water from the anchored ships near the shore water and its spread along the nearby beaches. Here, there is a scope to monitor the quantity of oily ballast water discharged and its impact on marine life. It is observed that steps have been taken to mitigate the abuse of shore waters by anchored ships.

BASF INDIA LIMITED is a new industry located in Bala village near Mangalore. This industry is commissioned recently in March 96 and manufactures Dyes and Dye intermediates. In the process it generates about 500 M^3/day of effluent and the treated effluent is discharged into the sea at a distance of 1.4 km from shore through a submarine pipeline.

Effluents Discharge

As per the quality of the effluents, it is inferred that the outflow of ammonia in the discharge is below the prescribed limit. Interestingly, the same level of ammonia, if let into fresh water, will have an adverse effect and will cause unsustainable damage to the fish life within. It has been presumed that the saline water neutralises and breaks down the effect of ammonia to a large extent through the contents of ammonia in the effluent discharges may sometimes reach above the normal.

The presence of ammonia in the sea is said to be very nutritive, since it helps to increase the biomass of the ecosystem but on the contrary it decreases the diversity of phytoplankton living within it, which in turn supports a wide variety of zooplankton. These planktons together and independently are very essential for the presence and existence of big populations and other marine fauna. Decrease in the diversity of phytoplanktons will therefore disrupt zooplanktons variety and quality which could affect the fish and other marine life within the sea. Therefore, if the same amount of MCF effluents are let into the sea over the years it would cause great damage to the ecosystem. Moreover, as far as the present state of the environment around Tannirbhavi and its vicinity is concernea, the study has been able to make definite assessments (Consultancy input to DANIDA-Report). A caustic soda plant of Ballapur Industries Limited is engaged in the manufacture of Caustic Soda, Phosphoric Acid, Dicalcium Phosphate, Kestra moulded equipment and discharges 18.00,000 lts/day of effluents into Arabian Sea at a distance of 800 mts. from the shore through a pipeline. It has been disposing its effluents into the sea from the Binaga shore, near Karwar. It was decided to lay the pipeline carrying the effluents, beyond 3.5 kms near Anjadiv islands from the Binaga shore. It was found that the occurrence of large scale mortality of fish was a phenomenon recurring often in the nallas. This is feared to be due to the residual chlorine in the water, which was beyond the permissible limits.

The other category of aerial pollutants as a result of industrial development are carbon monoxide from incomplete combustion of carbon in fuels, gases from the internal combustion engines, emissions of sulphur-dioxide, ammonia, dust and coal particulates, nitrous oxides from automobiles. fertilizer factories create sufficient air pollution problems. The emission through a stack from Mangalorezonal complex has maximum pollutants. Dust emissions from KIOCL is significant and a very high air-borne particulate has been recorded. Stack heights have been increased to 44 mtrs in respect of MCF to reduce their density downwards.

Some of the industries have provided necessary air pollution control measures and generally the emissions are meeting the standards. Since the pollution is in concomitant with most industrial activity, methods of monitoring is imperative.

Mangalore Power Company is going to establish a Thermal Power Plant at Nandikur in partnership with Cogentrix, USA and the China Light and Power (International) Ltd., Hong Kong. It is obvious in a project of this type environmental detailed impact assessment in respect of transportation of coal and its storage at plant, site, impact of flue gas emissions on ambient air quality with special reference to sociosphere, intake and disposal of cooling waters, fly ash disposal/utilisation plant and rehabilitation of oustees are the main issues. However, the Cogentrix has produced leaflets indicating commitments in respect of environment hazards, namely ash management, protection of air resources, responsible social programmes, environmental monitoring of air resources, water resources from plant operations, acid deposition, stock gas emissions from plant operations, and noise emissions around the surrounding land uses and help to establish environmental record programme by way of research issues, scholarships etc., in collaboration with research institutions. However, the said issues have been addressed in details in a of action and management over the years to adopt remedial measures as and when required for clean environment. However, the, rapid EIA report proposed by MPC is not clear on many of the likely hazards during the plant operation as indicated by NEERI, Nagpur in their report.

Granite blocks from inland quarries are stacked near Mangalore port for stevedoring at Mangalore. Its presence along a considerable portion of the backshore which has the potentiality of developing into a dark system should be looked into from a bio-esthetic point of view. Even though their presence has not created any ecological problem, however it is certain that their storage and further lifting by cranes would cause considerable disturbance to the adjacent developing urban life.

It is observed that there are many medium and small scale industries located in and outside industrial estates. There is no wastewater disposal system in such areas to analyse their impact on the environment. The small scale fishing industry along the beach areas produce hazardous wastes. It is clear that their waste products are not regulate through a proper channel. The fish oil plants along the beaches, channel their effluents directly into the sea and utilise the entire beach area for drying and storing of fish products. It is very clear that their activity is unregulated and harmful to the quality of beach environment (Rao 1990, 1991).

A few developments along the banks of rivers have increased the urban waste discharge to an alarming extent. The obtaining situation is a threat to the quality of river water used for drinking purposes. The absence of sewage systems in many of the medium and small scale industries would be a health hazard to a certain extent. These factor will certainly increase and create a significant impact on the beach developmental plans. It is imperative to establish a monitoring system to check and record the existing and possible pollution levels in the coastal zone to plan, pollution abatement measures in the integrated environmental planning and management.

References

Gupta, T.R.C., Katti, R.J., Hariharan, V., Satish. M., Gowda, G. and Shetty, H.P.C., 1988. Assessment of the impact of the effluent from the MCB on the zooplanktons in the First Indian Fisheries Forum, (Ed.) M. Mohan Joseph. *Proc. Indian Fisheries Soc.*, Indian Branch, Mangalore, pp. 287–295.

Menon, N.R., Gupta, T.R.C., Hariharan, V., Katti, R.J. and Shetty, H.P.C., 1977. *Marine Plankton of Mangalore Waters: A Pre-Pollution Assessment.* Zoological Spl. Pub., UNESCO NIO, pp. 274–283.

Rao, T.A., 1991. On the environmental master plan for the management of beach and its environs of the Karnataka Coast. Final Report, Department of Forests, Ecology and Environment. Government of Karnataka, pp. 1–700 (Unpublished).

Rao, T.A., 1990. Coastal Ecosystem of the Karnataka Coast I. Mangroves; II. Beaches. Final Report,

Department of Forest, Ecology and Environment, Government of Karnataka, pp. 1–400, pp. 1–625 (Unpublished).

Setty, T.M.R. *et al.*, 1995. Consultancy input to Danida–EMPS: A special study on coastal water quality off Dakshina Kannada. Report by College of Fisheries, Mangalore, pp. 1–33.

Chapter 13

Concentration of Nine Heavy Metals in Sardines and Mackerels from Gopalpur Coast, Orissa

Lakshman Nayak

P.G. Department of Marine Sciences, Berhampur University,
Berhampur – 760 007, Orissa

ABSTRACT

Two fishes namely, *Sardinella fimbriata* and *Rastrelliger kanagurta* were collected from Gopalpur Coast for analysis of nine heavy metals such as Copper, Nickel, Zinc, Iron, Lead, Manganese, Cadmium, Chromium and Cobalt, during the period January 1997 to December 1997 *Sardinella fimbriata* was found abundantly throughout the year. *Rastrelliger kanagurta* was found abundantly during winter. The samples were collected by the help of the local fishermen. The total length and weight of *Sardinella fimbriata* ranged from 9.5 cm to 18 cm and 9.66 gram to 43 gram respectively. The total length and weight of *Rastrelliger kanagurta* ranged from to cm to 24 cm and 9.42 gram to 130.65 gram respectively. Then the samples of these species were analysed for heavy metal concentration in liver, gills and muscle. In *Sardinella fimbriata* the high concentration of copper was found to be 55 ppm in liver and followed by gills being 14.3 ppm and muscle being 9.8 ppm. Nickel concentration was found to be highest 155 ppm in gills followed by liver being 155 ppm and muscle being 120 ppm.

Zinc concentration was recorded 367 ppm in liver followed by gills being 355 ppm and muscle being 346 ppm. Iron concentration was found to be highest in liver being 1,087 ppm followed by gills 1,080 ppm and muscle being 980 ppm. Lead concentration was found to be highest in liver being 95.3 ppm followed by gills being 50 ppm and muscle being 42 ppm Manganese concentration was found to be highest in liver being 100 ppm followed by gills being 65 ppm and muscle being 50 ppm. Chromium concentration was found to be highest in liver being 47 ppm followed by gills being 23 ppm and muscle being 17.5 ppm Cadmiumwas found to be highest in liver being 10 ppm followed by gills being 5 ppm and muscle being 4 ppm. Cobalt concentration was found to be highest in liver being 35 ppm followed by gills being 10.1 ppm and muscle being 9.5 ppm. In *Rastrelliger kanagurta* Copper concentration was observed to be highest in liver being 48.4 ppm followed by gills being 23 ppm and muscle 11 ppm. Nickel concentration was found to be highest in liver being 179 ppm followed by gills being 145 ppm and muscle being 120 ppm. Zinc concentration was found to be highest in liver being 390 ppm followed by muscle being 376 ppm and gills 300 ppm. Iron was found to be highest in liver being 1,083 ppm followed by gills 1,000 ppm and muscle being 930 ppm. Lead was found to be highest in liver being 59.6 followed by gills 53.32 ppm and muscle being 38.09 ppm. Manganese concentration was highest in liver being 120 ppm followed by gills 74 ppm and muscle being 53 ppm. Chromium concentration was highest in liver being 53 ppm followed by gills 25.54 ppm and muscle being 19 ppm. Cadmium was found highest in gills being 41.1 ppm followed by liver 11.2 ppm and muscle being 5 ppm. High concentration of Cobalt was recorded in liver being 40 ppm followed by gills being 11.6 ppm and muscle being 7 ppm. The trend of heavy metal concentration was found to be liver > gills > muscle.

Keywords: Heavy metals, Concentration, Marine fishes, Gopalpur, Orissa.

Introduction

The world population is rapidly increasing and man with the help of technological aids at his command, has exploited a major part of the resources available on land Now the man also exploited sea resources to provide sufficient food for the increasing population because sea foods are rich in protein, which fills the protein gap in human food supply in many parts of the world. Ocean occupies three-fourth of the area of world which is an important environmental compartment significance to mankind. They constitute an inexhaustible reservoir of food, chemicals, minerals, oils and natural gas. Still the marine environment is frequently treated with the greatest carelessness in terms of pollution. According to the Group of Experts on the Scientific Aspects of Marine Pollution (GESAMP) marine pollution is defined as "introduction by man, directly or indirectly of substances or energy to the marine environment resulting in deleterious effects such as hazards to human health; hindrance of marine activities, including fishing, impairment of the quality for the use of sea water; and reduction of amenities" (Clark, 1989). Most people thought that oceans have enormous capacity for uptake of pollutants without causing noticeable hazards.

Metals are natural constituent of sea water. Many metals, occur in the marine environment, at (race or ultra-trace level which is expressed as parts per million (ppm) to parts per billion (ppb) in certain sediments and organisms, down to parts per trillion (ppt) level in sea water. In open ocean, metals are present in very low concentration. But in coastal oceans and their associated embankments such as estuaries, lagoons, bays, creeks, swamps, etc. the concentrations of metals are relatively high. The increased level of metals in coastal ecosystem have been attributed to the entry of contaminants originated due to anthropogenic activities either on land or in the ocean.

Certain metals are essential to normal growth and development of organisms. A metal can be regarded as essential when the organism can neither grow nor complete its life cycle in the absence of the element; the element cannot be replaced by any other element and the element exerts a direct influence on organism and its metabolism. The essential metals are helpful in the physiological mechanism of the living beings by its presence. As for example, Iron occur in the haemoglobin of many vertebrates and invertebrates which help for respiratory mechanism. Copper and Vanadium are also occur in the respiratory pigment of molluscs, crustaceans and tunicates respectively. Many enzymes contain Zinc and Vitamin B_{12} contain Cobalt. The essential metals are also toxic to the living organisms after a certain level.

Similarly an element can be regarded as toxic if that element injures growth or metabolism of an organism when supplied above a certain concentration. Metals of biological concern may be of three types. The first one is light metal which generally transported as mobile actions in transitional solutions. This include Sodium, Potassium, Calcium and so on. The second one is transitional metal including Iron, Copper, Cobalt and Manganese. These are essential in low concentration but may be toxic in high concentration. The last one is metalloid which are normally not required for any metabolic activity but are toxic to the cell al quite low concentration.

These include Mercury, Lead, Tin, Selenium and Asrsenic. Transitional metals and metalloids are usually termed as heavy metals (Clark 1989).

The term heavy metal is a loose one including transitional metals like Chromium, Cobalt, Nickel, Copper, Zinc, Calcium, Mercury, Lead. Arsenic, Antimony and Bismuth. These heavy metals are also known as conservative pollutants due to its non-degradable characteristic. The most toxic metallic pollutants are Mercury, Lead, Zinc and Copper. Though heavy metals occur naturally in the marine environment, these are also enter into the environment by direct discharges via industrial and urban effluents, surface run off and indirectly from aerial fall out. Metals can also find their way into the marine environment from mining, smelting and refinery releases, combustion of fossil fuels. Metals can enter into the marine environment both in organic and inorganic forms and the toxicity is depend upon the form of the metal in the medium. From pollution view point, organic forms of metals deserve special attention, because the toxic characteristics of metal can be determined by the form of organometallic compound.

Toxic metals enter into the sea mainly from rivers and the atmospheric emission. It is difficult to measure the global influx of contaminants along either of these pathways. So global estimates of the relative contributions from different sources are rough. The United Nations Group of Experts on Scientific Aspects of Marine Pollution (GESAMP, 1990) estimates that 44 per cent of all pollution entering the world ocean is from run-off and discharges from land (mainly through rivers), 33 per cent is from atmospheric deposition, 12 per cent from maritime transportation (spills and operational discharges), 10 per cent from deliberate clumping of waters and 1 per cent is from the off-shore development of mineral resources (Norse, 1993).

When heavy metals reach in the sea, they enter into the food chain of the marine organisms. The common feature of these metals are that they are all relatively toxic even at fairly low concentrations and are readily concentrated by aquatic organisms and plants. The plants and animals are vary widely in their ability to regulate their metal content. Metal cannot be excreted as these are remain in the body in an unchanged state and are continuously added during the life of the organism which is known as bioaccumulation. Animals feeding on bioaccumulators have a diet enriched in these conservative materials and if they too are unable to excrete them or do so on slowly they in turn acquire an even greater body burden of the substance which is known as biomagnification (Clark, 1989). Through the bioaccumulation and biomagnification processes, heavy metals enter into the trophic level.

In marine environment, invertebrates have a particularly high capability for concentrating metals along with other foreign materials when they filter plankton during feeding. Fishes apparently can accumulate metals either directly from sea water or indirectly through food chain. Due to the complex formation with organic substances metals have a tendency to be fixed in the tissue and not to be excreted. This is known as biological half-time. Many marine fishes and shell fishes have got the capacity to concentrate trace metals in their tissues lo much higher level than they are in other organisms and also they concentrate them several hundred times more than the concentration of metals in sea water.

Toxicity of a metal and its effect on organisms will depend upon the chemical form of metal, presence of other metals, physiological status of organisms and environmental physico-chemical parameters like salinity, temperature, dissolved oxygen and pH of sea water. Entry of heavy metals into organisms depend upon the rate of absorption of a metal by animals which is affected by factors like salinity, pH, presence of other metals, complexing agents, temperature, size difference, maturity and starvation of the animal. The environmental parameters of water effect toxicity of the metal either by influencing physiology of organism or by altering chemical form of the metal in water. Generally metals are less toxic at lower temperatures and high salinity that at high temperature and low salinity. Toxicity of a metal is generally depends upon residence lime of the metal, Heavy metals from marine environment enter into the human body by consumption of fish.

Though the coastal oceans comprise less than 10 per cent of the total marine hydrosophere, they provide about 87 per cent of the total marine fish catch and almost the entire hulk of shellfish catch. Al the same time they are under the direct influence of the growing anthropogenic pressure. Man is continuously introducing materials into coastal ocean through river and coastal discharges, atmospheric emission, dumping of waters directly into the sea and through mariculture and other uses of coastal ecosystem. The more important routes of metal transport are through adsorption onto the particulate matter and accumulation of organic residues at various trophic levels of the food chain.

Fishes are the most important food among the marine organisms. Because these are contained 12 per cent animal protein which helps in the building material of the body and make good the wear and tear of the tissues. Though the ocean covers three fourth of the world, the fishery potential of marine fishes are used by only a few countries. There is no doubt about the fact that the fishes can supply a much greater amount of food than they provide now if advance biological and technological know-how are applied in their management and harvest. Among the animal foods, including fish, egg and meat, from which man gets proteins, fish is the best food because it is unexcelled, palatable and easily digestible. So it is said that for all around dictic quality and value, per unit of effort or cost to produce, no food product known to man exceeds or even equals the fatter varieties of ocean fish.

Sardines and mackerals are most important commercial pelagic fishes of Indian seas. These fishes are plankton feeders. Sardines are benthic and mackerls are surface feeders. When feeds are not available in surface water, mackereds go down-ward for feeding. The fishes feed on a variety of phytoplanktonic and zooplanktonic organisms. Lesser sardines are found in this coast. They belong to family Clupeidae and Dussumieridac. Indian mackerel belongs to the genus, Rastrelliger and family Scombridae. These fishes have high table value in India and outside India.

Marine pollution is known to us since a very long time. The problem of heavy metal pollution came to light only in 1930s, when

Goldwater (1936), traced the toxic effects of mercury on organisms. After that, two fatal incidents such as Minamata tragedy (1950's) and Niigata tragedy (1964) of Japan had created global awareness about the evil consequence of mercury poisoning. After the Minamata disaster in 1950's, Cadmium achieved notoriety when it was suspected of being responsible for an outbreak of "itai-itai" disease in a Japanese village on the Jintsu river. Cadmium and its compounds along with mercury and some other dangerous metals included in (lie "black list" which by international agreement may not be discharged or dumped into the sea Sewage sludge dumping grounds may be expected to contain high lead concentrations. Several work have been carried out in different countries on distribution of heavy metals on sea water, sediments, and metal concentration in marine organisms and their effects on human being. In world, reports on effects of heavy metals other than mercury on marine and estuarine organisms was given by Bryan (1971). The distribution of mercury in fish and its form of occurrence was studied by Krenkel (1979). Eisler (1981) has studied the trace metal concentration in different marine organisms. Andersen and Bower (1978) have studied the patterns of trace metal accumulations in Cray fish population. Kendall (1977) has given an account of acute effects of methyl mercury toxicity in channel catfish (*Ictalurus panctatus*). Hernadez *et al.* (1990) have estimated the heavy metal concentration in some marine organisms from the Mediterranean sea. Much work has not been done on heavy metal concentration of Indian fisheries. Heavy metal pollution studies in India have started late and more importance has been given on mercury pollution. Mercury concentration in thirty commercial species of fish have been analysed by Tejam and Halder (1975) from the Bombay and Thana region. Qasim and Gupta (1980) while reviewing the existing knowledge on the incidence and implication of heavy metal toxicity in fishes of the Indian ocean have emphasized the importance of carrying out studies on bioaccumulation of heavy metals in various tissues of fishes inhabiting the open sea, coastal areas and estuaries. Shaw *et al.* (1985) have studied the mercury distribution in sediments and some biotic components of Rushikulya estuary. Nayak *et al.* (1993) have studied heavy metals in fishes from Gopalpur coast, Orissa. Kureishy *et al.* (1981) has analysed heavy metals such as Copper, Manganese, Zinc, Iron, Nickel and Cobalt in muscle, liver, gills and hearts of several fishes and zooplanktons from the Andaman sea, west coast of India, Krishnakumar *et al.* (1990) have measured the bioaccumulation of trace metals by marine flora and fauna, near a Caustic soda plant (Karwar, India). Zingde *et al.* (1976) had analysed Arsenic, Copper, Zinc and Manganese in the marine flora and fauna of coastal and estuarine waters around Goa. Madhupratap *et al.* (1981) studied on the heavy metal concentration in marine zooplanktons and Invertebrates. Few workers such as Dious and Kasinathan (1992), Kureishy *et al.* (1983), Nair *et al.* (1997), Muralidharan and Raja (1997), Rajendran *et al.* (1988), Sankaranarayanan *et al.* (1978) and Sanzgiry and Brangaca (1981) have studied the heavy metal concentrations in commercial important fin and shell fishes from Indian waters.

In India some work on heavy metal concentrations in marine organism have been done mainly on Arabian Sea, West coast of India. Much work have not been done on Bay of Bengal, east coast. A little work has been done on the heavy metal concentration in fishes from the Bay of Bengal. Therefore the present work is an attempt to study the concentration of nine heavy metals in Sardines and Mackerels from Gopalpur Coast.

Materials and Methods

Description of the Study Area

Orissa is one of the four maritime states on the east coast of India. It is situated between latitude 17°49′–22°34′N and longitude 81°27′–87°29′E. The coastal area of Orissa is 480 km and the continental shelf area is about 24,000 sq.km. The coastal plain of the state comprises of fertile and alluvial soil. The coastal belt is more thickly populated than other areas of the state. The coastline of the state spread from Digha on its North to Bahuda on its South. The state is rifled with reach endowments of land, water, mineral, forest and animal resources. The study area is located along the South Orissa coast, North eastern part of Bay of Bengal.

The fishes were collected from Gopalpur coast for the experiment. It is situated 15 km away from the Berhampur city and 6 km from Berhampur University. Gopalpur is located between 19°16′ N latitude and 84°55′E longitude (Figure 13.1). It is a small fishing village and well known beach resort on lie South Orissa coast. The coastline at Gopalpur is fairly straight having an orientation of 48°E of North. It is represented by a completely sandy stretch for several

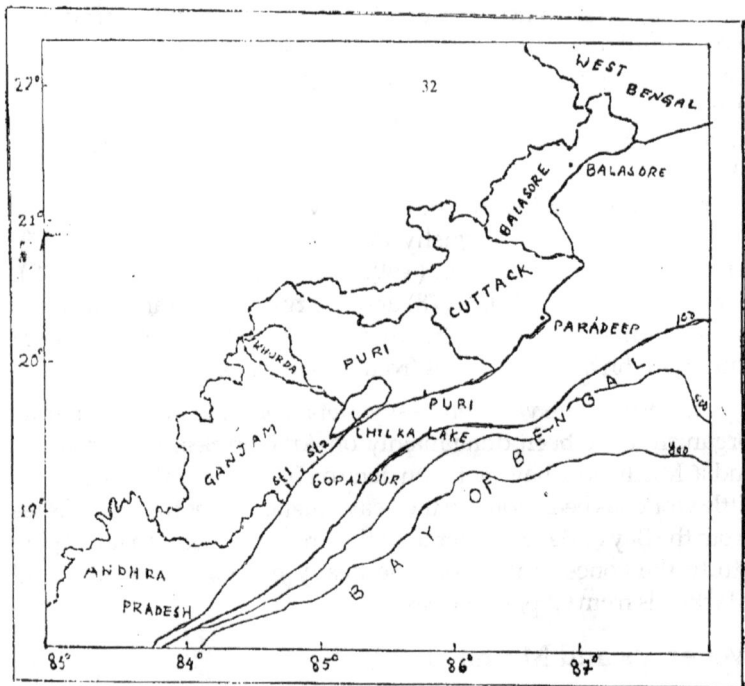

Figure 13.1: Map Showing Location of Gopalpur Coast and its Two Landing Centres (Stations I and II)

kilometers, on its North and South. The climate of this locality is semiarid type. According to Indian Meteorological Departments climatological chart for the period of 1993–1994, the annual average rainfall of Gopalpur is 1534.3 mm. This rain is mainly cine to the South-West monsoon.

Collection of Materials

The fishes were collected from local fishermen which they caught by using shore seines, beach seines, boat seines, gills nets, cast nets etc. For mackerel fishery they use boat scine and beach seines. Usually most of the fishermen used gill nets for sardine fishery which is locally known as 'BADAML'. The samples were collected from two landing centers namely Gopalpur station I and station II. Sardines are pelagic and mackerels are epipelagic in habitat. The samples were collected once in every month for a period of one year *i.e.,* January

1997 to December 1997. After collection, the fishes were brought to the laboratory and identified.

Identification

Sardine

At Gopalpur coast, out of four species of sardines only one species *i.e.*, *Sardinella fimbriata* is available plentiful amount and throughout the year. Other three species such as *Sardinella longiceps*, *Sardinella albella* and *Sardinella gibosa* were available in small amount only in few months. The collected fishes were identified using FAO identification sheets, Day (1978), Bal and Rao (1984), Munro (1955).

Synonyms

Sardinella fimbriata (Cuvier and Valenciennes); *Sardinella fimbriata* (Cuvier and Valenciennes), 1847

Clupea fimbriata (Part) Day 1878; *Clupea* (Harengula) *fimbriate* (Part) Weber and Beaufort, 1913

Sardinella fimbriata Regan 1917; Munro, 1955, Nair 1973; Flower, 1941; Herra, 1953; Whitehead, 1973; *Sardinella jussieu* Mishra, 1976.

Common Names

English: Fringscale sardine; *Tamil*: Sudoi or Choodi; *Telugu*: Kavallu; *Malayalam*: Chalmathi; *Marathi*: Waslii; *Kannada*: Pedi, Erabi; *Hindi*: Charreaddee; *Oriya*: Kahalii.

Characteristics

Sardines belong to the family clupeidae. General characteristics of this family: small silvery fishes, usually with fusiform, sub-

Figure 13.2: Lateral View of *Sardinella fimbriata*

cylindrical bodied but sometimes quite strongly compressed; abdominal scutes usually present. Lower jaw short but deep, giving typical clupeidae mouth shape. Teeth small or absent. This fish belongs to sub-family clupeinae. Body usually compressed laterally, with scutes on lower edge of belly. Mouth terminal, with feeble or no teeth; two supra maxillae present. Anal fin short with less than 30 rays. Pelvic fin with 7 to 9 rays. Caudal fin forked. Scales thin and deciduous. All the sardines were noted for their oily skins.

This lesser sardine belongs to the Genus Sardinella and species fimbriata. This genus is characterised by are compressed body, keeled or rounded belly, terminal mouth, operculum smooth, without radiating grooves. Dorsal fin origin before pelvic-fin origin; last dorsal ray normal, not prolonged. Anal fin comparatively short, with less than 24 rays last two anal rays more extensively branched than anterior ones and somewhat elongated so that the lip of anal fin slightly produced. Vertical scale striae usually interrupted at center of scale at least in anterior scales. Gill rackers fine, numerous body silvery. Species (fimbriata) is characterised by fusiform, compressed body, body depth 3 to 3.5 cm, head length 4 cm. Snout as long as diameter of the eye. Belly sharp with keeled scutes, pre-pelvic 17 or 18, post-pelvic 12 to 14. Pseudobranch short, flat; gill rackers on lower arm of firs arch 53 to 82. Dorsal fin with 14 to 16 branched rays, its origin slightly before mid point of body. Anal fin with 16 to 10 branched rays, lying far behind dorsal fin. Scales cycloid, thin, deciduous; anterior scales perforated and fimbriated at posterior margin, posterior scales with 4 or 5 vertical striae, interrupted at centers of scale; pre-dorsal median ridge covered by adjacent sides of two longitudinal series of scales.

Colour of the Fish

Bluish green above, silvery below, black dots and blotch at base of anterior rays of dorsal.

These fishes are distributed in India, Indo-Malayan Archipelago to the Philippines, New Guinea and possibly to Northern tip of Australia. These are inhibit in pelagic region the fishes feed on a variety of phyto and zooplanktons. *Sardinella fimbriata* is caught along both the coast of India, but it occurs in abundance, along the north-eastern and south-western coastal regions.

Mackerels

Indian mackerel belongs to the genus *Rastrelliger* (Jordan and Stark, 1908). Three species belong to the genus, namely kanagurta (Guvier, 1817), bracliyyoma (Bleeker, 1851) and faugni (Matsui, 1967), The bulk of the commercial landings of mackerel are contributed by a single species *Rastrelliger kanagurta*. *Rastrelliger branchysoma* enters the fishery in the Andaman waters, but its contribution is extremely limited. Recently *Rastrelliger fanghni* has been icporled to occur along the Madras Coast (Gnanamuthu, 1971).

Synonyms

Scomber kanagurta Cuvier, 1817; Rupell, 1928; Cuvier and Valenciennes, 18311; *Scomber macrolepidotus* Rupell, 1835; Day, 1958.

Scomber reain Day, 1870

Rastrelliger macrolepidotus, Banard, 1917.

Rastrelliger kanagurta Jordan and Stark 1917; Flower, 1935; Jones and Silao, 1962; Jones and Rosa Jr., 1965; Rao, K.V. 1970.

Macrolepidotus Barnard, 1917; *Rastrelliger kanagurta* Jordan and Stark, 1917; Flower, 1917, Jones and Silao, 1962; Jones and Rosa, Jr. 1965; Rao, K.V., 1970.

Local Names

English: Indian mackerel, Rake-gilled mackerel; *Hindi*: Bangdi; *Marathi*: Kaulagedar, Bangda; *Tamil*: Kumln, Kannageluthi, Ailae; *Telugu*: Kannyurta or Kannagadatha; *Sindhi*: Obigedar; *Malyalam*: Ayila or Ayla; *Oriya*: Karankita.

Characteristics

Mackerel belongs to the family *Scombridae*. Torpedo shaped

Figure 13.3: Lateral View of *Rastrelliger kanagurta*

powerful body, mostly metallic or blue green above. Month rather large; teeth in jaws usually feebly; roof of mouth and lounge may be toothed. Gills membranes free from isthmus. Two dorsal fins, with finlets behind second dorsal and oval fins, first dorsal fin welt behind head; pectoral fin inserted high on body; pelvic fin with 6 lays, placed beneath the pectoral fin. Lateral line simple or branched. Body either uniformly covered with small cycloid scales or restricted to a corselet around the front part of the body. The distinguishing character is that the caudal fin rays completely cover the hypural plate. It is due to the adaptation for high-speed swimming. The other character is the presence of a pair of oblique keels near the end of caudal fin which function to direct and accelerate a current of water over the fork of the tail. This is hydro-dynamically important and allows fishes to swim-faster. This family is epipelagic in habitat and is important for both commercial and recreational fisheries all over the world. The flesh is mostly oily, rich and tasty.

The Indian mackerels belong to the genus Rastrelliger Body fusiform and moderately compressed. Teeth small in jaws; no teeth in vomer and palatines. Body and cheeks with minute scales. Spinous dorsal fin separated from the soft-rayed second dorsal fin by a distance equal to length of base of spinous dorsal fin; and fin spine thin, rudimentary, 5 (rarely 6) finlets present behind each dorsal and anal fins. Caudal fin deeply forked; two small keels on either side of caudal penduncle. Tills fish is greenish above with horizontal rows of spots on each side of back. The fish belongs to the species kanagurta. Head length equal to height of body, head longer than broad, body laterally compressed, snout pointed eyes with thick adipose eyelids; mouth large, oblique with deep cleft; teeth small in a single row in both jaws, but usually own out with age, vomer and palatine without teeth; gill rackers numerous, 17–24/33–45 respectively on upper/lower limbs of the first gill arch, long and feather like and distinctly visible when the mouth is opened, first dorsal spinous and its first spine shorter than second spine; second dorsal soft rayed, dorsal and anal finlets five or six, pectoral soft rayed and pelvics with one spine and five soft ray, caudal deeply forked with pointed lobes. Body colour is bluish green tending towards grey above and yellow on belly at sides, with about three greyish longitudinal stripes along upper half of body, two rows of small dark spots on sides of dorsal fin bases, narrow dark longitudinal spot on body near lower margin of pectoral fin. Dorsal fin yellowish, other fins dusky.

This is widely distributed in tropical regions of the Indian and Pacific Ocean between 30°N–30'S latitude and 30°E–160'N longitude. It occurs along the coastal waters of Eastern Africa, north of Durban, Seychelles, Malagasy, Mauritius and Reunion Islands, with countries boardering the Red sea and tile Persian Gulf, Pakistan, India including the Union territory of Andaman and Nicobar Islands, Sri Lanka, Burma, Malaysia, Thailand, Cambodia, Indonesia, Philippines etc.

The Indian mackerel is a surface feeder. It feeds on phyto and zooplanktons, In Orissa mackerel Fishes are caught by masula boat (Ber.), beach seine and boat seine which is locally known as "BER JAL and IRAGHAI VALA" respectively.

Methods

After identification fishes were washed with double distilled water. The surface moisture of the fish is removed by using blotting papers. Length of the fish were measured by using scale. The total length was measured from the tip of the snout to the caudal lobe of the fish. The length was recorded in centimeter. Weight of the fishes were weighed by singlepan balance; The unit expressed in grams.

After taking length and weight of the fishes these were prepared for dissection. Stainless steel cissor, forcep and needle were used for dissection. Before their use, these were washed with contaminant free double distilled water. Petridishes were also washed with distilled water. The fishes were dissected for muscle, liver and gills and the tissues were prepared from five individuals of each species. The fishes were different in sizes depending on the availability. Different tissues were taken in different petridishes which were cleaned. The samples were dried in hot air over at 60°C for 8 hours. After this the dried tissues were ground to fine powdered form and they redried until a constant weight was observed. After getting a constant weight, the powder was prepared for digestion. Glass bottles with covered mouth were first washed with double distilled water and then kept with concentrated nitric acid (HNO_3) for about 6–12 hr. After this the powdered samples were weighted to 0.5 gm of each samples and were taken in separate bottles. Then 20 ml of concentrated nitric acid were added in each bottle these mixture were digested for 12 hrs or more at 100°C until the formation of clear solution. If necessary extra content rated nitric acid were also added. If the solution being charred. 2–3 drops of perchloric acid ($HClO_4$) was also added. Then the samples were cooled at room temperature.

For volume make-up, measuring flasks were first cleaned with double distilled water and then with concentrated nitric acid. Then the solution were made upto 100 ml with double distilled water in the acid washed measuring flask. After that these samples were ready for determination of heavy metals. The analysis were done in Flame Atomic Absorption Spectorphotometry. The values were expressed in dry weight ppm.

Results

During the experimental period from January 1997 to December 1997, the length, weight and heavy metal concentration for Copper, Nickel, Zinc, Iron, Lead, Manganese, Chromium and Cobalt in gills, liver and muscle of lesser sardine, *Sardinella fimbriata* and mackerel *Rastrelliger kanagurta* were studied.

The length and weight of *Sardinella fimbriata* were measured from January, 97 to December 97 and represented in Table 13.1. The length was expressed in centimeter (cm) and weight in gram (gm).

In *January*, the highest and lowest length of Sardines were 18 cm and 15 cm respectively. The highest weight was 43 gm and the lowest weight was 28 gm. The average length and weight were 17 cm, and 37.63 gm respectively.

In *February*, the highest length was 16.1 cm, and the lowest length was 14.7 cm. The highest and lowest weight were 29.5 gm and 27.0 gm respectively. The average weight and length of the fish were 28.74 gm and 15.64 cm respectively.

The highest length and weight were 15 cm and 30 gm in the month of March. The lowest length and weight were 9.8 cm, and 18.5 gm respectively. The average length and weight of the fish were 12.4 cm and 24.3 gm respectively.

The highest length of the fish were 15 cm and lowest length 135 cm in the month of *April*. The highest and lowest weight of the fish were 31.8 gm and 28.0 gm and the average length and weight of the fish were 14.2 cm, and 29.8 gm respectively.

**Table 13.1: Length and Weight of *Sardinolla fimbriata*,
During the Month of January, 1997 to December, 1997
from Gopalpur Coast, Orissa**

Months and Year, 1997	Length in cm	Weight in gm	Months and Year, 1997	Length in cm	Weight in gm
January	15.0	28.0	July	12.0	23.962
	16.5	36.8		13.0	25.0
	17.0	39.0		14.0	27.87
	18.0	43.0		12.5	24.0
	18.0	42.650		12.9	25.0
Average	17.0	37.63	Average	12.88	25.16
February	15.9	29.0	August	9.5	12.0
	16.1	29.5		10.0	15.5
	15.5	28.0		11.2	16.0
	16.0	29.2		10.5	16.5
	14.7	27.0		10.3	14.962
Average	15.64	28.74	Average	10.3	14.99
March	13.9	26.262	September	12.0	24.642
	12.0	26.0		13.0	25.0
	15.0	30.0		11.8	24.0
	9.8	18.5		12.5	24.750
	11.3	20.782		14.5	26.0
Average	12.4	24.2	Average	12.76	24.8
April	14.0	30-0	October	12.5	16.464
	13.5	29.792		10.7	9.666
	15.0	31.872		14.0	18.582
	14.6	29.562		13.5	17.762
	13.9	28.0		15.0	19.0
Average	14.2	29.84	Average	13.14	16.29
May	15.9	29.0	November	15.5	30.341
	16.5	30.65		17.1	35.701
	14.5	23.0		16.5	31.755
	14.5	23.0		16.0	31.777
	15.0	28.0		16.4	32.704
Average	15.28	26.73	Average	16.22	32.45

Contd...

Table 13.1–Contd...

Months and Year, 1997	Length in cm	Weight in gm	Months and Year, 1997	Length in cm	Weight in gm
June	14.5	29.0	December	16.0	31.178
	13.0	27.0		15.0	29.372
	12.7	25.262		15.8	30.291
	12.5	25.5		15.7	32.672
	12.0	24.962		15.8	35.98
Average	12.94	26.34	Average	15.66	31.899

In the month of *May* the highest length and weight of the fishes were 16.5 cm and 30.65 gm. The lowest length and weight of fishes were 14.5 cm and 23 gm respectively. The average length and weight of the fish were 12.94 cm and 26.34 gm respectively,

In the month of *July,* the highest length was 14 cm and the lowest was 12 cm. The highest and lowest weight of the fishes were 27.87 gm and 23.962 gm respectively. The average length and weight of the fish were 12.8 cm and 25.16 gm respectively.

In *August*, the highest length and weight of the fishes were 11.2 cm and 16.5 gm respectively. The lowest length and weight were 9.5 cm and 12 gm respectively. The average length and weight of the fish were 10.3 cm and 14.99 gm respectively.

The highest length and weight of the fishes in September, were 14.5 cm and 25 gm respectively. The lowest length and weight of the fishes were 11.8 cm and 24 gm respectively. The average length and weight of the fishes were 12.76 cm and 24.8 gm respectively.

In *October* the highest length was 15 cm and lowest was 10.7 cm. The highest weight was 19 gm and lowest was 9.66 gm. The average length and weight of the fish were 13.1 cm and 16.29 gm respectively.

In *November*, the highest weight was 35.70 gm and lowest weight were 30.341 gm. The highest and lowest length were 16.7 cm and 15 cm respectively. The average length and weight of the fishes were 16.22 cm and 32.45 gm respectively.

In December, the highest length and weight of the fishes were 16 cm and 35.98 gm. The lowest length and weight were 15 cm and

29.37 gm respectively. The average length and weight of the fish were 15.66 cm and 31.89 gm respectively.

The length varied from 9.5 cm to 18 cm, and weight varied from 9.66 gm to 43 gm during January 97 to December 97.

The heavy metal concentration (Copper, Nickel, Zinc, Iron, Lead, Manganese, Chromium, Cadmium and Cobalt) in gills, liver and muscle of *Sardinella fimbriata* were analysed from January'97 to December'97 (Table 13.2–13.4). The values were expressed in ppm dry weight.

Table 13.2: Heavy Metal Concentration (in ppm dry wt.) in Liver of *Sardinella fimbriata* from Gopalpur Coast During January 1997 to December 1997

Month	Cu	Ni	Zn	Fe	Pb	Mn	Cr	Cd	Co
Jan	40.69	67.00	135.00	290.00	37.00	53.00	16.00	5.00	9.00
Feb	23.00	104.00	134.00	250.00	22.00	63.00	19.00	600	5.00
Mar	10.00	69.00	153.00	210.00	14.00	30.00	9.00	2.90	4.90
Apr	20.00	15.00	123.00	230.00	22.00	22.00	9.00	4.00	5.00
May	13.00	26.00	80.00	267.00	46.00	15.00	Tr	4.30	5.10
June	22.00	33.00	288.00	300.00	37.00	35.00	2.90	5.00	5.00
July	13.00	151.00	367.00	347.00	20.00	100.00	8.00	10.00	35.00
Aug	55.00	73.00	172.00	1087.00	38.00	49.00	47.00	8.00	15.00
Sept	45.00	19.12	155.00	700.00	25.00	68.00	45.00	9.80	10.00
Oct	10.00	54.09	250.00	497.00	27.00	10.61	4.24	5.00	7.00
Nov	13.50	53.00	125.00	667.00	27.00	45.09	18.00	1.20	4.00
Dec	43.83	15.00	132.00	312.00	95.35	28.15	17.30	2.90	4.00

Tr: Trace amount.

In *January*, the highest concentration of Copper was studied in liver being 40.69 ppm followed by gills being 7.87 ppm and muscle being 7.58 ppm, The same trend (Liver > gill > muscle) was observed by Nickel, Zinc, Manganese, Chromium Cadmium and Cobalt, Iron was found highest in gill being 300 ppm followed by liver being 290 ppm the lowest concentration found in muscle being 242 ppm. Lead was also followed this trend gills > liver > muscle.

Table 13.3: Heavy Metal Concentration (in ppm dry wt.) in Gills of *Sardinella fimbriata* from Gopalpur Coast During January 1997 to December 1997

Month	Cu	Ni	Zn	Fe	Pb	Mn	Cr	Cd	Co
Jan	7.87	64.93	123.00	300.00	39.37	23.00	10.50	2.00	5.00
Feb	6.00	100.00	130.00	225.00	17.00	26.00	9.70	3.00	3.80
Mar	4.70	65.00	165.00	200.00	6.00	11.00	3.50	1.50	3.30
Apr	9.50	14.50	120.00	228.00	19.00	10.50	5.00	2.00	2.00
May	8.00	17.90	98.00	270.00	50.00	8.50	2.60	2.70	4.00
June	10.00	29.00	280.00	276.00	35.00	23.00	1.80	2.90	3.50
July	11.50	155.00	355.00	298.00	25.00	65.00	4.00	5.00	10.10
Aug	12.00	69.80	170.00	1080.00	40.00	35.00	23.00	4.00	10.00
Sept	10.00	18.13	150.00	697.00	23.00	53.00	19.00	3.90	5.70
Oct	7.50	66.82	253.00	450.00	25.00	10.87	3.61	3.00	4.50
Nov	7.41	61.02	123.00	665.00	27.84	51.96	5.56	1.84	3.70
Dec	14.30	11.32	130.00	322.00	47.66	64.35	15.00	1.53	2.90

Table 13.4: Heavy Metal Concentration (in ppm dry wt.) in Muscle of *Sardinella fimbriata* from Gopalpur Coast During January 1997 to December 1997

Month	Cu	Ni	Zn	Fe	Pb	Mn	Cr	Cd	Co
Jan	7.58	60.44	117.00	242.00	30.34	16.15	8.97	1.50	3.00
Feb	5.00	90.00	122.00	198.00	9.00	19.00	8.00	2.50	2.90
Mar	4.50	59.00	150.00	190.00	4000	7.00	Tr	Tr	1.50
Apr	6.30	4.20	100.00	199.00	12.00	9000	2.50	1.90	0.51
May	7.90	15.30	75.00	210.00	42.00	5.00	Tr	1.90	2.30
June	9.50	22.00	269.00	250.00	23.00	10.90	2.00	1.30	1.01
July	6.70	122.00	346.00	300.00	9.00	50.00	2.00	4.00	9.50
Aug	9.80	54.00	150.00	980.00	23.00	20.00	17.50	3.00	7.00
Sept	9.50	20.16	110.00	670.00	10.50	35.00	17.00	2.30	4.00
Oct	6.00	11.02	242.00	425.00	18.00	5076	4.32	1.50	4.30
Nov	3.78	25.00	121.00	650.00	14.62	29.71	5.51	1.01	1.01
Dec	8.64	7.90	128.00	275.00	28.82	10.50	6.19	1.40	2.00

Tr: Trace amount.

In *February*, the highest concentration of Copper was studied in Liver, being 23 ppm, followed by gills being 6 ppm. The lowest concentration was recorded in muscle being 5 ppm, the same trend was followed by Nickel, Zinc, Iron, Lead, Manganese, Chromium, Cadmium and Cobalt.

In the month of *March*, the highest concentration of Copper was found in liver being 10 ppm followed by gills being 4.7 ppm and muscle being 4.5 ppm. The same trend was followed by Nickel, Iron, Lead, Manganese. Chromium, Cadmium and Cobalt. The highest concentration of Zinc was found in gills being 165 ppm followed by liver being 153 ppm and muscle being 150 ppm. This trend: gills > liver > muscle, was obtained by Cadmium. Chromium was found trace in muscle, highest in liver being 9 ppm followed by gills being 3.5 ppm.

In *April*, the highest concentration of Copper was found in liver being 20 ppm followed by gills, being 9.5 ppm. The lowest concentration was found in muscle being 6.3 ppm. All other metals *i.e.,* Nickel, Zinc, Iron, Lead, Manganese, Chromium, Cadmium, Cobalt were also followed the same trend liver > gills > muscle.

In the month of *May*, the highest concentration of Copper was found in liver being 13 ppm followed by gill being 8 ppm and muscle 7.9 ppm. Nickel, Manganese, Cadmium and Cobalt obtained the same trend; liver > gills > muscle. Zinc was found highest in gills being 98 ppm followed by liver 80 ppm. The lowest concentration was found in muscle being 75 ppm and Lead was followed this pattern. Chromium was present in trace amount in liver and muscle whereas it was presents 2.6 ppm in gills.

In the month of June, Copper was found highest in liver being 22 ppm to followed by gills being 10 ppm. Muscle contains lowest concentration being 9.5 ppm. All other heavy metals, Nickel, Zinc, Iron, Lead, Manganese, Chromium, Cadmium, Cobalt were also followed the same trend liver > gills > muscle.

The highest concentration of Copper was found in liver being 13 ppm followed by gills being 11.5 ppm and muscle being 6.7 ppm in the month of *July*. Zinc, Iron, Manganese, Chromium, Cadmium and Cobalt were found the same pattern. Iron was found highest in liver being 347 ppm followed by muscle being 300 ppm. The lowest concentration was found in gills being 298 ppm. Lead was found highest in gills being 25 ppm followed by liver being 20 ppm. The lowest concentration was found in muscle being 9 ppm.

During *August* copper was found highest in liver being 55 ppm followed by gills being 12 ppm. The lowest concentration was found in muscle being 9.8 ppm. Other metals Nickel, Zinc, Iron, Manganese, Chromium, Cadmium, Cobalt were followed the same trend liver > gills > muscle. All metals except Lead followed the trend. The concentration of lead was found to be highest in gills being 40 ppm followed by liver being 38 ppm. Lowest concentration was followed by muscle being 23 ppm.

In *September*, the highest concentration of Copper was found in liver being 45 ppm followed by gills being 10 ppm and muscle being 9.5 ppm. The saline trend was observed in all other metals.

In the month of *October*, Copper was found highest in liver being 10 ppm followed by gills being 7.5 ppm. The lowest concentration was found in muscle being 6.0 ppm. Iron, Lead, Chromium, Cadmium mid Cobalt followed the pattern. Nickel was found to be highest in gills being 66.82 ppm followed by liver being 54.09 ppm, lowest concentration was found to be 11.2 ppm in muscle. Chromium concentration was found to be highest in muscle being 4.32 ppm followed by liver being 4.2 ppm and gills being 3.6 1 ppm.

In *November* the highest concentration of Copper was found in liver being 13.5 ppm followed by gills being 7.41 ppm and muscle being 3.78 ppm. Zinc, Iron, Chromium, cobalt followed the same pattern. The highest concentration of Nickel was found in gills being 61.02 ppm followed by muscle being 25 ppm. The lowest concentration was found in liver being 15 ppm. Lead concentration was found highest in gills being 27.84 ppm, followed by liver being 27 ppm and muscle being 14.62 ppm. The same trend was followed by Manganese and Cadmium.

In *December*, the highest concentration of Copper was found in liver being 48.83 ppm followed by gills being 14.3 ppm, The lowest concentration was found in muscle being 8.64 ppm. Nickel, Zinc, Lead, Chromium, Cadmium and Cobalt followed the same pattern. Iron concentration was found to be highest in gills being 322 ppm followed by liver being 312 ppm. and muscle being 275 ppm. Manganese followed the same trend.

**Table 13.5: Length and Weight of *Rastrilliger kanagurta*
During the Month of January, 1997 to December, 1997 from
Gopalpur Coast, Orissa**

Months and Year, 1997	Length in cm	Weight in gm	Months and Year, 1997	Length in cm	Weight in gm
January	23.0	129.0	July	12.3	22.0
	22.0	128.0		11.9	23.0
	23.5	130.25		13.0	26.762
	24.0	130.65		14.0	27.0
	23.0	130.0		14.0	27.560
Average	23.1	129.58	Average	13.04	25.26
February	21.5	104.0	August	10.0	19.0
	14.7	39.25		10.5	19.25
	15.2	41.786		11.0	22.0
	22.7	109.876		11.2	20.0
	19.4	102.582		12.0	21.562
Average	8.7	79.49	Average	10.94	19.96
March	12.9	35.65	September	11.0	9.422
	13.0	37.0		12.0	12.0
	11.5	35.0		11.5	9.752
	14.5	40.262		14.0	13.0
	15.0	42.0		13.5	12.652
Average	13.38	37.98	Average	12.4	10.96
April	22.3	102.0	October	13.7	23.892
	9.8	101.0		14.0	24.0
	9.0	100.0		12.5	22.582
	20.0	101.5		11.5	22.364
	21.0	103.0		10.8	20.780
Average	20.02	101.5	Average	12.5	22.74
May	15.9	29.0	November	22.0	115.112
	16.5	30.65		19.5	110.567
	14.5	23.0		20.0	112.0
	14.5	23.0		24.0	125.0
	15.0	28.0		21.7	114.786
Average	15.28	26.73	Average	21.44	115.493

Contd...

Table 13.5–Contd...

Months and Year, 1997	Length in cm	Weight in gm	Months and Year, 1997	Length in cm	Weight in gm
June	18.0	70.0	December	23.0	124.566
	17.0	69.5		24.0	125.0
	16.5	65.0		22.0	125.262
	17.2	69.0		22.6	123.670
	16.0	63.982		21.5	122.0
Average	16.94	67.49	Average	22.22	123.09

The length and weight of *Rastrilliger kanagurta* were measured from January, 97 to December, 97 and represented in Table 13.5. The length was expressed in centimeter (cm) and weight in grain (gm).

In *January*, the highest length and weight of fishes were 24 cm and 130.65 gm respectively. The lowest length was 22 cm and weight was 129 gm. The average weight and length were 129.58 gm and 23.1 cm respectively.

In *February*, the highest and lowest length were 22.7 cm and 14.7 cm. The highest and lowest weight were observed 109.876 gm and 39.252 gm respectively. The average length and weight of the fish were 18.7 cm and 79.49 gm respectively.

The highest length was observed to be 15 cm and lowest length 11.5 cm in the month of *March*. The highest and lowest weight of fishes were studied 42 gm and 35 gm respectively. The average length and weight were 13.38 cm and 37.98 gm respectively,

In *April*, the highest length and weight of the fish were 21 cm and 103 gm and the lowest length and weight were 19 cm and 100 cm respectively. The average length and weight of the fish were 20.02 cm and 101.5 gm respectively.

In the month of *May* the highest length and weight of the fish were 16.5 and 30.65 gm and the lowest length and weight of fishes were 14.5 cm and 23 gm respectively. The average length and weight were 15.28 cm and 26.73 gm respectively.

In *June*, the highest and lowest length were 18 cm and 16 cm and the highest and lowest weight were 70 gm and 63.98 gm respectively. The average length and weight were 16.94 cm and 76.49 gm respectively.

The highest length and weight of the fishes were studied 14 cm and 27.56 gm and lowest length and weight of fishes were 11.9 cm and 22 gm respectively. The average length and weight were 13.04 cm and 25.26 gm respectively in the month of *July*.

In *August*, the highest weight was 21.562 gm, and lowest was 19 gm. The highest length was 12 cm and lowest was 10 cm. The average length find weight were 10.94 cm and 19.96 gm respectively.

In *September*, the highest length was 14 cm and lowest length was 11 cm. The highest weight was 13 gm, and lowest weight was 9.422 gm. The average length and weight were 22.4 cm and 10.96 gm respectively.

In *October*, the highest length and weight were 14 cm, and 24 gm. The lowest length and weight were 10.8 cm and 20.78 gm respectively. The average length and weight were 12 5 cm and 22.72 gm respectively.

In *November*, the highest and lowest length of fishes were 24 cm and 19.5 cm respectively. The highest weight was 125 gm and lowest weight was 110.50 gm. The average length and weight were 2 1.44 cm and 115.493 gm respectively. In December, the highest length was 24 cm and the lowest was 21.5 cm. The highest and lowest weight of fishes were observed to be 125 gm and 122 gm respectively. The average length and weight of the fishes were 22.22 cm and 123.09 gm respectively.

The length of the fish varied from 10 cm to 24 cm and the weight varied from 9.42 gm to 130.65 gm during January, 97 to December, 97.

The heavy metal concentration (Copper, Nickel, Zinc, Iron, Lead, Manganese, Chromium, Cadmium and Cobalt) in gill, liver and muscle of *Rastrelliger kanagurta* was analysed from January, 97 to December, 97 and represented in Tables 13.6–13.8. The values were expressed in ppm dry weight.

In *January*, the highest concentration of Copper was found in liver being 8.16 ppm followed by gills being 7.81 ppm and muscle being 7 ppm. The same trend was observed by Nickel, Zinc, Iron, Manganese, Chromium, Cadmium and Cobalt. Highest concentration of Lead was found in gills being 40.8 ppm followed by liver 39.06 ppm and muscle being 38.09 ppm.

Table 13.6: Heavy Metal Concentration (in ppm dry wt.) in Liver of *Rastrelliger kanagurta* from Gopalpur Coast During January 1997 to December 1997

Month	Cu	Ni	Zn	Fe	Pb	Mn	Cr	Cd	Co
Jan	8.16	68.11	220.00	198.00	39.06	90.00	27.00	4.90	8.70
Feb	20.00	95.00	150.00	250.00	20.00	65.00	18.70	7.00	5.00
Mar	14.00	110.00	201.00	198.00	18.00	35.00	8.50	3.00	5.30
Apr	12.00	17.50	250.00	200.00	14.00	25.00	4.60	4.80	6.00
May	12.30	23.00	298.00	342.00	40.00	14.00	6.00	4.90	6.30
June	14.00	38.00	300.00	350.00	33.00	40.00	4.00	6.00	4.00
July	15.90	123.00	390.00	496.00	30.00	120.00	9.00	11.20	40.00
Aug	65.00	179.00	130.00	1083.00	29.00	50.00	53.00	7.00	18.00
Sept	27.00	28.56	214.00	988.00	12.00	98.00	42.00	5.60	12.00
Oct	20.10	92.40	197.00	338.73	25.00	10:20	4.08	6.00	6.30
Nov	21.00	25.09	138.00	940.76	59.62	93.00	4.20	4.18	4.18
Dec	48.40	13.50	125.00	930.00	51.40	60.50	9.00	3.00	5.00

Table 13.7: Heavy Metal Concentration (in ppm dry wt.) in Gills of *Rastrelliger kanagurta* from Gopalpur Coast During January 1997 to December 1997

Month	Cu	Ni	Zn	Fe	Pb	Mn	Cr	Cd	Co
Jan	7.81	50.65	210.00	116.00	40.80	62.00	25.54	3.30	4.00
Feb	9.70	93.00	147.00	200.00	12.00	29.00	10.00	3.00	4.00
Mar	7.90	100.00	200.00	120.00	9.00	9.00	4.00	1-90	2.70
Apr	6.00	16.70	243.00	150.00	11.00	8.60	3.90	2.00	3.00
May	10.00	19.80	250.00	290.00	36.00	9.00	3.00	2.70	5.00
June	12.30	35.00	267.00	300.00	40.00	25.00	2.90	2.90	2.20
July	9.00	120.00	300.00	400.00	22.00	70.00	5.00	41.10	11.60
Aug	10.00	145.00	120.00	1000.00	39.00	33.00	25.00	6.30	9.00
Sept	23.00	21.16	200.00	750.00	10.00	70.00	20.00	7.00	7.00
Oct	8.00	64.00	95.21	117.95	23.00	9.06	7.50	2.00	3.20
Nov	14.00	16.32	134.00	716.00	28.00	74.00	2.00	6.00	4.00
Dec	10.65	10.00	128.00	831.60	53.32	10.65	14.23	2.00	2.00

Table 13.8: Heavy Metal Concentration (in ppm dry wt.) in Muscle of *Rastrelliger kanagurta* from Gopalpur Coast During January 1997 to December 1997

Month	Cu	Ni	Zn	Fe	Pb	Mn	Cr	Cd	Co
Jan	7.00	30.55	209.00	100.00	38.09	26.69	10.26	2.00	1.90
Feb	8.00	8400	130.00	123.00	6.00	15.00	8.00	1.90	2.00
Mar	7.00	93.00	187.00	140.00	5.00	5.00	2.50	0.89	1.70
Apr	5.30	5.20	192.00	143.00	6.00	7.00	Tr	Tr	1.50
May	6.00	9.00	253.00	230.00	30.00	4.00	1.00	1.80	2.80
June	11.00	25.00	270.00	267.00	29.00	15.00	1.50	2.50	1.00
July	7.00	110.00	376.00	397.00	11.00	53.00	3.00	4.30	7.00
Aug	9.00	120.00	195.00	935.00	22.00	19.00	19.00	5.00	5.90
Sept	8.00	1506	150.00	700.00	Tr	29.00	14.20	4.00	2.00
Oct	7.30	32.58	90.00	155.00	11.50	6.02	7.53	1.98	2.70
Nov	4.61	13.67	129.00	416.63	15.30	19.41	1.42	1.52	3.26
Dec	349	6.00	120.00	723.75	23.30	5.34	5.34	1.00	1.01

Tr: Trace amount.

In *February*, the highest concentration of Copper was found in liver being 20 ppm followed by gills being 9.7 ppm and muscle being 8 ppm. The same trends were observed in Lead, Nickel, Zinc, Iron, Manganese, Chromium, Cadmium and Cobalt.

In *March*, the copper concentration was found highest in liver being 14 ppm followed by gills being 7.9 ppm. The lowest concentration was observed in muscle being 7 ppm. Other metals Lead, Nickel, Zinc, Manganese, Chromium, Cadmium and Cobalt were followed the same pattern. Iron concentration was found to be highest in muscle being 140 ppm followed by liver being 198 ppm and gills being 120 ppm.

In *April*, Copper concentration was found to be highest in liver being 12 ppm followed by gills being 6 ppm and the lowest concentration was found in muscle being 5.3 ppm. The same trend were observed in Lead, Nickel, Zinc, Iron, Manganese, Chromium, Cadmium and Cobalt.

In the month of *May*, the highest concentration of Copper was found highest in liver being 12.3 ppm followed by gills being 10 ppm and muscle being 6 ppm. Other metals except zinc were followed the same pattern liver > gills > muscle. Zinc concentration was found highest in liver being 298 ppm followed by muscle being 253 ppm. Lowest concentration was found in gills being 250 ppm.

In the month of *June,* Copper concentration was found to be highest in liver being 14 ppm followed by gills 12 ppm. The lowest concentration was found in muscle being 11 ppm. Nickel, Iron, Manganese, Cadmium and Cobalt were followed the pattern liver > gills > muscle. The highest concentration of Zinc was present in liver being 300 ppm followed by muscle being 270 ppm and gills being 267 ppm. The highest concentration of Lead was found in gills being 40 ppm followed by liver being 33 ppm and muscle 29 ppm. Other metal concentration followed the trend liver > gills > muscle.

In the month of *July,* the Copper concentration was found to be highest in liver being 15.9 ppm followed by gills being 9 ppm. The lowest concentration was found in muscle being 7 ppm. Other metals except Cadmium were followed the pattern of liver > gills > muscle. Cadmium concentration was found to be highest in gills being 41.1 ppm followed by liver being 11.2 ppm and muscle being 1.8 ppm.

During *August* highest concentration of Copper was found highest in liver being 65 ppm followed by gills being 10 ppm and muscle being 9 ppm. Other metals Nickel, Zinc, Iron, Manganese, Chromium, Cadmium and Cobalt were followed the same pattern. Lead concentration was found to be highest in gills being 39 ppm, followed by liver being 29 ppm. The lowest concentration of Lead found in muscle being 22 ppm.

In *September,* Copper concentration was found to be highest in liver being 27 ppm followed by gills being 23 ppm and muscle 8 ppm. The same trend liver > gills > muscle was followed by Nickel, Zinc, Iron, Manganese, Chromium, Cobalt, and Lead. The highest concentration of Cadmium was present in gills being 7 ppm followed by liver being 5.6ppm and muscle being 4 ppm.

In *October,* Iron was found highest in liver being 338.73 ppm followed by muscle being 155 ppm and gills being 117.95 ppm. The higher concentration of Chromium was found in muscle being 7.53 ppm followed by gills being 75 ppm and liver being 4.08 ppm. The highest concentration of other metals were found in liver followed by gills and muscle.

In *November*, the highest concentration of Cadmium was observed in gills being 0 ppm followed by liver being 4.18 ppm and muscle being 1.52 ppm Copper concentration was found to be highest in liver being 21 ppm followed by gills being 14 ppm. The lowest concentration of copper was found in muscle being 4.61 ppm. Nickel, Zinc, Iron, Lead, Managanese, Chromium and Cobalt were followed the same pattern liver > gills > muscle.

In the month of *December*, the highest concentration of Zinc was found in gills being 128 ppm followed by liver being 125 ppm and muscle being 120 ppm. Lead and Chromium were followed the pattern of, gills > liver > muscle. Copper was found to be highest in liver being 48.4 ppm followed by gills being 10.6 ppm and muscle being 3.49 ppm. Other metals like Nickel, Iron, Manganese, Cadmium, and Cobalt followed the same pattern of liver > gills > muscle

Discussion

The metals having atomic number between 22 to 92, of all the group of the periodic group from 3 to 7 are called heavy metals. Out of all the heavy metals 11 are more important. These are Iron Copper, Zinc, Manganese, Chromium, Molybdenum, Vanadium, Nickel and Strontium. Some heavy metals are essential for living organisms. Metals like Iron, occur in haemoglobin of many vertebrates and invertebrates which help for respiratory mechanisms. Copper and Vanadium are also occurring in the respiratory pigments of mollusca, crustaceans and tunicates respectively. Copper is one of the trace element essential to many plants, animals and human life. On the other hand, even in small quantities of copper can be highly toxic to aquatic; organisms and lower member of food chain. It is also toxic to mammals and human being. In sea water 1.25 parts per billion (ppb) of copper is present. Adult human body contain about 100 mg of copper, one third of it in muscle tissues. Liver and brain are rich in copper and the concentration of copper in human fetus is ten times higher in the liver of adults (Khangrot and Ray, 1988).

The concentration of all the nine metals *i.e.,* Copper, Nickel, Zinc, Iron, Lead, Manganese, Cadmium, Chromium and Cobalt analysed for both *Sardinella fimbriata* and *Rastrelliger kanagurta* appear to vary widely. Concentration of metals in an organism mainly depends on the season, environmental conditions like salinity, temperature, concentration of metals in their locality and the quality of food available. A fresh water organism contain high concentration of heavy metals than the marine organisms. During the rainy season marine organisms show high concentration of heavy metals. It is due to the mixing of fresh water in the sea water, by which the salinity decreases. Lowest concentration of heavy metals in marine organisms are found in summer because of the increasing values of the salinity in sea water due to evaporation. In the present study highest concentration and lowest concentration of heavy metals in the tissues of *Sardinella fimbriata* and *Rastrelliger kanagurta* were recorded mainly during the rainy season and summer season respectively. It is also confirmed earlier by Dious and Kasinathem (1992) that the low saline media has higher capacity to maintain metals either in the form of solution or in suspension, during summer, the salinity of biotope increased and the metal uptake decreased.

The concentration of heavy metals was to be found highest in the liver and lowest in muscle, the edible portion of the fish, observed by Kureishy *et al.* (1981) from Andaman Sea. The highest concentration of analysed heavy metals during January, 97 to December, 97 of *Sardinella fimbriata* and *Rastrelliger kanagurta* were as follows.

Copper

The highest concentration of copper was recorded in liver being 55 ppm in the month of August followed by gills being 14.3 ppm in the month of December and being 9.5 ppm in muscle in the month of August studied in *Sardinella fimbriata*. In *Rastrelliger kanagurta* highest concentration was recorded in liver being 48.4 ppm in the month of December followed by gills being 23 ppm (September) and muscle being 11 ppm (June). The lowest concentration was to be 10 ppm in liver (March and October) followed by 4.7 ppm in gills (March) and 3.78 ppm in muscle (November) *Sardinella fimbriata* and *Rastrelliger kanagurta* in. In *Rastrelliger kanagurta* the lowest concentration in liver was 8.16 ppm (June) followed by gills 6 ppm (Apr) and 3.49 ppm in muscle (December).

The highest and lowest concentration of copper was followed the pattern of liver > gills > muscle. Here the values of copper were high as compared to the values obtained by other workers (Nair *et al.*, 1997). Jaffer and Ashraf (1988) also found that copper

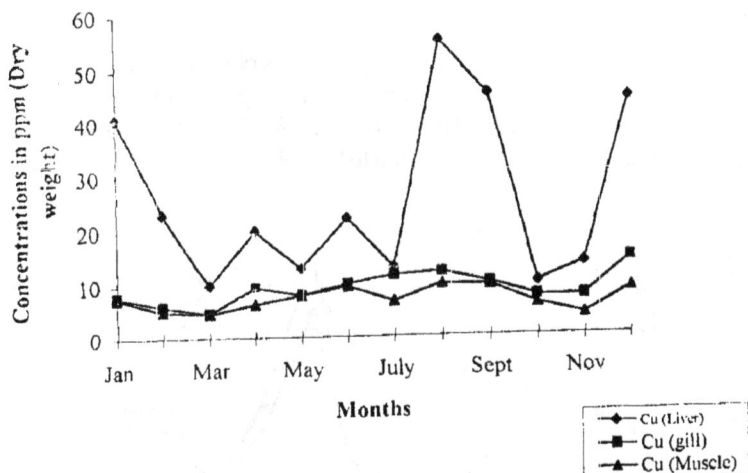

Figure 13.4: Copper Concentration in Liver, Gills and Muscle of
***Sardinella fimbriata* During January 1997 to December 1997**

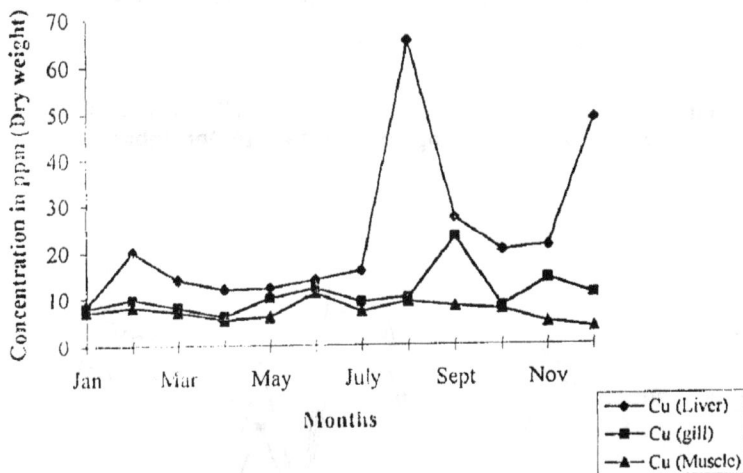

Figure 13.5: Copper Concentration in Liver, Gills and Muscle of
***Rastrelliger kanagurta* During January 1997 to December 1997**

concentration in fishes from Pakistan was higher in liver than the
muscle. The present finding is in agreement with the results of Nayak
et al. (1993).

Nickel

The highest concentration of Nickel was observed in gills being 155 ppm (July) followed by liver being 151 ppm and muscle being 122 ppm (July) in *Sardinella fimbriata*. In *Rastrelliger kanagurta* the highest concentration was observed to be 179 ppm in liver (September)

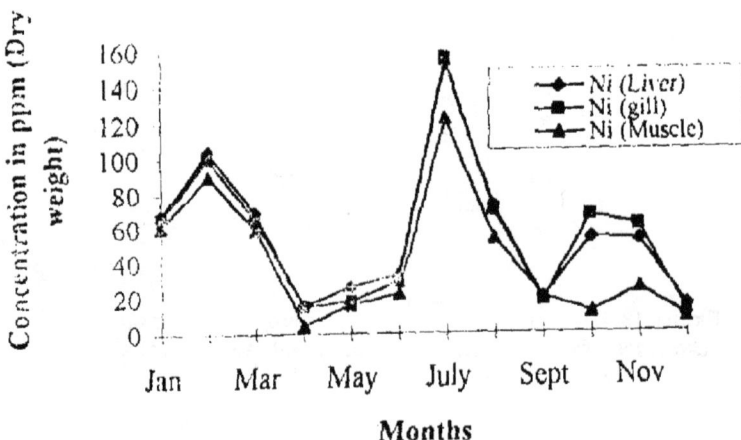

Figure 13.6: Nickel Concentration in Liver, Gills and Muscle of *Sardinella fimbriata* During January 1997 to December 1997

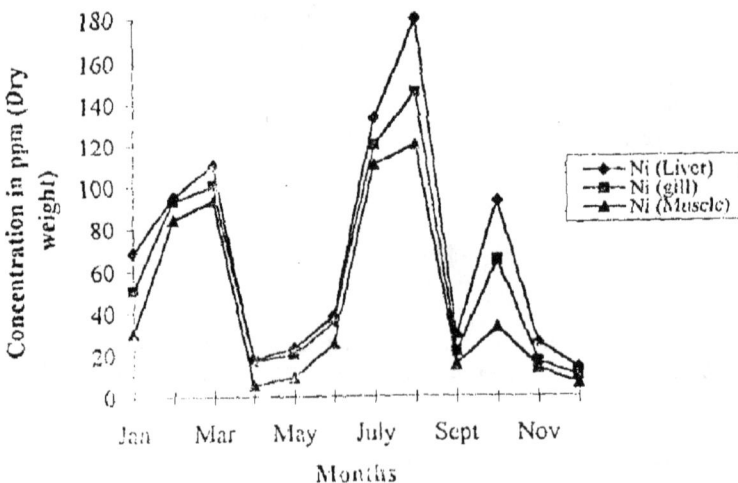

Figure 13.7: Nickel Concentration in Liver, Gills and Muscle of *Rastrelliger kanagurta* During January 1997 to December 1997

followed by gills being 145 ppm and 120 ppm in muscle (August). The lowest concentration was recorded in the liver of *Sardinella fimbriata* was 15 ppm (April) followed by gills being 11.32 ppm (December) and muscle 4.2 ppm (April). In *Rastrelliger kanagurta* the lowest concentration was recorded being 8.16 ppm in liver (June) followed by 6 ppm in gills (April) and 3.49 ppm in muscle (December). The Nickel concentration was followed the pattern of gills > liver > muscle in case of *Sardinella fimbriata* and liver > gill> muscle in case of *Rastrelliger kanagurta*. In *Sardinella fimbriata* the highest concentration was found in gills than in liver. This may be due to the absorption of the metal by gills during uptake of water. The pattern was observed by Kureishy *et al.* (1981) before.

Zinc

The highest concentration of Zinc in *Sardinella fimbriata* was recorded to be as 307 ppm in liver during July. The highest concentration of the metal in gills and muscle was recorded to be 355 ppm and 346 ppm respectively in the month of July. In *Rastrelliger kanagurta* the highest concentration in liver was 390 ppm followed by muscle being 376 ppm and gills being 300 ppm in the month of

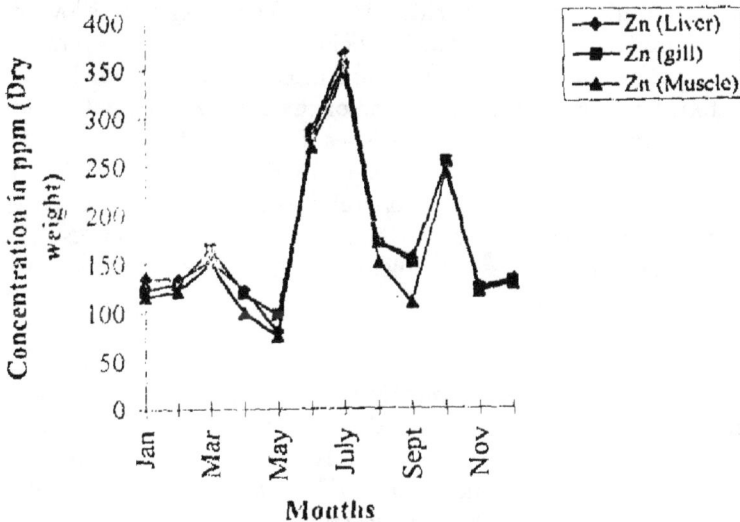

Figure 13.8: Zinc Concentration in Liver, Gills and Muscle of
***Sardinella fimbriata* During January 1997 to December 1997**

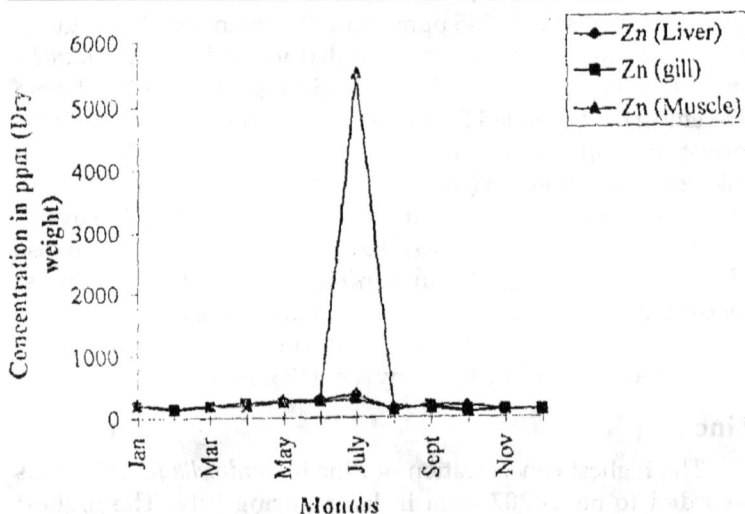

Figure 13.9: Zinc Concentration in Liver, Gills and Muscle of
***Rastrelliger kanagurta* During January 1997 to December 1997**

December. The lowest concentration was found 98 ppm in gills
followed by 8.0 ppm in liver and 75 ppm m muscle in *Sardinella
fimbriata* during the month of May. In *Rastrelliger kanagurta* the lowest
concentration was recorded in liver being 125 ppm during December
followed by gills being 95.21 ppm and muscle being 90 ppm during
October. The highest concentration of zinc followed the pattern of
liver > gills > muscle in both the fishes. The present study has little
variation than the values observed by De Solizb and Naqvi (1979)
and Nayak *et al.* (1993). It is generally believed that fish activity
regulates zinc concentrations in their muscle tissue and as a result it
does not reflect changes in ambient available levels of this element
in the environment (Cross *et al.*, 1973).

Iron

The highest concentration of iron found in liver, gills and muscle
of *Sardinella fimbriata* were recorded as 1087 ppm, 1080 ppm and
980 ppm during the month of August respectively. The highest
concentrations were found in *Rastrelliger kanagurta* liver being 1083
ppm followed by gills 1000 ppm and muscle being 935 ppm during
the month of August. The lowest concentration were recorded in

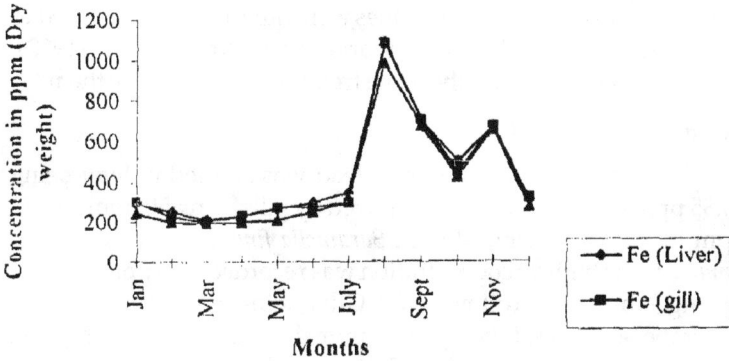

Figure 13.10: Iron Concentration in Liver, Gills and Muscle of
***Sardinella fimbriata* During January 1997 to December 1997**

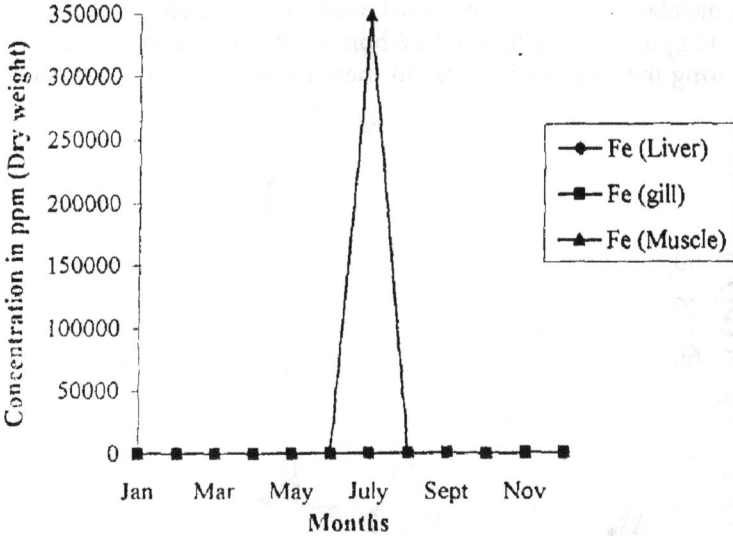

Figure 13.11: Iron Concentration in Liver, Gills and Muscle of
***Rastrelliger kanagurta* During January 1997 to December 1997**

Sardinella fimbriata liver being 210 ppm followed by gills 200 ppm and muscle 190 ppm during the month of March. In *Rastrelliger kanagurta* it was 198 ppm in liver followed by 116 ppm in gills and muscle 100 ppm during the month of January, The distribution of this metal of these two fishes followed the pattern of liver > gills >

muscle. In present study iron values were quite high as compared to the values obtained by other workers like Nayak *et al.* (1993), Kureishy *et al.* (1981) and the same trend was observed by them.

Lead

The highest concentration of lead was studied in liver being 95.35 ppm during December followed by gills being 50 ppm and 42 ppm in muscle during May in *Sardinella fimbriata*. In *Rastrelliger kanagurta* the highest concentration was recorded 59.6 ppm in Liver during November followed by Gills being 53.32 ppm during December and muscle being 38.09 ppm during January. The lowest concentration was studied in *Rastrelliger kanagurta* as 12 ppm in liver during September followed by 9 ppm in gills during March. A trace amount of Lead concentration was found in muscle during September, In *Sardinella fimbriata* lowest concentration was recorded as 14 ppm in liver followed by 6 ppm in gills and 4 ppm in muscles during the month of March. In these to fishes the concentration

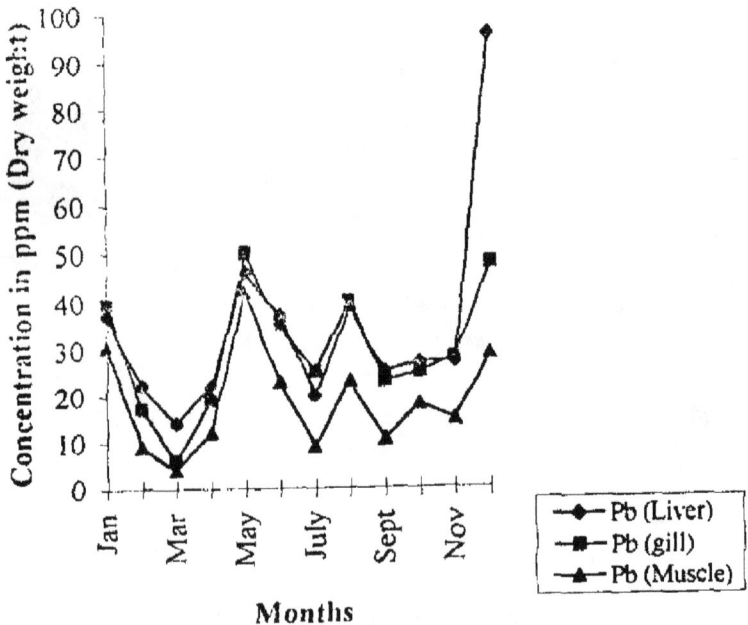

Figure 13.12: Lead Concentration in Liver, Gills and Muscle of *Sardinella fimbriata* During January 1997 to December 1997

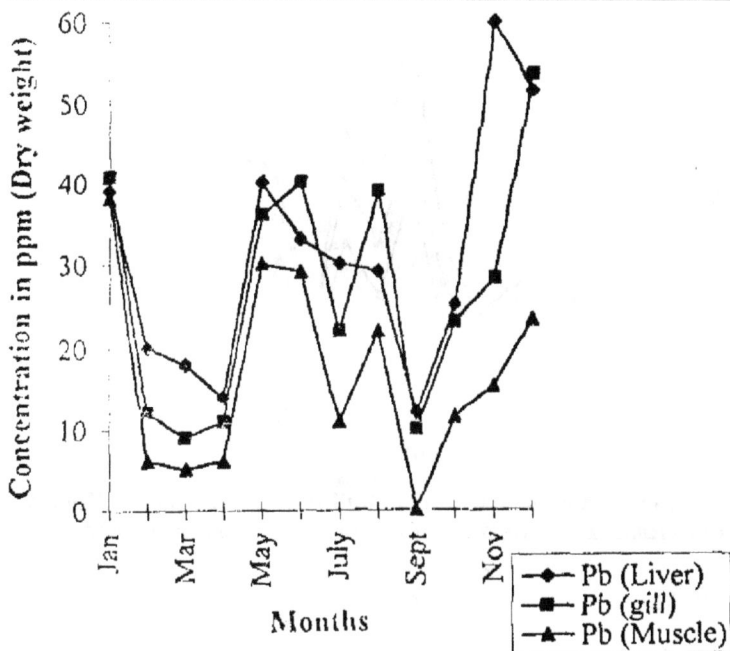

Figure 13.13: Lead Concentration in Liver, Gills and Muscle of
Rastrelliger kanagurta **During January 1997 to December 1997**

distribution followed the pattern of liver > gills > muscle. Lead is highly accumulated by liver is well as by gills Hernandez *et al.* (1990) studied an important accumulation of metals in liver and gills for lead and chromium in some in marine organisms from Mediterranean sea.

Manganese

The highest concentration of Manganese was observed in liver being 100 ppm followed by gills being 65 ppm and muscle being 50 ppm in the month of July in *Sardinella fimbriata*. In case of *Rastrelliger kanagurta* the highest concentration was 120 ppm in liver during July followed by 74 ppm in gills during November and 53 ppm in muscle during July. The lowest concentration was recorded as 10.2 ppm in liver during October followed by gills 8.6 ppm during April and muscle being 4 ppm in May. In *Sardinella fimbriata* the lowest

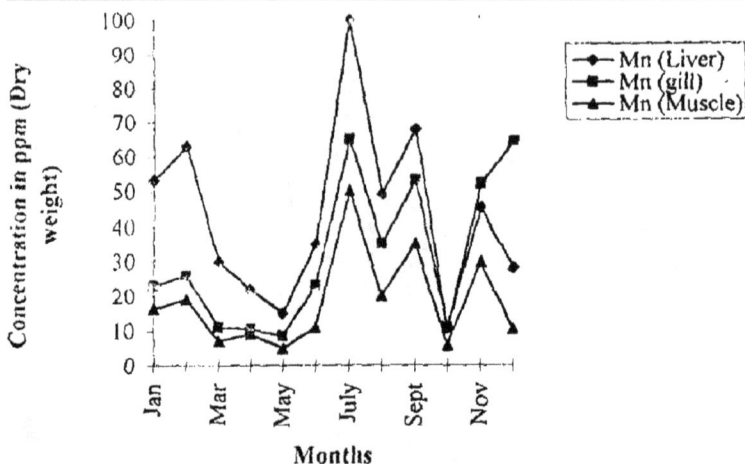

Figure 13.14: Manganese Concentration in Liver, Gills and Muscle of *Sardinella fimbriata* During January 1997 to December 1997

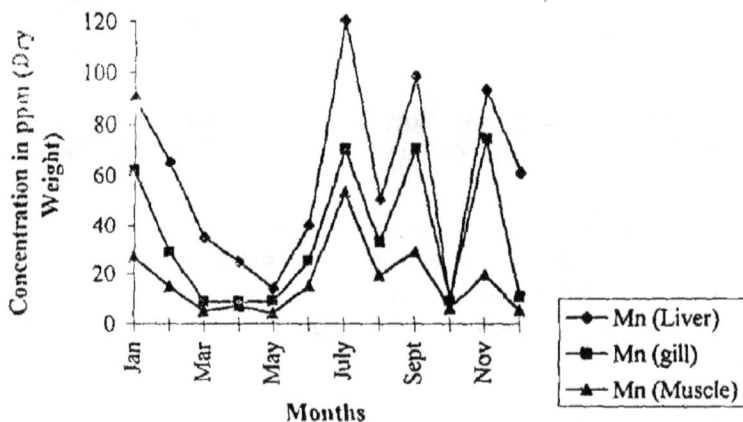

Figure 13.15: Manganese Concentration in Liver, Gills and Muscle of *Rastrelliger kanagurta* During January 1997 to December 1997

concentration was recorded in liver being 10.6 ppm during October followed by gills being 8.5 ppm and muscle being 5 ppm during May. The manganese concentration of these two fish tissues followed the pattern of liver > gills > muscle like other metals.

Chromium

The highest concentration of chromium in *Sardinella fimbriata* was studied in liver being 47 ppm followed by 23 ppm in gills and

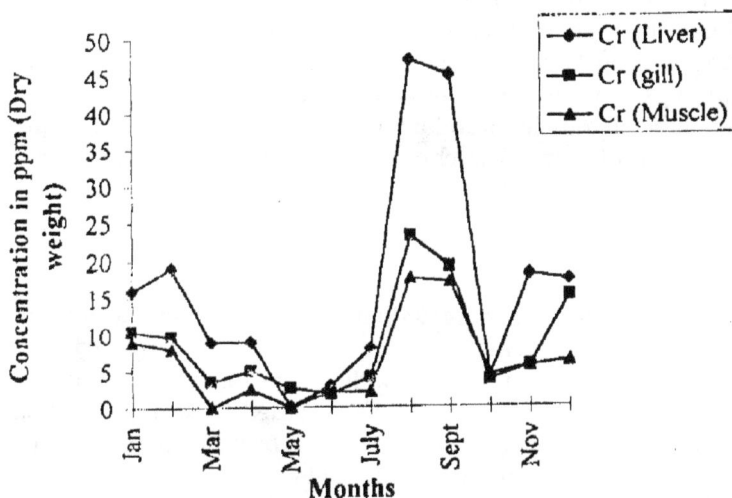

Figure 13.16: Chromium Concentration in Liver, Gills and Muscle of *Sardinella fimbriata* During January 1997 to December 1997

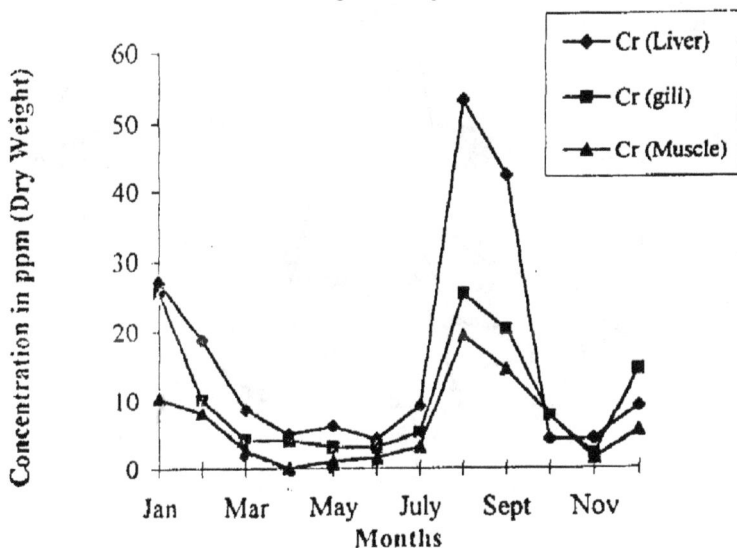

Figure 13.17: Chromium Concentration in Liver, Gills and Muscle of *Rastrelliger kanagurta* During January 1997 to December 1997

17.5 ppm in muscle during August. In case of lowest concentration, during May trace amount of chromium was found in liver and muscle where 1.8 ppm was recorded in gills. In *Rastrelliger kanagurta* the highest concentration in liver was recorded to be 53 ppm during August followed by 25.4 ppm in gills during January and 19 ppm in muscle during August. The lowest concentration was recorded being 4 ppm in liver during June followed by 2 ppm in gills during November in muscle chromium was found in trace amount during the month of April. The chromium concentration followed the pattern of liver > gills > muscle. The toxicity of chromium depends upon its valancy.

Cadmium

In *Rastrelliger kanagurta* Cadmium concentration was found 10 be highest in gills being 41.1 ppm during July followed by liver being 11.22 ppm in the same month and 5 ppm in muscle during August. The lowest concentration was recorded 3 ppm in liver during February and December followed by gills being 1.9 ppm in March During April trace amount of Cadmium was recorded in muscle of *Rastrelliger kanagurta*. In *Sardinella fimbriata* the highest concentration

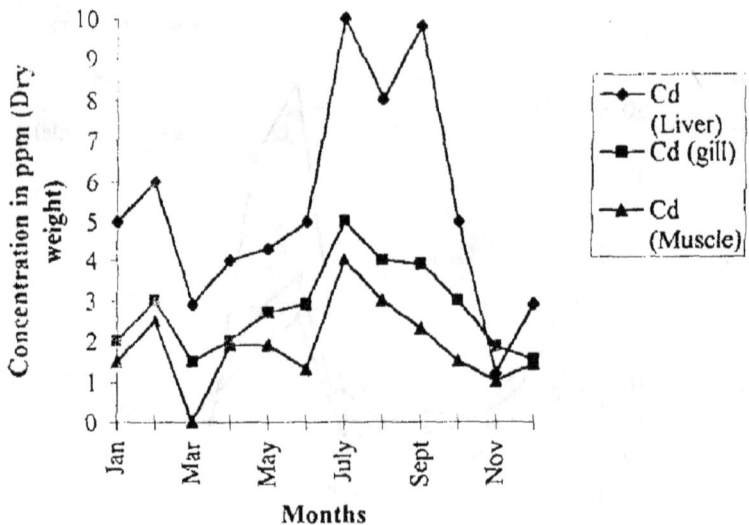

Figure 13.18: Cadmium Concentration in Liver, Gills and Muscle of
***Sardinella fimbriata* During January 1997 to December 1997**

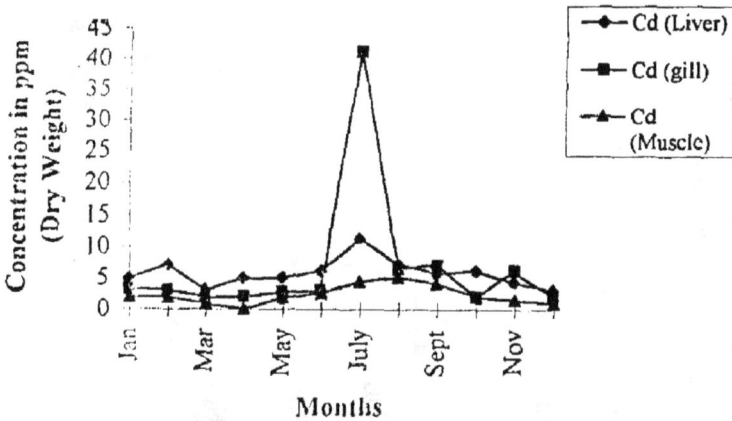

Figure 13.19: Cadmium Concentration in Liver, Gills and Muscle of Rastrelliger kanagurta During January 1997 to December 1997

was recorded to be 10 ppm in liver followed by gills being 5 ppm and muscle being 4 ppm during July. The lowest concentration was recorded 1.2 ppm in liver during November and 1.5 ppm during March. Trace amount of Cadmium was found in muscle during March. The highest concentration of metal was followed the pattern of gills > liver > muscle in *Rastrelliger kanagurta* and liver > gills > muscle in *Sardinella fimbriata*.

Cobalt

Cobalt concentration was recorded highest 35 ppm in liver followed by 10.1 ppm gills and 9.5 ppm in muscle during the month July in *Sardinella fimbriata*. The lowest concentration was recorded 4 ppm in liver during November and December followed by 2 ppm in gills during April and 1.01 ppm in muscle during June and November. In *Rastrelliger kanagurta* the highest concentration was recorded 40 ppm in liver during July followed by 11.6 ppm in gills and 7 ppm in music during July. The lowest concentration was recorded 4 ppm in liver during June followed by 2 ppm in gills during December and 1 ppm in muscle during June. The concentration was followed the pattern of liver > gills > muscle in these fishes.

The sea water contains lowest amount of heavy metals. These metals enter through various environmental compartments (Figure 13.22) to the marine environment. The natural sources of metals in

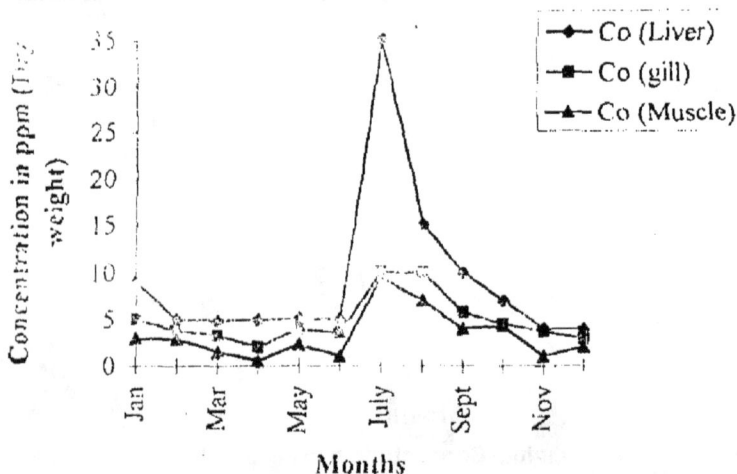

Figure 13.20: Cobalt Concentration in Liver, Gills and Muscle of
***Sardinella fimbriata* During January 1997 to December 1997**

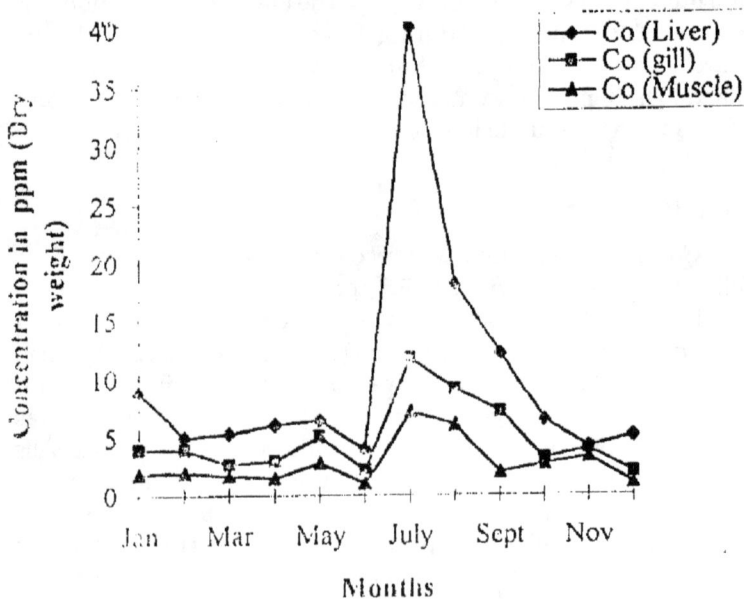

Figure 13.21: Cobalt Concentration in Liver, Gills and Muscle of
***Rastrelliger kanagurta* During January 1997 to December 1997**

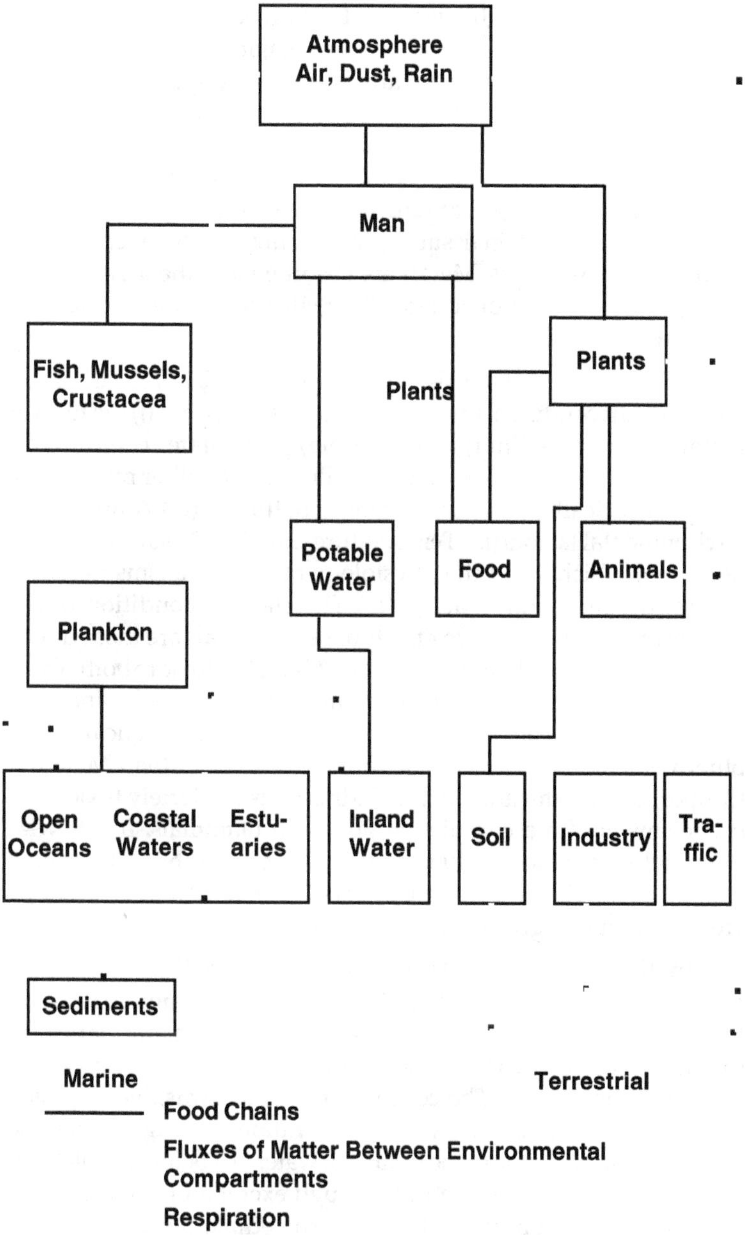

Figure 13.22: Pathways of Toxic Metals from Various
Environmental Compartments

coastal water are through river run-off. The mechanical and chemical weathering of rocks serves as another major source. In addition, components washed from the atmosphere through rainfall, wind blown dust, forest fire and volcanic particles also add to this. These heavy metals enter into organisms directly through water uptake and indirectly through food chain. Heavy metals are concentrated by all organisms. Some organisms show exceptional powers of accumulation in certain tissue *e.g.* Vanadium in ascidians, mercury in tuna and shell fishes. Metals are absorbed into the tissues either through body surface or across gills or cilliary feeding mechanisms or through food,

The toxic effect of metals depend upon chemical, biological and environment condition. Important amongest them are (*i*) the form of metal *i.e.*, organic/inorganic, soluble/particulate, complexed/chelated; colloidal/absorbed etc. (*ii*) Presence of other metals may act synergistically, antagonistically or naturally (*iii*) A number of environmental factors (*i.e.* Temperature, Salinity, Dissolved oxygen, light, pH) which affect the physiology of the organisms may also determine the metal toxicity. (*iv*) The state or condition of the organisms. The best known effects of heavy metals are those due to Mercury, Cobalt and Cadmium. Cobalt is highly toxic to both plants and animals The toxicity of Chromium depends on its valancy and this metal is known to act synergetically and antogonistically with other metals. Lead is less toxic to marine animals than Mercury, Copper, Chromium and Nickel.Cadmium is modertely toxic to all organisms and in a cumulative poison in mammals. It acts with other substances to produce synergestic effects. Nickel in small amounts said to be dangerous to oysters. Fate map of metals in marine environment was given in Figure 13.23.

In the present study it was observed that the metal concentrations were found to be highest mainly in rainy season and lowest in summer due to increasing and decreasing nature in salinity of sea water. The highest concentration was round in liver and lowest in muscle. The concentration trend was liver > gills > muscle. In some cases also will concentration was higher than the liver which may be due to the uptake of water just before its catching. Though metals are coming out through excretata from the body, these are also accumulated by liver. The edible portion (muscle) contain a low amount of metals.

Aquatic Environment

```
                              |
        ┌─────────────────────┼─────────────────────┐
   Dispersed            Concentrated          Transported (Biota,
   (atmospheric              |                Current suspended
   exchange, currents        |                    particle)
   and biota)                |
                  ┌──────────┴──────────┐
           Biological Process      Physico-chemical
                  |                      |
          ┌───────┴──────┐     ┌─────────┼─────────────┐
     Uptake and      Biota-   Absorption  Ion-      Precipitation
     accumulation    Biota               Exchange
```

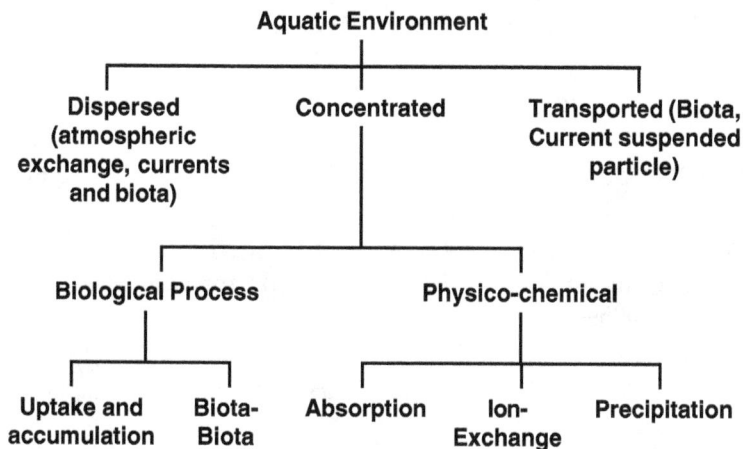

Figure 13.23: Fate Map of Metals in Marine Environment

The principle of uptake seems to be tile absorption of these elements through the gills filaments as all the samples of gills analysed showed appreciable con cent rat ion of practically all the metals. There was no correlation between the concentration of different tissues and the size of the fish. Correlation of metal concentration to body length, weight and age were obtained by Honda *et al.* (1983). They indicated that the metabolic turnover was more important than age or exposure lime in determining the levels of Iron, Manganese, Zinc and Copper. The age or exposure time is a dominant factor for the accumulation of lead. Nickel, Cadmium and Mercury in marine animals. However there is no established fact in this regard. The heavy metal concentration depends on the environmental characteristics of the area of catch (Kureishy *et al.*, 1983).

References

Anderson, R.V. and Brower, J.E., 1978. Pattern of trace metal accumulation in crayfish population. *Bull. Conta. Aq. Toxicology*, 20(1): 217–221.

Bal, D.V. and Rao, R.V., 1984. *Marine Fisheries*. Tata Mc-Graw Hill Publishing Company Limited, New Delhi, p. 271–282.

Bryan, G.W., 1971. The effects of heavy metals other than Hg on marine and estuarine organisms. *Proceeding R. Society*, London, B.177: 389–410.

Clark, B.R., 1989. *Marine pollution,* 3rd edn. Oxford University Press, p. 1–172.

Cross, F.A., Hardy, L.H., Joney, N.Y. and Barber, R.T., 1973. *J. Fish. Res. Board.,* Canada, 30: 1287.

Day, F., 1978. *Fisheries of India,* Vol. I: Text, Vol. II: Atlas. 198 Pis. Taylor and Francis, London.

DeSouza, S.N. and Naqvi, S.W.A., 1979. Metal concentrations in the grey mullet (*Mugil cephalus*) from Visakhapatnam. *Mahasagar,* 12(4): 259–264.

Dious, S.R.J. and Kasinathan, R., 1992. Concentration of Fe, Mn, Zn and Cu in cephalopod *Sepiella inermis* (Mollusca: Decapoda). *IJMS,* 21: 224–225.

Eisler, R., 1981. *Trace Metal Concentrations in Marine Organisms.* Pergamon Press, New York, NY, USA.

FAO, 1974. FAO species identification sheets for fishery purposes, Eastern Indian Ocean and Western Central Pacific. Food and Agriculture Organisation of United Nations, Rome, I–IV.

Goldwater, L.J., 1936. From Hippocrates to Ramazzin: Early history of industrial medicine. *Ann. Med. Hist.,* 8–27.

Gnanamuthu, J.C., 1971. *Indian J. Fish,* 18: 170–173.

Hernandez, J., Medina, J. Ansuagui, M. Conesa. 1990. Heavy metal concentration in some marine organisms from the Mediterranean Sea (Castellon, Spin): Metal accumulation in different tissues. *Sci. Mar.,* 54(2): 113–129.

Honda, K.R., Tatsukawa, K., Itano, N., Miyazaki and Fujiyama, T., 1983. Heavy metal concentration in muscle, liver and kidney tissues of striped dolphin *Slenella coeruleoalba* and their variations with body length, weight, age and sex. *Agric. Biol. Chem.,* 47: 1219–1228.

Jaffar, M. and Asharf, M., 1988. Selected trace metal concentration in different tissue of fish from coastal waters of Pakistan. *IJMS,* 17(3): 213–224.

Kendall, M.W., 1977. Acute effect of methyl mercury toxicity in channel catfish (*Ictalurus punctatuus*), liver. *Bull. Env. Cont. and Toxic.,* 18(2): 253–258.

Khangarot, B.S. and Ray, P.K., 1988. Environmental copper and human health. *Science Reporter*, 25(6): 352–353.

Krenkel, P., 1979. *Journal of Water Pollution Control Federation*, 51: 2168–2188.

Krishnakumar, P.K., Pillai, Y.K. and Vasala, K.K., 1990. Bioaccumulation of trace metals in marine flora and fauna near a caustic soda plant (Karwar, India). *Indian J. Fish*, 37(2): 129–137.

Kureishy, T.W., Sanzgiry, S. and Braganca, A., 1981. Some heavy metals in fishes from the Andaman Sea. *IJMS*, 10: 303–307.

Kureishy, T.W., Sanzgiry, S. and George, M.D., 1983. Hg, Cd and Pb in different tissues of fishes and zooplankton from the Andaman Sea. *IJMS*, 21(1): 60–63.

Madhupratap, M., Achuthakutty, C.T. and Sreekumaran Nair, S.R., 1981. Toxicity of some heavy metals to copepods *Acartica spinicauda* and *Tortamus forcipatus*. *Indian Journal of Marine Sciences*, 10(4): 382–383.

Munro, I.S.R., 1955. *The Marine and Freshwater Fishes of Ceylon*. Ext. Affairs, Canberra, pp. 349.

Muralidharan. G. and Vivek Raja, P., 1997. Trace element concentration in the meat of the edible calm *Marcia recens* (Chemnitz) (Pelecypoda : Veneridae). *IJMS*, 26: 383–385.

Nair, M., Balachandra, K.K., Sankaranarayanan, V.N. and Joseph, T., 1997. Heavy metals in fishes from coastal waters of cochin, South-West Coast of India. *IJMS*, 26: 98–100.

Nayak, L., Sahu, K.C. and Sahu, D.K., 1993. Heavy metals in some commercial fishes from Gopalpur coast, India. *Bull. Env. Sci.*, 11:33–35.

Norse. Elliot, A., 1993. *Global Marine Biological Diversity*. Island Press, Washington, D.C. pp. 1–383.

Qasim, S.Z. and Gupta, R. Sen, 1980. Present status of marine pollution studies in India. In: *Management of Environment*, (Ed.) B. Patel. Wiley Eastern, New Delhi, p. 310–329.

Rajendra, N., Tagore, J. and Kasinathan, R., 1988. Heavy metal concentration in Oyster, *Crassostrea madrasensis* of Cuddalore backwaters, South-East coast of India. *IJMS*, 17(2): 174–175.

Sankaranarayan, V.N., Purushan, K.S. and Rao, T.S.S., 1978. Concentration of some of the heavy metals in the Oyster *Crassoslrea madrasensis* (Preston) from the Cochin Region. *Indian Journal of Marine Sciences*, 7(2): 130–131.

Sanzgiry, S. and Braganca, A., 1981. Trace metals in Andaman sea. *Indian Journal of Marine Sciences*, 10: 238–240.

Shaw, B.P., Sahoo, A. and Panigrahi, A.K., 1985. Residual mercury concentrations in brain, liver and muscle of contaminated fish collected from an estuary near caustic chlorine industry. *Current Science*, 54: 810–812.

Sunda, W.G., Tester, P.A. and Huntsman, S.A., 1990. Toxicity of trace metal to *Acartia tonsa* in the Elizabeth river and southern Chesapeake Bay. *Estua. Coast. Shelf. Sci.*, 30(3): 207–221.

Tejam, B.M. and Halder, B.C., 1975. Mercury concentration in the thirty commercial fishes from Bombay and Thana region. *Indian J. Environ. Health*, 17: 9–16.

Zingde, M.D., Singhal, S.Y.S., Moraes, C.F. and Reddy, C.V.G., 1976. Arsenic, copper, zinc and manganese in the marine flora and fauna of coastal and estuarine waters around Goa. *Indian Journal of Marine Sciences*, 5: 212–217.

Chapter 14

Impact of Physico-Chemical Factors on Primary Productivity of Phytoplankton in Papnash Pond, Bidar (Karnataka)

N. Shiddamallayya* & Pratima Mathad

*Phycology Laboratory,
Department of P.G. Studies and Research in Botany,
Gulbarga University, Gulbarga – 585 106, Karnataka*

ABSTRACT

The study was undertaken to note an influence of physico-chemical nature of water on phytoplankton primary productivity in Papnash pond, Bidar. Physico-chemical nature of the water was analysed by using standard methods and the productivity by light and dark bottle method. The study revealed that temperature, pH, alkalinity, hardness, dissolved oxygen, free carbon dioxide, chloride, nitrite, phosphate, sulphates were involved in the variation of gross primary productivity, net primary productivity and community respiration. The physico-chemical nature of water body

* Corresponding Author: E-mail: siddamallayya_matapati@rediffmail.com.

regulated the primary productivity of the studied water body of Bidar, Karnataka.

Keywords: CR, GPP, NPP, phytoplankton, physico-chemical.

Introduction

Primary productivity of an aquatic ecosystem gives the details regarding energy fixation and availability to support bio-activity of the total system. Phytoplakton primary productivity in association with limno-chemical qualities could be fruitfully utilized as indices of trophic status of tropical reservoirs. This has been well documented by Ravindra Singh (1990). Janakiram and Jayaraj (1996), Bais *et al.* (1997) Singh (1998) and Arun and Dilip (1999). Keeping this in view, the Papnash pond has been selected for the study. Which is situated in Bidar district of Karnataka and lies at 17°55' N latitude ad 77°32' E longitude, at an altitude of 551m above MSL. The pond covers an area of 0.25 square kilometer with an average depth of 2.1 m.

The Papnash pond is found near the famous temple of Papnash. This temple is considered to be an important tourist and pilgrim place of Bidar. A perennial spring originates near the temple, where pilgrims take bath and washes their clothes. This polluted water enters in to the Papnash pond. In addition to this, disposal of domestic sewage and washing of cattle was also noticed. Taking these things in to consideration an attempt has been made in the present study to understand the impact of physico-chemical factors on phytoplankton primary productivity in Papnash pond, Bidar.

Materials and Methods

The present study has been carried out from October 1999 to September 2000. Using standard methods APHA (1995) did the physico-chemical analysis of water samples. The primary productivity of phytoplankton was estimated by following light and dark bottle method of Gaarder and Gran (1927). A set of bottles was incubated at 0.5m depth for 3 hours. (9-30 a.m. to 12-30 p.m.). The dissolved Oxygen values were converted into carbon values by multiplying with a factor of 0.375. Considering a 12 hours photoperiod as a day, values were presented as g $C/m^3/day$.

However, macrophytes were not taken in to consideration for assessing the total carbon production.

Results

The range of variation of physico-chemical parameters during the study period in the pond has been tabulated in Table 14.1. The phytoplankton primary productivity of Papnash pond (Table 14.2) showed seasonal variation. The grass primary productivity (GPP) ranged from 2.99 g C/m³/day in August to 5.49 g C/m³/day in October. The net primary productivity (NPP) varied from 1.50 g C/ m³/day in January and August to 3.99 g C/m³/day in April and October. Finally the community respiration (CR) showed its minimum value of 0.99 g C/m³/day in May and maximum value 2.25 g C/m³/day in January. The ratio of NPP to GPP ranged between 0.39 to 0.83 and the percentage of CR in GPP varied between 16.88 to 60.16 in selected water body.

Table 14.1. Physico-chemical Factors of Papnash Pond, Bidar (October–1999 to September–2000)

Factors	Range of Variation	Average
Water temperature (°C)	19.00-26.5	22.8
pH	7.2–7.8	7.46
Dissolved Oxygen	5.6–11.4	7.95
Free Carbondioxide	0.20–1.10	0.56
Total alkalinity	345–750	569
Total hardness	58–270	162.5
Chloride	5–23	14.5
Nitrite	0.015–0.182	0.06
Phosphate	0.010–0.464	0.128
Sulphate	1.2–16.3	7.4

All the values are expressed in mg l⁻¹ except temperature and pH.

Discussion

The GPP, NPP and CR do not show similarity throughout the study period, except in the month of August, in which CR is similar to NPP. The data obtained in the present study supports that temperature, dissolve O_2, free CO_2, sulphates, alkalinity, Nitrites and Phosphates were sufficient in quantity for the phytoplankton.

Table 14.2: Average Phytoplankton Primary Productivity of Papnash Pond Bidar (October–1999 to September–2000)

Month	GPP g C/m³/ day	NPP g C/m³/ day	CR g C/m³/ day	NPP : GPP	CR (%) GPP
October	5.49	3.99	1.50	0.72	27.32
November	4.99	2.99	2.00	0.60	40.08
December	4.24	2.74	1.50	0.64	35.37
January	3.74	1.49	2.25	0.39	60.16
February	4.24	2.99	1.25	0.70	29.48
March	3.99	2.25	1.74	0.56	43.60
April	4.99	3.99	1.00	0.79	20.04
May	3.24	2.25	0.99	0.69	30.55
June	3.49	1.99	1.50	0.57	42.97
July	4.50	2.74	1.76	0.60	39.11
August	2.99	1.50	1.50	0.49	50.16
September	4.50	3.74	0.76	0.83	16.88

The other workers have also co-related phytoplankton primary productivity with physico-chemical factors. Among various workers Ketchum (1969) and Schindler and Fee (1975) opined that Nitrogen and Phosphates have played an important role in primary productivity. Ayyappan and Gupta (1985), Prasad (1990), Bais *et al.* (1997), recorded high GPP due to the presence of high temperature, Carbon and Phosphate, which also favours maximum growth of phytoplakton. Similarly Verma and Mohanty (1995) and Gujarathi and Kanhere (1998), have also noted in addition to that pH and dissolved oxygen to support the growth of phytoplankton.

The factors such as alkalinity, Chloride, sulphate, free CO_2, hardness were also noted in an enhancement of primary productivity by Hedge and Bharati (1984), Rao *et al.* (1993), Mukharjee and Pankajakshi (1995), Arun and Dilip (1999), and Prakasham and Joseph (2000). CR fluctuation has shown less compared to NPP due to dominance of phytoplankton. In the month of January CR was high compare to NPP, which shows that there may be increase in zooplankton population in relation to other season. Similarly Das (2002), noted that NPP was surpassed by the dominance of zooplankton in the selected reservoirs of Andhra Pradesh.

The ratio of NPP to GPP varied from 0.39:1 to 0.83:1 and CR expressed as percentage of GPP ranged from 60.16 to 16.88, both in the month of September and January respectively. It indicates that NPP and CR have a quite apposite relationship with one another.

The average phytoplankton GPP increased during Winter (4.62 g C/m^3/day) followed by Summer (4.12 g C/m^3/day) and Rainy season (3.87 g C/m^3/day) in the pond.

The correlation co-efficient (r) between physico-chemical factors and phytoplakton primary productivity was computed and presented in Table 14.3. The CR showed significant negative correlation with Chloride (r=0.5 178, P>0.1).

Table 14.3: Correlation Co-efficient (r) Value Between Physico-chemical Parameters and Phytoplankton Primary Productivity of Papnash Pond, Bidar

Factors	GPP	NPP	CR
Water temperature (°C)	−0.1509	0.0981	−0.4616
pH	0.1231	0.1033	0.0056
Dissolved Oxygen	0.4119	0.2289	0.2548
Free Carbondioxide	−0.2165	−0.0069	−0.2373
Total alkalinity	0.3511	0.1322	0.2954
Total hardness	0.0924	0.0229	0.1148
Chloride	−0.1564	0.1252	−0.5178*
Nitrite	−0.0448	−0.2362	0.4001
Phosphate	−0.0975	−0.1544	0.1427
Sulphate	−0.0376	−0.3029	−0.0425

Values are Significant at P > 0.1 (*)

The present study clearly indicates that the phytoplankton primary productivity in the Papnash pond was governed by many physico-chemical factors of the water body.

Acknowledgements

The authors are indebted to late Dr.S.B. Angadi, Reader, Dept. of Botany, for his valuable guidance. We also thank the Head of the Dept. of Botany, Gulbarga University, Gulbarga for providing the facilities.

References

APHA, AWWA, WPCF, 1995. *Standard Methods for the Examination of Water and Wastewater,* 19th edn. American Public Health Association, Washington, D.C.

Ayyappan, S. and Gupta, T.R.C., 1985. Limnology of Ramsamudra tank primary production. *Bull. Bot. Soc.,* 32: 82–88.

Bais, V.S., Agrawal, N.C. and Arasta, Tazeen, 1997. Seasonal changes in phytoplankton productivity due to artificial enrichment of nutrients: A situ experiment. *J. Environ. Biol.,* 18(3): 249–255.

Das, A.K., 2002. Phytoplankton primary production in some selected reservoirs of Andra Pradesh. *Geobios,* 29: 52–57.

Gaarder, T. and Gran, H.H., 1927. Investigations of the production of plankton in the Oslo Fjord. *Rapp. et proc. Verb., Cons. Internat. Explor. Mez.,* 42: 1–48.

Gujarathi, A.S. and Kanhere, R.R., 1998. Seasonal dynamics of phytoplankton population in relation to biotic factors of a fresh water pond at Barwani (M.P.). *Poll. Res.,* 17(2): 133–136.

Hegde, G.R. and Bharati, S.G., 1984. Ecological studies in ponds and lakes of Dharwad: Trophic status. *Phykos,* 23(1 and 2): 71–74.

Jankiram, K. and Jayaraj, Y.M., 1996. Effect of organophosphorus insecticides on primary productivity of stabilisation pond. *Geobios,* 23(4): 219–222.

Ketchum, B.H., 1969. The development and restoration of deficiencies in phosphorus and nitrogen content of unicellular plants. *J. Cellular. Comp. Physiol.,* 5: 55–74.

Mukherjee, B. and Pankajakshi, G.V.N., 1995. The impact of detergents on plankton diversity in freshwaters. *J. Environ. Biol.,* 16(3): 211–218.

Prakasham, V.R. and Joseph, M.L., 2000. Water quality of Sasthamcotta lake, Kerala (India) in relation to primary productivity and pollution from anthropogenic sources. *J. Environ. Biol.*, 21(4): 305–307.

Prasad, D.Y., 1990. Primary productivity and energy flow in upper lake. Bhopal. *Ind. J. Environ. Hlth.*, 32(2): 123–139.

Rao, V.N.R., Mohan, R., Hariprasad, V. and Ramasubramanian, 1993. Seasonal dynamics of physico-chemical factors in a tropical high altitude lake: An assessment in relation to phytoplankton. *J. Environ. Biol.*, 14(1): 63–75.

Schindler, D.W. and Fee, E.J., 1975. The role of nutrient cycling and radiant energy in aquatic comminutes. *I.B.P. Symposium on Photosynthesis and Production in Different Environment*, (Ed.) J.P. Cooper, p. 323–344.

Singh, Arun K. and Singh, Dilip K., 1999. A comparative study on the phytoplanktonic primary production of river Ganga and pond of Patna (Bihar), India. *J. Environ. Biol.*, 20(3): 263–270.

Singh, H.P., 1998. Studies on primary production in Gobindsagar reservoir, Himachal Pradesh. *J. Environ. Biol.*, 19(2): 167–170.

Singh, Ravindra, 1990. Correlation between certain physico-chemical parameters and primary production of phytoplankton at Jamalpur, Munger. *Geobios*, 17(5–6): 229–234.

Verma, J.P. and Mohanty, R.C., 1995. Phytoplankton and its correlation with certain physico-chemical parameters of Danmukundpur pond. *Poll. Res.*, 14(2): 233–242.

Chapter 15

Evaluation of Drinking Water Quality Status of Kanale Tank with Reference to Physico-chemical Factors, Sagara Taluk, Karnataka

R. *Purushothama,* B.R. *Kiran,*
J. *Narayana &* E.T. *Puttaiah*

Department of Environmental Science, Kuvempu University,
Jnana Sahyadri, Shankaraghatta – 577 451, Karnataka

ABSTRACT

The present study deals with water quality status of the Kanale tank water with reference to physico-chemical factors to assess the potability of water from January 2004 to December 2004. Kanale tank exhibits total alkalinity 15–60mg/L, with pH ranging from 6.5 to 8.3, total hardness ranged between 12–26.66 mg/L, phosphate fluctuates between 0.1–0.21 mg/L, BOD deviates between 0.81–6.08 mg/L. From the present observation it can be known that the Kanale tank is free from pollution and suitable for drinking purposes. The possible factors responsible for variation have been discussed in details.

Keywords: Water quality, Physico-chemical factors, Kanale tank.

Introduction

Quantity of potable water is as important as its quality. Various physico-chemical and biological factors governing the quality of water (Sunkad and Patil). In India there are enormous numbers of natural and man made water bodies used for various purposes, mainly for drinking and agriculture. However, in recent years due to rapid urbanization, industrialization, and modern agricultural activities, the quality of water bodies deteriorated causing environmental hazards. Due to direct or indirect interference of hazardous chemical pollutant causing an adverse impact on human health and aquatic life as well.

Studies have been made on the limnobiological status of natural and man made water bodies in India mainly with an intention to asses the water quality (Singh, 2000; Shastri and Pendse, 2001). Hence, the present work has been undertaken to study the water quality parameters monthly during January to December 2004. The results obtained are discussed in light of available literature.

Study Area

Kanale tank (Kanale kere) is one of the important water resources located in the Kanale village at Sagara taluk. It is a perennial water body as it receives the water by rainfall. It lies between14° 12' latitude and 74° 84' longitude, catchment area of the tank is 1.30 Sq/km, water spread area is 22.1 Hectare.

Methods

Water samples were collected in black polythene cans (2-Litre capacity) with well-fitted caps for a period of one year from January 2004 to December 2004. The parameter such as Water temperature, pH and Dissolved oxygen were analyzed on the spot. The samples were transported to the laboratory in iceboxes and standard methods of preservation and analysis were used as per APHA (1998).

The parameters estimated and the methods used are given in Table 15.1

Results and Discussion

The results of the water quality analysis is presented in Table 15.2. The water temperature ranged between 24°C and 29.5 °C during the study period. The variation in temperature was varies between summer and rainy seasons. Similarly, seasonal variation of temperature also reported by Jain *et al.* (1999). The hydrogen ion concentration of water in all months slightly acidic to alkaline (6.5–8.3). Maximum pH of 8.3 was observed in the month of November and minimum of 6.5 in the month of October, due to heavy rainfall and surface runoff. The fluctuation in pH have been found to follow changes in sunshine (Lind, 1938) and higher pH values were considered to be indicative of higher rate of productivity. The present study revealed a similar relationship in the tank. pH and dissolved oxygen varied together in the water body confirming earlier observations of Bharathi and Kore(1975) and Armugam and Furtado (1980).

Table 15.1

Parameters	Method
Temperature	Direct measurement
pH	pH meter
Conductivity	Systronics conductivity meter, Model 304
Dissolved Oxygen	Modified Winklers method
Biochemical Oxygen Demand	Modified Winklers method
Chlorides	Argentometric method
Total Hardness	EDTA titrimetric method
Sulfate	EDTA titrimetric method
Calcium	EDTA titrimetric method
Magnesium	EDTA titrimetric method
Phosphate	UV/Visible spectrometer
Sodium	Systronic flame photometer
Potassium	Systronic flame photometer
Nitrate	Brucine-sulfate method spectrophotometer
Nitrite	Phenol disulphonic acid method spectrophotometer

Turbidity values ranged from 15.3 to 35.5 NTU. Maximum turbidity was observed during rainy months due to surface runoff. Minimum values of turbidity observed in Feb, it may be due to higher temperature, which increases the decomposition of organic matter. Electrical conductivity values fluctuated from 33 to 135 mmhos/cm.

Table 15.2: Physico-chemcial Factors of Kanale Tank,

Parameter	Jan	Feb	Mar	Apr	May	Jun	July	Aug	Sept	Oct	Nov	Dec	Range
A. T. °C	30	33.5	33	32	31	27.5	25	29	32	32	30	31.5	25–33.5
W.T. °C	28	28.5	29.5	30	28.5	26.5	24	28	30	28.5	28	29	24–29.5
pH	7.3	8.0	8.0	7.3	8.2	6.6	7.4	7.1	7.4	6.5	8.3	6.9	6.5–8.3
E.C	40	53	62	98	135	58	47	45	47	35	33	37	33–135
TDS	25.6	33.92	39.68	62.72	89.1	37.12	30.08	28.8	30.08	22.4	21.1	23.58	21.1–89.1
Turbidity	12	15.3	35.5	23.3	28.5	31.0	31.3	20.1	25.5	19.0	30	32	15.3–35.5
DO	9.16	8.8	6.48	1.21	8.1	6.89	4.86	2.02	6.75	14.19	16.62	4.86	1.21–16.62
Free CO_2	4.4	8.8	8.8	17.6	8.8	4.4	11.0	13.2	13.2	13.2	4.4	17.6	4.4–17.6
Cl_2	8.5	15.60	18.43	21.27	17	17.0	14.18	9.92	14.18	14.18	12.76	12.76	12.76–21.27
Ca^{++}	2.52	4.20	5.05	3.36	5.89	2.77	4.2	4.2	4.2	5.05	4.20	2.52	2.52–5.89
Mg^{++}	1.1	4.0	3.43	0.87	2.26	1.39	1.68	0.3	0.36	1.36	2.31	2.2	0.87–4.0
TH	14.0	20.6	26.66	12	24	12.66	26	12	12	14.0	20.0	16.0	12–26.66
BOD	2.71	3.6	2.88	2.5	3.64	4.46	2.8	4.2	2.4	6.08	14.18	0.81	0.81–2.16
T alk	30	60	50	40	63	26	30	30	20	15	30	23	15–63
PO_4	0.2	0.21	0.16	0.12	0.14	0.04	0.002	0.1	0.2	0.001	0.001	0.001	0.001–0.21
NO_2	0.014	0.01	0.01	0.015	0.01	0.01	0.03	0.02	0.01	0.01	0.01	0.002	0.002–0.03
NO_3	0.12	0.13	0.11	0.14	0.18	0.26	0.19	0.21	0.11	0.15	0.21	0.18	0.11–0.26
SO_4	8.6	7.87	14.4	9.36	10.36	1.78	4.6	9.6	4.9	14.9	5.76	1.92	1.92–14.9
Na^+	4.3	4.9	3.6	3.1	3.6	4.1	4.0	4.1	3.6	19	2.1	1.6	1.6–4.9
K^+	3.8	4.2	1.1	2.9	3.2	3.1	3.2	3.4	3.2	1.5	1.4	1.1	1.1–4.2

Note: All the parameters are in mg/L except pH, Temperature (°C), Turbidity (NTU) and Electrical conductivity (mmhos/cm).

Dissolved oxygen showed maximum level 16.2 mg/L in the month of November, while the minimum DO level of 1.21 mg/L was recorded at this tank in April. Solubility of oxygen decreases with increases in temperature (Sabata and Nayar, 1995). Similarly increase in DO is obviously related to decrease in temperature, However similar observation was noticed in the present study also. Highest free CO_2 was recorded during April and December (17.6 mg/L), and lower content in the months of June and November *i.e.* 4.4 mg/L, higher concentration of free CO_2 is due to respiration of organisms and absence of photosynthesis. Phenolphthalein alkalinity was absent in all the months. The maximum value of total alkalinity (63mg/L) was recorded in the month of May and lower concentration of 15 mg/L in October. Philipose, (1959) suggested that a water body with alkalinity value higher than 100 mg/L is nutritionally rich. By this standard tank water is oligotrophic. The lowest value of 20 mg/L was observed in the month of May, Kannan (1991) has classified water on the basis of hardness values in the following manner; 0-60mg/L soft, 60-120 moderately hard, 120-160mg/L hard, 180 and above is very hard. On the basis of hardness values, this tank water can be included in a soft category.

The amount of total dissolved solids (TDS) found in experimental sampling is 21.1 mg/L to 89.1 mg/L. TDS in water is mainly due to the presence of bicarbonates, sulfates, chlorides of calcium, magnesium and sodium. In the present study the water body contain lower concentration of TDS.

Calcium is very important element influencing flora of ecosystem, which plays potential role in metabolism and growth. The range of it varied from 2.52 to 5.89 mg/L. The maximum calcium concentration 5.89 mg/L was recorded in the month of May and minimum of 2.52 mg/L in December. The concentration of magnesium varied between 0.87 to 4.0 mg/L. Normally these ions are not problematic but at higher concentration increase total hardness of water.

The higher amount of chloride 21.27 mg/L was recorded at this tank in the month of April. Minimum concentration of 12.76 mg/L was noticed in November and December. Thresh, *et al.* (1944) also pointed out that high concentration of chloride is due to large quantity of organic matter in it, which was further supported by Adoni, (1985). However, such a condition is not observed in the present water body.

BOD values in water samples are ranged between 0.81 to 2.16 mg/L. These values are within permissible limit. Phosphate concentration was low in the water. The maximum of 0.23 mg/l was recorded in the month of February, low values of phosphate was recorded in months of July, October, November and December. Phosphate is considered amongst the primary limiting nutrients in ponds and lakes (Schindler, 1971). Low values of phosphate have been reported from various Himalayan lakes and reservoirs (Raina and Peter, 1999)

The nitrate is one of the most oxidisable forms of nitrogen and is an essential for plant nutrient. NO_3 concentration is associated with rainwater runoff and sullage discharge. The nitrate ranged from 0.011–0.26 mg/L in the tank water. Sugunan (1995) reported much lower level of NO_3 in the Govind sagar reservoir. The highest concentrations of the nutrient were recorded after the on set of rains, probably by the transport of nutrients from the watershed areas with the runoff water. From this aspect, the tank water is healthy for drinking purpose. Pant *et al.* (1985) reported on the rising level of nitrogen in the lake Nainital due to increasing of human interference.

Out of the four cations analyzed (Na^+, K^+, Ca^{++}, Mg^{++}) except for calcium others behaved identically without much variation. Sodium and potassium ions varied from 1.6 to 4.9 and 1.1 to 4.2 mg/L respectively. On the basis of above-mentioned physico–chemical parameters, it may be concluded that the tank water is very soft recommended for drinking purpose, based on the physico-chemical values.

Remedial Measures

After studying these it is felt that with proper care even now many of these water bodies can be saved from dying. Some of the remedial measures suggested are

1. Survey of the tanks should be immediately undertaken and boundary lines should be clearly demarked.

2. Pitching of tank boundaries and plantation to prevent soil erosion.

3. In future, washing of clothes, bathing, washing of domestic animals and human activities should be strictly prohibited.

4. Harvesting of aquatic macrophytes can be assigned to private agencies to make use of waterbodies and protecting them from eutrophication.

5. Regular water quality monitoring by authorized agencies.

6. Establishing educational centres for creating awareness about the importance of an Eco-friendly heritage and also as a source of income.

7. Recycling of the wastewater from nallas and use for non-potable purpose.

8. Protect the water body from human encroachment.

Acknowledgement

The first author is thankful to Kuvempu University, Shankaraghatta and UGC, New Delhi for providing facilities and fellowship.

References

Adoni, A.D., 1985. *Workbook on Limnology*. India MBA Committee. Department of Environment, Govt. of India, pp. 216.

APHA, 1998. *Standard Methods of Examination of Water and Wastewater*, 18th edn. Washington, U.S.A.

Armugam, P.T. and Furtado, J.I., 1980. Enrichment and recovery of a Malasian reservoir. *Developments in Hydrobiology*, 2: 281–285.

Bharathi, S.G. and Kore, A.P., 1975. Limnological studies in ponds and lakes of Dharwad. Diuranal variation in sayadapurkere pond (part I). The Karnataka University. *J. Science*, 20: 157–167.

Kannan, K., 1991. *Fundamental of Environmental Pollution*. S. Chand and Company Ltd., New Delhi.

Lind, A., Rai, S.C., Pal, J. and Sharma, E., 1999. Hydrology and nutrient dynamics of sacred lake in Sikkim Himalaya. *Hydrobiologia*, 416: 13–22.

Pant, M.C., Sharma, P.C. and Sharma, A.P., 1985. Physico-chemical limnology of lake Nainital Kumaun Himalaya (Uttarakhand) India. *Acta Hydrochem. Hydrobiol.*, 13(3): 331–349.

Philipose, M.T., 1959. Freshwaters phytoplankton in land fisheries. In: *Proc. Symp. Algalogy*, ICAR, New Delhi, p. 272–291

Raina, H.S. and Peter, T., 1999. Coldwater fish and fisheries in the Indian Himalayas lakes and reservoirs. In: *Fish and Fisheries at Higher Altitudes Asia,* (Ed.) T. Peter. FAO Fish. Tech. Paper No. 385. Rome, pp. 64–88.

Sabata, B.C. and Nayar, M.P., 1995. *River Pollution in India: A Case Study of Ganga River.* APH Publishing Corp., New Delhi, pp. 223.

Schindler, D.L., 1971. Carbon, nitrogen, phosphate and eutrophication in freshwater lakes. *J. Phycol.,* 17: 321–329.

Shastri and Pendse, D.C., 2001. Hydrobiological study of Dahikhuta reservoir. *J. Environ. Bio.,* 22(1): 67–70.

Singh, D.N. 2000. Seasonal variation of zooplankton in a tropical lake. *Geobios,* 27: 92–100.

Sugunan, V.V., 1995. *Reservoir Fisheries of India.* FAO Fish. Tech. Paper No. 345. FAO, Rome.

Sukunda, B.N. and Patil, H.S., 2004. Seasonal dynamics of phytoplankton in relation to physico-chemical factors of fort lake, Belgaum (North Karnataka). *J. Environment and Ecology,* 22(2): 337–347.

Thresh, J.C., Sacking, E.V. and Beale, J.F., 1944. *The Examination of Water Supplies,* (Ed.) E.W. Taylor.

Chapter 16

Delineation and Management of Salt Water Contaminated Zones in Venkatagiri Taluq, Nellore District, Andhra Pradesh

S.V. Lingeswara Rao

Department of Geology, Sri Venkateswara University,
Tirupati – 517 502, Andhra Pradesh

ABSTRACT

Salt water contamination zones in the groundwater of Venkatagiri Taluq, Nellore District, Andhra Pradesh, have been delineated basing on hydrogeochemical parameters such as Ca/Mg, $Cl\text{-}CO_3+HCO_3$, TA/TH and Base Exchange Index and Mg/Ca versus Cl. In order to delineate salt water contaminated zones, 77 groundwater samples have been collected for various hydrogeochemical and chemical parameters. It is found that salinity in groundwaters is due to local environmental conditions such as drainage, application of fertilizers in crop fields, water logging, domestic sewage and solid waste disposal sites.

Keywords: Groundwater, Salt water contaminated zones, Delineation and management.

Introduction

Demand for groundwater is increasing incessantly day by day owing to population explosion, increasing industrial demands, greed and white revolutions and improving technologies. The study area is located in the south-west corner of Nellore district of Andhra Pradesh, India and lies between N. Lat.13° 49'–14° 12' and E. Long.79° 26'–79° 49' (Figure 16.1). The total geographical area of the study area is 944.2 sq. km. Out of 944.2 sq km, 21.12 per cent of the area is occupied by Velikonda Hill ranges and 23.37 per cent by forest land. The topography of the area is undulating with an altitude ranging from 30 to 80m above MSL gently sloping towards the north-east. Velikonda Hill Ranges form the western boundary of the area ranging

Figure 16.1: Location Map of Well Inventoried in the Study Area

an altitude from 152 to 650 m. The study area enjoys tropical climate. The average annual rainfall of the area is 1035 mm, of which monsoon rainfall is 820 mm. The area is mainly occupied by the geological formations quartzites, mica hornblende schists, granites and gneisses, phyllites and alluvium. Groundwater in the area generally occurs under water table conditions. The present study aims at visualizing the genesis of salinity in groundwaters on the basis of their hydrogeochemistry. Several people used (Ravi Prakash and Chandu, 1993 and Sambasiva Rao and Janardhan Raju, 1997) have applied hydrogeochemical parameters for delineation of salt-water contamination zones.

Methodology

Seventy seven groundwater samples have been collected covering all the geological formation of the study area and analysed for various chemical constituents using the methods described in Rainwater and Thatcher (1960) and Brown *et al.* (1970). The sample locations are presented in Figure 16.2.

Hydrogeochemical Parameters

Total dissolved solids and constituent ions play a pivotal role in the salinity of groundwater. Chemical constituents such as Ca, Mg, Na, K, CO_3, HCO_3, Cl, SO_4 and NO_3 involve in the aqueous system of groundwater. Ion exchange affects and modifies composition of water. Saline water zones have been delineated in the study area by using the hydrogeochemical parameters such as Ca/Mg, Cl/CO_3 + HCO_3, TA/TH (Total alkalinity/Total Hardness) ratios and base exchange index and Mg/Ca versus Cl. Table 16.1 shows the chemical ratios computed from chemical data.

Results and Discussion

Ca/Mg Ratio

Seawater contamination is indicated by Ca/Mg ratio (Mandel and Shiftan, 1981). Magnesium is present in seawater in much greater concentration than calcium (Hem, 1975). Majority of saline water show significantly lower Ca/Mg ratio than non-saline waters (Muthuraman *et al.*, 1992). The maximum and minimum values of Ca/Mg ratio are 1.284 (S. No.47) and 0.080 (S.No.62) respectively. Ca/Mg ratio values than 0.20 indicate saline zones (Figure 16.2)

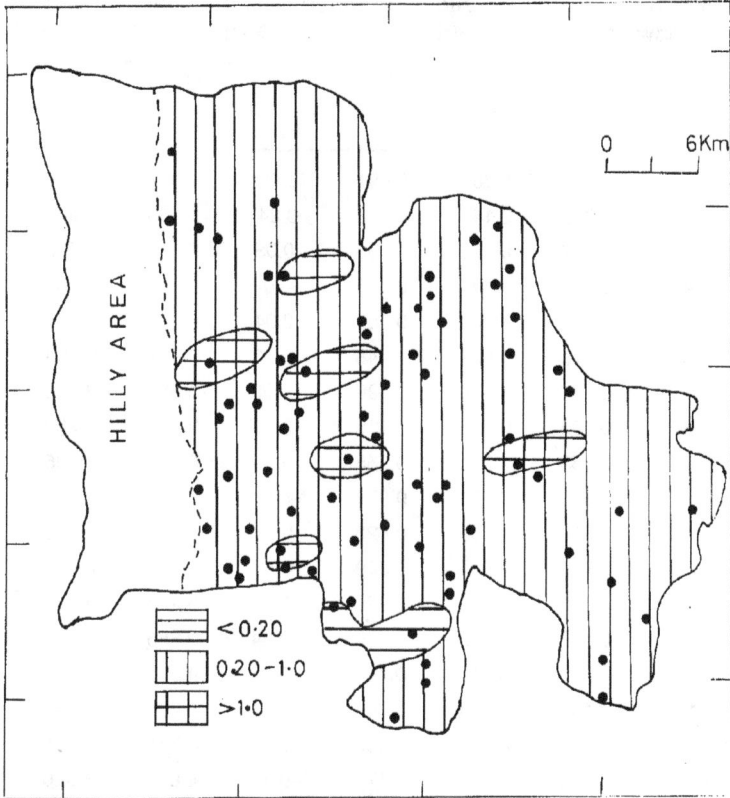

**Figure 16.2: Zones of Salt Water Contamination on
the Basis of Ca/Mg Ratio**

between 0.20 and 1.00 suggest slightly saline water zones and
exceeding 1.00 have been considered as fresh water zones. Ca/Mg
values lower than 0.20 (S. Nos. 9, 16,59 and 62) have been recorded
in isolated patch in the southern portion. Ca/Mg values between
0.20 and 1.00 have been distributed throughout the area. Values
exceeding 1.00 have been distributed in isolated patches in northern,
eastern, central, eastern and southwestern portions of the study area.
Since Ca/Mg ratio for seawater is 0.18 such low values are suggestive
of salt-water contamination.

Table 16.1: Chemical Ratios Computed from the Chemical Data of Groundwater of Venkatagiri Taluq, Nellore District, Andhra Pradesh

Sl.No.	Ca/Mg	Cl/CO$_3$+ HCO$_3$	TA/TH	Base Exchange Index	Mg/Ca	Cl
1.	0.250	0.285	0.691	0.02	4.001	0.959
2.	0.766	2.141	0.583	0.24	1.305	7.447
3.	0.312	0.796	1.053	0.03	3.206	6.629
4.	0.839	0.498	0.868	−0.04	1.190	2.087
5.	0.223	4.804	0.183	2.34	4.476	28.661
6.	1.078	1.045	1.246	−0.03	0.927	4.0796
7.	0.277	0.878	0.796	−0.08	3.612	4.598
8.	0.303	0.801	0.710	0.13	3.297	4.259
9.	0.153	1.224	0.755	0.20	6.511	8.125
10.	0.277	0.578	0.978	−0.07	3.607	2.652
11.	0.219	1.330	0.623	0.26	4.565	8.125
12.	0.868	1.634	0.519	0.46	1.152	12.230
13.	0.303	0.738	0.320	−0.12	3.297	4.400
14.	0.346	0.518	0.225	−0.34	2.886	3.535
15.	0.232	0.217	0.786	−0.04	4.305	1.403
16.	0.159	1.048	0.859	0.12	6.264	7.701
17.	0.570	0.303	3.409	−0.28	1.751	2.031
18.	0.206	0.460	1.171	−0.10	4.854	3.526
19.	1.010	0.246	1.653	−0.39	0.989	2.510
20.	0.367	0.376	2.465	−0.29	2.720	4.344
21.	1.280	0.841	0.987	0.06	0.765	4.626
22.	0.480	0.812	0.976	0.03	2.082	5.163
23.	0.223	0.414	1.130	−0.16	4.472	2.200
24.	0.653	0.559	1.073	−0.13	1.531	2.679
25.	0.490	0.979	1.033	0.02	2.039	8.350
26.	0.460	0.412	1.189	−0.19	2.169	3.385
27.	0.888	1.099	0.588	0.37	1.125	6.121
28.	0.369	1.236	0.559	0.30	2.708	5.755
29.	0.512	0.879	0.957	0.05	1.952	5.416
30.	0.825	0.681	2.380	−0.33	1.211	6.827
31.	0.992	0.135	1.550	−2.25	1.007	0.789

Contd...

Table 16.1–Contd..

Sl.No.	Ca/Mg	Cl/CO$_3$+ HCO$_3$	TA/TH	Base Exchange Index	Mg/Ca	Cl
32.	0.279	0.683	1.608	−0.82	3.573	7.249
33.	0.929	0.461	1.462	−0.49	1.075	3.893
34.	0.346	0.205	1.899	−1.36	2.884	2.962
35.	0.653	0.162	7.113	−3.48	1.529	1.794
36.	0.257	0.228	1.739	−1.130	3.886	2.990
37.	0.957	0.184	2.784	−2.06	0.347	1.354
38.	0.798	0.784	1.284	−0.06	1.253	6.883
39.	0.510	0.422	2.974	−0.94	1.958	3.554
40.	1.079	0.299	2.967	−1.33	0.926	2.078
41.	0.347	0.735	1.322	−0.19	2.884	5.726
42.	0.582	0.274	2.082	−1.53	1.717	3.442
43.	0.996	0.979	1.926	−0.25	1.003	7.108
44.	0.491	0.121	2.515	−3.37	2.033	1.298
45.	0.513	0.215	2.969	−2.22	1.939	2.200
46.	0.279	0.274	2.699	−1.40	3.572	3.183
47.	1.284	0.296	1.497	−0.67	0.788	2.934
48.	0.834	0.264	1.231	−0.85	1.198	1.890
49.	0.441	0.294	4.061	−1.39	2.266	4.232
50.	0.657	0.503	2.280	−0.73	0.521	4.457
51.	0.466	0.240	2.259	−1.73	2.143	2.934
52.	1.079	0.066	3.844	−8.31	0.923	0.903
53.	0.488	0.105	1.933	−5.34	2.033	0.903
54.	0.685	0.123	3.706	−3.81	1.458	1.354
55.	0.536	0.192	1.216	−1.26	1.865	1.410
56.	0.320	1.254	0.540	0.56	3.118	10.522
57.	1.021	0.074	2.562	−5.71	0.978	0.677
58.	0.448	0.090	3.135	−5.94	2.231	0.874
59.	0.144	0.023	3.246	−4.19	6.906	0.451
60.	0.684	0.269	2.866	−1.54	1.460	4.062
61.	0.212	0.277	3.790	−2.53	4.706	4.005
62.	0.080	0.315	1.048	−0.09	12.366	3.949

Contd...

Table 16.1–Contd..

Sl.No.	Ca/Mg	Cl/CO_3+ HCO_3	TA/TH	Base Exchange Index	Mg/Ca	Cl
63.	0.308	0.125	2.202	–3.09	3.238	1.577
64.	0.265	0.531	0.730	0.28	3.759	3.272
65.	0.789	0.089	3.727	–5.62	1.269	0.959
66.	0.219	0.258	1.377	–1.08	0.219	1.354
67.	0.379	0.632	0.769	0.17	2.637	3.272
68.	0.579	0.403	1.433	–0.29	1.726	3.385
69.	0.479	0.744	0.984	–0.06	2.084	6.009
70.	0.664	0.303	2.479	–1.12	1.505	4.344
71.	0.787	0.350	2.981	–1.14	1.270	5.416
72.	0.233	0.406	1.850	–0.53	4.301	5.473
73.	0.242	0.557	1.268	–0.55	4.123	4.908
74.	0.829	0.168	2.965	–3.02	1.205	1.354
75.	0.653	0.420	0.954	–0.16	1.530	2.595
76.	0.570	0.355	0.954	–0.63	1.474	1.749
77.	0.370	0.707	0.805	–0.02	2.692	3.949

Cl/CO_3+HCO_3 Ratio

The degree of salt-water contamination can also determined on the basis of Cl/CO_3+HCO_3 ratio (Revelle, 1941). Chloride is most abundant anion in seawater and bicarbonate is predominant in fresh groundwater. On the basis of $Cl/CO_3 + HCO_3$ ratio, Simpson (1946) classified the following classification of groundwater.

Range of Cl/CO_3+HCO_3 Ratio	Description
< 0.05	Normally fresh groundwater
0.05–1.30	Slightly contaminated groundwater
1.30–2.80	Moderately contaminated groundwater
2.80–6.60	Injuriously contaminated groundwater
6.60–15.50	Highly contaminated groundwater
> 200.00	Seawater

Gamble and Fisher (1966) suggested that substantial contribution of chloride is erosion of crystalline rocks. Saline tract is characterized by the addition of Na and Cl ions in the waters during the process of acquition of the dissolved phase (Hem, 1975). Carbonate has an important role in attacking primary silicates and in contributing dissolved solids to waters. The anion produced during the attack is bicarbonate.

Moderately contaminated groundwater is observed as isolated patches in north-east and south-central part of the study area (Figure 16.3). Injuriously contaminated groundwater is observed in the almost central portion of the study area as an isolated patch.

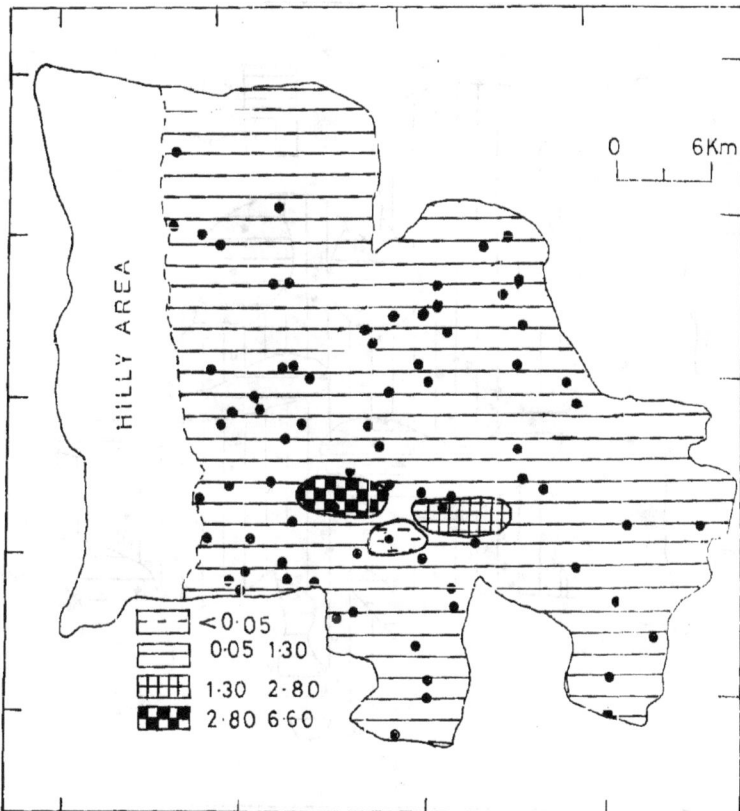

Figure 16.3: Zones of Salt Water Contamination on the Basis of Cl/CO$_3$+HCO$_3$ Ratio

TA/TH Ratio

Total Alkalinity/Total Hardness ratio is another parameter employed to ascertain the degree of saltwater contamination (Hem, 1975). Sodium bicarbonate is present if excess of total alkalinity over total hardness is observed.

Most of the study area reveals an excess of total alkalinity over total hardness indication non-contamination of salt-water zones (Figure 16.4). Only in few isolated zones at north, north-east, eastern, southeastern, southern, southwestern and western portions of the study area. The TA/TH is less than one. Some of the packets showing

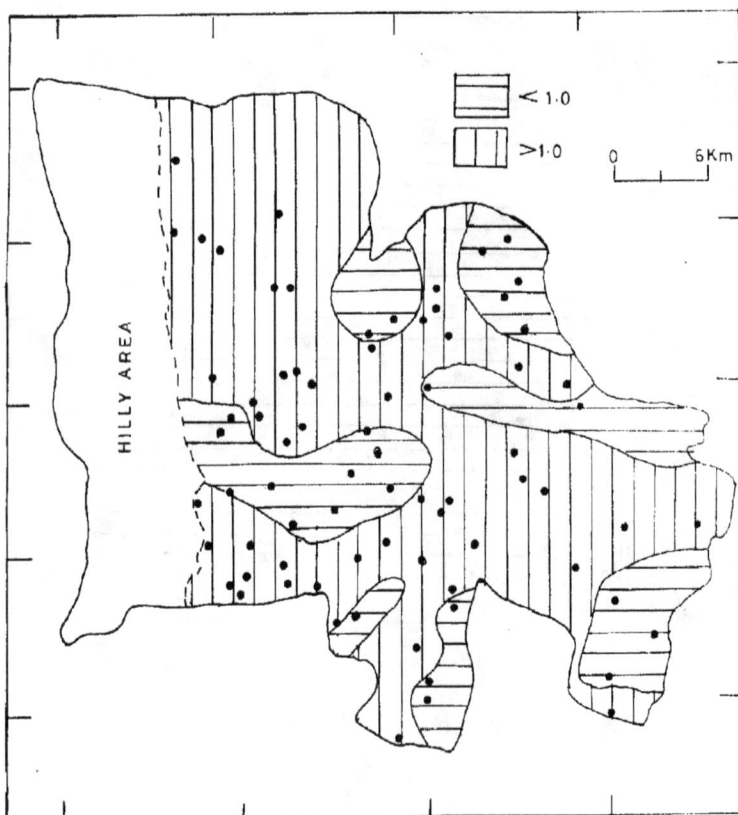

Figure 16.4: Zones of Salt Water Contamination on the Basis of TA/TH Ratio

less values (less than one) of TA/TH also show less values of Ca/ Mg (0.20) indicating salt-water contamination in these zones.

Base Exchange Index

Base Exchange Index was derived by Cl–Na+K/Cl (Schoeller, 1967). Negative values indicate no contamination but positive values indicate salt-water contamination of groundwater. Majority of the samples (50) in the study area show negative base index values and only a few samples (27) show positive base index values. It shows that major portion of this water is meteoric in origin and cannot owe its source to seawater. Schoeller concept of water types if related to the evolution of groundwater with respect to chemistry.

Mg/Ca Verses Cl

The extent of salt-water contamination is often indicated by Mg/Ca ratio plotted against Cl (Nair *et al.*, 1981) as shown in Figure 16.5. A straight line shows a simple mixing of salt water and groundwater. Any deviation from the line would indicate the

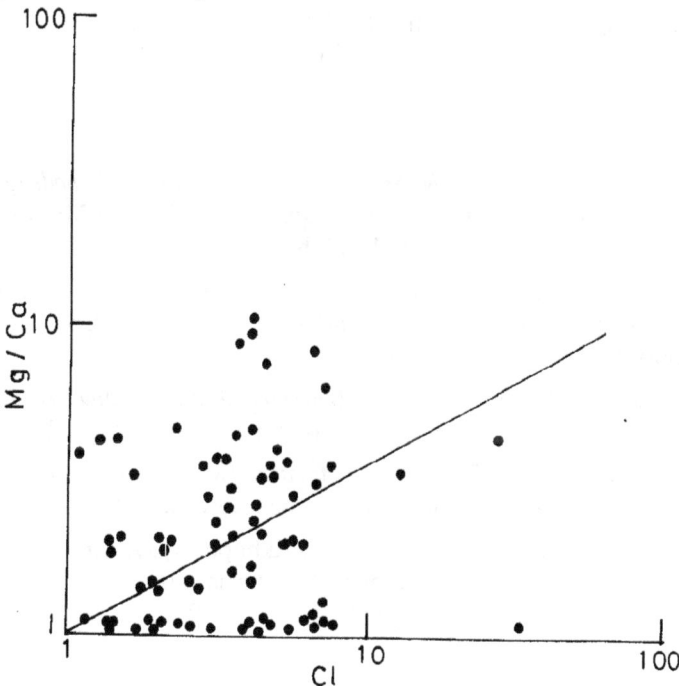

Figure 16.5: Mg/Ca Versus in Groundwater

influence of aquifer materials on the mixed water by way of exchange process. The plotted points for the area under study are higher in the upper part and lower in the lower part of the sea water value suggesting an increase in the quantity of magnesium in groundwater or depletion of calcium precipitation of calcium seems to be an important factor influencing the trend of the data plots.

Conclusion

Salt water contaminated zones in Venkatagiri Taluq, Nellore District, AP., India have been delineated basing on certain hydrogeochemical parameters. In some places more than one parameter (*i.e.*, Ca/Mg, $Cl\text{-}CO_3+HCO_3$ and TA/TH) are indicating salt-water contamination. There are some exceptional cases where all the parameters are not coinciding for delineating salt-water contamination zones, which necessitates further study of local environmental conditions.

Acknowledgement

The first author is thankful to the University Grants Commission, New Delhi for providing financial assistance to carry out the work.

References

Brown, E., Skougslad, M.W. and Fishman, M.J., 1970. *Methods for Collection and Analysis of Water Samples for Dissolved Minerals and Gases*. Technique of W.R.I. of the USGS.

Gambell, A.W. and Fisher, D.W., 1975. *Chemical Composition of Rainfall in Eastern North Carolina and Southeastern Virginia*. US GS Water Supply Paper 1535–K, 41 p.

Hem, J.D., 1975. *Study and Interpretation of the Chemical Characteristics of Natural Water*. USGS Water Supply Paper 1459–A, p. 31.

Mandel, S. and Shiftan, Z.L., 1981. *Groundwater Resources Investigation and Development*. Academic Press, Inc., New York.

Muthuraman, K., Tiwari, M.P. and Mukhopadhyay, P.K., 1992. Salinity in the groundwater of the Purna basin: Its genesis. *Journal of Geological Society of India*, 39: 50–60.

Nair, V.N., Singhal, B.B. and Prakash, B., 1981. Salt water encroachment in the karstic miliolite limestones, coastal parts of Saurastra. *Indian Geohydrology*, 15(I–IV): 211–220.

Raviprakash, S. and Chandu, S.N., 1993. Delineation of salt water contamination zones by chemical parameters in Dakor area, Jalaun district, Uttar Pradesh. *Indian Journal of Geology*, 65(4): 282–289.

Revelle, R., 1941. Criteria for recognition of sea water in groundwater. *Trans Amer. Geophys. Union*, 22: 593–597.

Sambasiva Rao, T. and Janardhana Raju, N., 1995. Hydrogeochemical parameters for delineation of salt water contamination zones in upper Nagari river basin, Chittoor District, Andhra Pradesh, India. *Ecology*, 10(6): 34–38.

Schoeller, H., 1967. *Arid Zone Hydrology: Recent Developments*. UNESCO, Rev., Reicardilla.

Simpson, T.R., 1946. *Salinas Barin Investigation*. Bul. 52, Calif. Div. Water Resources, Sacramento, 230 p.

Rainwater, F.H. and Thatcher, L.L., 1960. Methods for collection and analysis of water samples. *U.S. Geological Survey Water Supply Paper* 1454, 301 p.

Chapter 17

Phytoremediation: A Plant Assisted Technology for Treatment of Industrial Wastewater

Yashpal Singh,Vikas Singhal & J.P.N.Rai*

Department of Environmental Sciences,
G.B.Pant University of Agriculture and Technology,
Pantnagar – 263 145

ABSTRACT

Phytoremediation is an environmental cleanup technology, which involves the use of green plants to remove, contain, inactivate and/or render environmentally toxic contaminants harmless. In this new emerging technology, phytoremediator plants work on the "Green liver concept" and act as solar driven natural detoxification systems, involving the phytoextraction, phytovolatalization, phytodegradation, phytostabilisation and rhizofilteration etc. The key factors involved are low cost (compared to other classical remediation techniques), low technology, high efficiency and aesthetic aspect, making phytoremediation suitable for simultaneous cleanup of multiple, mixed contaminants from industrial wastewater. The ultimate success of this technology relies on better understanding of complex interaction between

---

* Corresponding Author.

pollutants, soil, plant roots and rhizospheric microflora to increase the bioavailability of pollutant to fully exploit the detoxification capabilities of phytoremediator plants so as to make it acceptable for industries. The present paper deals with phytoremediation process along with phytoremediation potential of Water hyacinth (*Eichhornia crassipes*) against wastewater of pulp and paper mill, distillery, brass and electroplatting industries.

Introduction

Industrialization is the index of modernization, which leads to alteration in the physical, chemical and biological properties of environment. Since industrialization and pollution are complementary to each other, measures are to be adopted so that pollution can be lessened to biosphere. India has witnessed a very rapid industrial growth, especially in last four decades. Presently the country has developed a sound base in several industries like metal, chemicals, pulp and paper, petroleum, textile, fertilizers, distillery and others. According to one estimate, there are at present about 500 large and medium scale industries in the country which contribute about 70 per cent by volume of the total industrial water, the remaining 30 per cent contributed by small and cottage sector units. During the last century, the explosive development of chemical industries has produced a great variety of chemicals leading to the modernization of our lifestyle. The large-scale production of a variety of chemicals, however, caused a global deterioration of environmental quality (Iwamoto and Nasu, 2001). Contamination of soil, groundwater, sediments, surface water and air with hazardous and toxic chemicals are serious problem, which has been faced by world today (Boopathy, 2000). Undoubtedly, advances in technology have sustained our industrialized society, but the problem of pollution goes unrecognized, particularly due to very few studies and surveys. The inadequacy of such studies further enhances the toxicity hazards and thus our concern to take up such studies and surveys on large scale is very pertinent.

Moreover, no single industry discharges all types of pollutants. At the same time, the discharge of only one substance in sufficient quantity depends upon nature of industry, raw materials and processes involved in the industry. Even in a given industry, the qualitative and quantitative fluctuations in effluents take place from day to day even hour to hour (Mahabal, 1993). In general, the industrial effluents have a varying spectrum of organic and inorganic pollutants, which need to be treated for meeting the legal requirement and ecological and social acceptance of wastewater with no and/or minimum adverse effects. Otherwise, some large industries such as pulp and paper, distilleries and metal industries of the region with deadliest effluents will result into inhospitable environs of western Uttar Pradesh.

Several methods such as precipitation (hydroxide and sulphide), adsorption, electro-chemical treatment, solvent extraction and evaporation are well known for removing pollution load from a variety of industrial effluents (Eccles, 1999). However, these methods are quite expensive and have restricted applicability due to several short comings such as incomplete removal of pollutants, low-moderate selectivity, very low and high working land of contaminants and the production of sludge and other waste products that also need proper disposal (Bunke *et al.*, 1999; Eccles, 1999). At this juncture, investigation to develop a simple, cost effective and environment friendly phytoremediation technology for the treatment of various industrial wastes is highly warranted.

Phytoremediation

Phytoremediation offers such a technology which exploits the natural detoxification abilities of plants and their associated microorganisms to degrade, contain and to change the chemical nature with potential for the effective inexpensive clean up of a broad range of organic and inorganic wastes from contaminated soil and water systems (Cunninghum *et al.*, 1999; Ensley, 2000). In broad sense, phytoremediation involves the accumulation of the contaminants in the root tissues of plants growing in contaminated water/land and their transfer to harvestable plant part. The main processes involved in phytoremediation are:

Phytoextraction

In this process, a plant species known to accumulate inorganic (metals, metalloids or radio nuclides) and organic contaminant is grown in contaminated soil and water environment. The employed plant is harvested at regular interval and contaminants are recovered, if economically feasible (Brooks, 1998). Bioconcentration of heavy metals in plants is well established and today a large number of higher plants that hyper accumulate heavy metals are known (Brooks, 1998; Dahmani-Muller *et al,* 2001). Plants exhibiting such strategies have been classified as (*i*) *indicators*: plants in which uptake and translocation reflect soil metal concentration and show toxic symptoms (*ii*) *excluders*: plants that restrict the uptake to toxic metals into shoot over a wide range of background concentrations (*iii*) *accumulators*: plants in which uptake and translocation reflect background metal concentration without showing toxic symptoms and (*iv*) *hyper accumulators*: plants in which metal concentration is up to 1 per cent in dry matter. As such, plants growing in metal contaminated ecosystems, regularly take up toxic heavy metals and other organic pollutants and thus, help in environmental decontamination. To date, there are several commercial ventures gaining economic benefits from phytoextraction, not only by recovering extracted metals from plant biomass, but also using the biomass for energy (Glass, 1999).

Hyper accumulator plants could potentially recontaminate the environment, if not treated or processed appropriately. Any treatment or processing should result in a minimal volume of contaminants, requiring no further treatment. Two treatment methods, cofiring with coal/incineration and composting are common. The third, liquids extraction method involves burning of harvested phytomass, composting of other fraction of phytomass and liquid extraction of third fraction using chelating agents. A holistic treatment approach may be the recovery of left over energy in form of biogas from the phytoremediated biomass of plant as a result of methanogenesis.

Advantages

The plant biomass containing the extracted contaminants can be a resource. For example, biomass that contains selenium (Se), an essential nutrient, has been transported to areas that are deficient in Se and used for animal feed (Banuelos *et al,* 1997 a). In green house experiments, gold was harvested from plants (Anderson *et al,* 1998).

Limitations

1. Metal hyperaccumulators are generally slow-growing with a small biomass and shallow root systems.
2. Plant biomass must be harvested and removed, followed by metal reclamation of proper disposal of the biomass.

3. Metal may have a phytotoxic effect (Kumar *et al*, 1995).

4. Phytoextraction studies conducted using hydroponically-grown plants, with the contaminant added in solution, may not reflect actual conditions of soil.

Phytostabilization

Phytostabilization at some metal contaminated sites involves the use of plants especially roots and/or plant exudates to stabilize, demobilize and bind the contaminants in the soil matrix, there by reducing their bioavailabilty. This approach is suitable for metal contaminanted soils. Field trials for *in situ* decontamination of heavy metal polluted soils using metal accumulating crops was proved to be advantageous, particularly with fibre yielding crops (Felix, 1997; Prasad *et al*, 2003). Phytostabilization relies on plants, or compounds they secrete, to stabilize low levels of contaminant that are present in soils (*e.g.*, by absorption or precipitation), to prevent them from mobilizing or leaching in a manner that would endanger public health (Cunningham *et al.*, 1995). Possible mechanism might include sequestering the contaminant in or on cell wall lignin ("lignification"), absorption of contaminants to soil humus, via plant or microbial enzymes ("humification"), or other mechanism, whereby the contaminant is sequestered in the soil, *e.g.*, by binding to organic matter. Phytostabilization is primarily applicable to metal contamination, and might best be used near the end of remediation by traditional means, or after site closure, for those sites where it is acceptable under regulatory guidelines to leave a non-bioavailable portion of the contaminant remaining in the soil (*e.g.* a risk-based regulatory approach). The term also refers to the use of a vegetative cover, *e.g.*, at a landfill, to groundwater or surface waters (Glass, 1999). Phytostabilization reduces the mobility of the contaminant and prevents migration to the groundwater or air, and it reduces bioavailability for entry into the food chain. (Barman and Lal, 1994; Banuelos and Mayland, 2000; Bunzl *et al.*, 2001; Dudka *et al.*, 1996). Phytostabilization occurs through root zone microbiology and chemistry, and/or alteration of the soil environment or contaminant chemistry. Soil pH may be changed by plant root exudates or through the production of CO_2. Phytostabilization can change metal solubility and mobility or impact the dissociation of organic compounds. The plant affected soil environment can convert metals from a soluble to an insoluble oxidation state. Phytostabilization can occur through sorption, precipitation, complexation, or metal valence reduction. Plants can also be used to reduce the erosion of metal-contaminated soil.

Advantages

1. It circumvents the removal of soil.
2. It has a lower cost and is less disruptive than other more-vigorous soil remedial technologies.
3. Revegetation enhances ecosystem restoration.

Limitations

1. The contaminants remain in place.
2. The vegetation and soil may require long-term maintenance (Zorpas *et al.*, 1999).
3. Plant uptake of metals and translocation to the above ground portion must be avoided.
4. The root zone, root exudates, contaminants, and soil amendments must be monitored to prevent an increase in metal solubility and leaching.
5. It might be considered to only be an interim measure.

Phytovolatilization

This method can be used for those contaminants that are highly volatile. Mercury or selenium, once taken up by the plant roots, can be converted into non-toxic forms and volatilized into the atmosphere from the roots, shoots, or leaves. For example, Se can be taken up by plants of the *Brassica* genus and other wetland plants (floating aquatic macrophyte), and converted (*e.g.*, by methylation to the volatile dimethyl selenium) into non-toxic forms which are volatilized by the plants. A similar mechanism can be exploited for Hg, although there are no naturally occurring plants that can accomplish this: the goal here is to engineer bacterial genes for mercury reduction into plants, and here to laboratory experiments are highly encouraging (Bizily *et al.*, 1999; Heaton *et al.*, 1998). Phytovolatiliztion occurs as growing trees and other plants take up water and the contaminants. Some of these contaminants can pass through the plants to the leaves and volatilize into the atmosphere at comparatively low concentration. Mercury has been shown to move through a plant and into the air in a plant that was genetically altered to allow it to do so.

Advantages

1. Contaminants could be transformed to less toxic forms.
2. Contaminants or metabolites released to the atmosphere might be subjected to more effective or rapid natural degradation process such as phytodegradation.

Limitations

The contaminants or hazardous metabolites might accumulate in vegetation and be passed on in later products such as fruit or tubers.

Rhizofiltration

Rhizofiltration refers to the use of plant roots to sorb, concentrate, and precipitate metal contaminants from surface or groundwater. Roots of plants are capable of sorbing large quantities of lead and chromium from soil, water or from water that is passed through the root zone of densely growing vegetation. Shallow lagoons have been engineered as wetlands and maintained as facultative microbial systems with low dissolved oxygen in the sediment. Groundwater or wastewater is pumped through the system for the removal of contaminants by rhizofiltration. Wetlands have been used with great success in treating metals for many years (Mc Grath *et al.*, 2001; Wenzel *et al.*, 2003; Fitz and Wenzel, 2002). Long-term utilization of wetland plants and sulfate reducing conditions results in an increase in pH and a decrease in toxic metals concentration for treatment of acid mine drainage. Rhizofiltration is the use of plant roots to accumulate metals from water. Hydroponically cultivated plants rapidly remove heavy metals from water and concentrate them in the roots and shoots. Harvested plants containing heavy metals can be disposed of or treated to recycle the metal and metal recovery can be done in econimically feasible manner. Rhizofiltraction can be practiced *in situ* (*e.g.*, on surface water), but more likely would be operated in an above ground artificial flow-through reactor, akin to hydroponics, and could thus be applicable to groundwater or industrial waste streams.

Advantages

1. Either terrestrial or aquatic plants can be used.
2. This system can be either *in situ* (floating rafts on ponds) or *ex situ* (an engineered tank system).
3. An *ex situ* system can be placed anywhere because the treatment does not have to be at the original location of contamination.

Limitations

1. The pH of the influent solution may have to be continuously adjusted to obtain optimum pollutant uptake.
2. The chemical speciation and interaction of all species in the influent have to be understood and accounted for.

Plant Suitability for Phytoremediation

The term hyperaccumulator, first introduced by Prof. R. R. Brooks, Department of Soil Science, Massey University, North Newzeland in 1977, includes grasses, aquatic macrophytes and many crop plants must possess certain characteristics; Superior capacity of tolerance to prevailing contaminants, extensive root system, rapid growth, Ease of handling and established cultural practices and species preferably be indigenous to the region.

Owing to the above mentioned suitability criteria, phytoremediation potential of water hyacinth (*Eichhornia crassipes*) against different industrial effluent such as Pulp and Paper mill, distillery, textile, brass and electroplating industry has been assessed to develop an ecologically sound and economically viable remediation technology for the safe disposal of industrial wastes.

Present Status of Industries Involved in the Study

Pulp and Paper Mill

In India there are about 250 paper mills; 38 are large, producing more than 35 tones of paper/day and rest are medium or small scale. A small scale paper mill consume around 250 M^3 water/tonnes of paper while a large paper mill consumes about 250-360 M^3 water/ tones of paper and the entire quantity appears as effluent requiring proper treatment and disposal (Parekh, 1993).

Distillery

In India, there are about more than 250 distilleries which release annually 4×10^{12} KL of spent wash, as one of the most complex caramelized and cumbersome waste having very high BOD (40-50, 000 ppm) and COD (85-1,20, 000 ppm) content with other organic and inorganic contents (Mishra, 1993). The ultimate disposal of the untreated or inadequately treated effluent in to water/land ecosystem is gradually becoming a major threat to environment quality (Chandra, 1999).

Metal Industry

Several industries use metal salts as a raw material, including, pesticide, fertilizers, brass and stainless steel industry, electroplating, paint and pigment, batteries and others. Owing to rapid industrialization issue of heavy metal contamination and their bioaccumulation in higher trophic levels of the ecosystems is becoming increasingly common. Greater awareness of the ecological effects of toxic metals and their biomagnification through food chain as well as highly publicized episodes such as infamous *itai-itai* and *Minamata* diseases from Japan has propped an urgent demand for decontamination of heavy metals from the aquatic system. The release of heavy metals in biologically available forms may causes physiological and genetic disorders in various life forms and ultimately mammals. The industries like electroplating, brass and stainless steel manufacturing units are no way exception and generate huge amount of liquid and/or solid effluents containing heavy metals (Sharma and Pandey, 1998; Barman and Ray, 1999).

Materials and Methods

Effluents of Century Pulp and Paper mill Lalkuan(8 km away from Pantnagar),Keshar Enterprises(Distillery)Baheri(28 km S-W of Pantnagar) and the Brass and Electroplating industry, Moradabad were collected in clean plastic containers and mixed to form a composite sample for a given industry. Immediatly after collection, the effluent was analyzed for its physico-chemical properties and used for experiments.The phytoremediator plants of *Eichhornia crassipes* (Mart) solms, were collected from natural pond and stored in cemented tank for experimental use.

The effluent characteristics, *viz*, pH, EC, BOD, COD, TSS, TDS, TS, N, P, Na and K were assessed following standard methods described in APHA (1998) and Na, K, N and P contents were measured by methods of Tandon (1993). The heavy metals were determined by Atomic absorption spectrophotometer.

To assess to phytoremedition potential of *Eichhornia crassipes* against the various industrial effluent, two concentrations of pulp and paper mills and distillery effluent *i.e.*, 20 and 40 per cent (P 20, P40) were prepared with distilled water along with their corresponding control concentration *i.e.*, NP 20, NP 40. In case of brass and electroplating industry, four concentrations *i.e.* 15, 30, 45 and 60 per cent (P 15, P 30, P 45 and P 60) were made along with their corresponding control treatments *i.e.* NP15, NP 30, NP 45 and NP 60. All the treatments were kept in plastic tubs of 8 litre capacity with inner diameter of 42 cm. In each treatment, 2 plants/tub were kept and mechanical aeration was given by air pump 15 min/tub at 3 day interval. Physico-chemical analysis of phytoremediated effluent was done at regular interval.

Results

Effect of *E. crassipes* on Physico-chemical Characteristics of Pulp and Paper Mill Effluent

The effluent had a dark brown colour with alkaline pH (8.36)and high BOD, COD, TS, TSS, TDS, N, P, Na, K, etc. (Table 17.1). The effect of *E. crassipes* on physico-chemical characteristics of pulp and paper mill, showed significant reduction in all the selected parameters. Out of the four treatments, the maximum reduction in all the parameters of effluent was recorded at P 20 treatment. With increasing phytoremediation period, there was substantial decrease in values of selected physico-chemical parameters.

The EC, BOD, COD, TSS, TS, TDS were reduced up to 15.57, 18.79, 15.34, 22.13, 19.88 and 18.41 per cent in NP 20, 9.73, 11.73, 0.89, 9.24, 16.84 and 22.29 per cent in NP40 treatment, 73.01, 72.51, 76.00, 75.41, 76.92 and 77.89 per cent in P 20 and 70.31, 62.60, 70.74, 57.99, 65.26 and 70.48 per cent in P 40 treatment respectively in 45 days long phytoremediation period. Further, color unit, lignin, Na and K were reduced up to 18.23, 5.27, 3.11 and 2.28 per cent in NP 20.12.97, 3.47, 1.87 and 1.70 per cent in NP40. 80.93, 53.05, 38.34 and 43.40 per cent in P 20 and 74.96, 47.93, 29.87 and 34.46 per cent in P 40 treatment respectively in 45 days (Figure 17.1).

Table 16.1: Initial Physico-chemical Characteristics of Pulp and Paper Mill, Distillery and Brass and Electroplating Industry Effluent

Effluent Parameters	Pulp and Paper Mill Effluent	Distillery Effluent	Brass and Electroplating Industry Effluent
Colour	Dark brown	Brownish Yellow	Blackish grey
pH	8.36±0.01	5.54±0.02	6.89±0.01
EC (μ Scm⁻¹)	1937±7.20	2767±5.44	
BOD	1237±7.20	3705±43.01	–
COD	3211±11.52	10400±99.78	3723±16.33
TSS	1713±5.44	7313±138.99	1280±9.6
TS	4213±14.40	16640±183.55	4129±5.95
TDS	2500±18.86	9327±44.56	2770±5.34
Colour unit (hazen)	5347±100.06	9456±143.69	–
Lignin		1390±9.18	NA –
N	18.5±0.14	59.9±0.56	89.7±0.6
P	36.0±0.24	41.8±0.36	22.3±0.03
Na	86.7±1.96	216.0±1.89	210±3.34
K	188.0±0.94	336.0±1.89	810±6.10
Cu	–	–	5.660
Ni	–	–	0.367
Cd	–	–	0.023
Cr	–	–	0.280

*: All values of given in mgL⁻¹ except pH, colour, and EC

–: Indicates the non aplicablity.

Effect of *E. crassipes* on Physico-chemical Parametes of Distillery Effluent

Initial Physico-chemical characteristics of distillery effluent is given in Table 1. The EC, BOD. COD, TSS, TS and TDS were reduced by 71.68, 62.23, 79.30, 71.64, 70.35 and 69.34 per cent in P20 treatment while 64.78, 56.31, 76.18, 67.62, 63.67 and 60.60 per cent reduction in case of P40 respectively over a phytoremediation period of 45 days. Where as, colour unit, Na and K were found to decrease upto 80.50, 45.84 and 48.03 per cent in P20 treatment and 77.18, 4206 and

Contd...

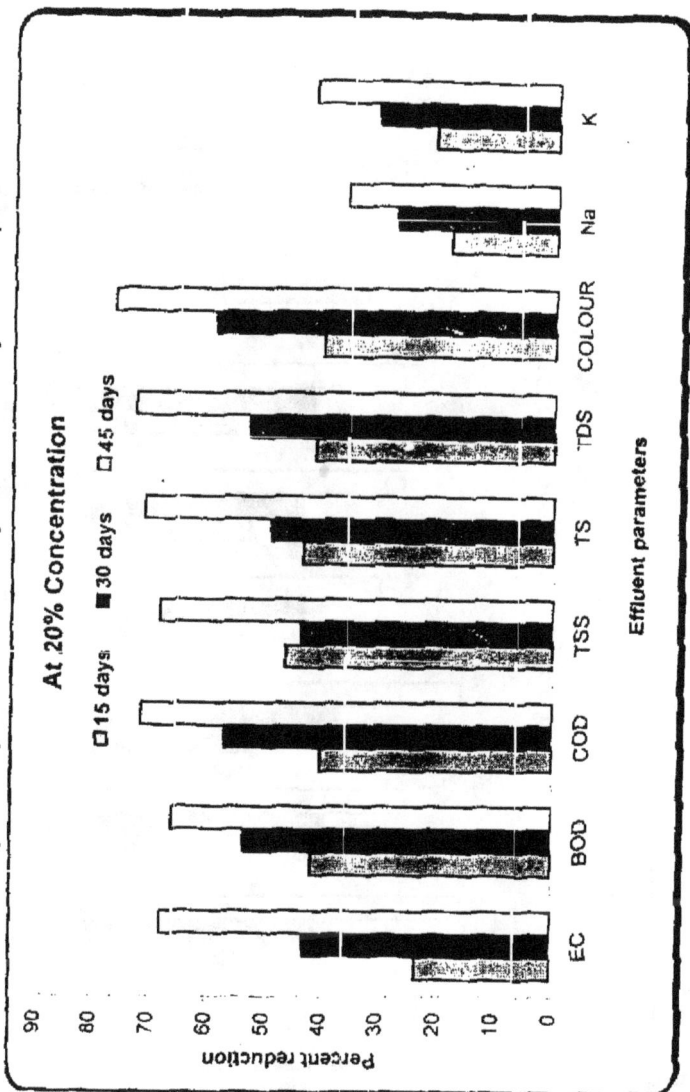

Figure 17.1: Per cent Reduction in Physico-chemical Characteristiscs of Pulp and Paper Mill Effluent Phytoremediated by *E. crassipes*

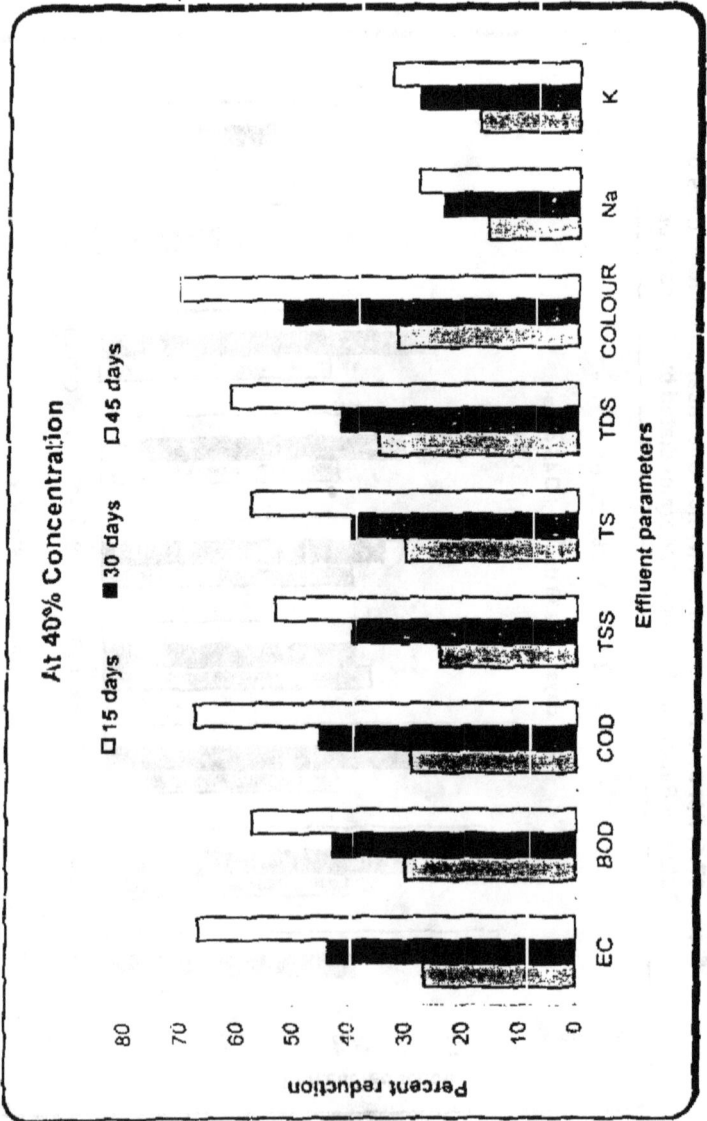

At 40% Concentration

☐ 15 days ■ 30 days ☐ 45 days

Figure 17.1–Contd...

45.55 per cent in P 40 treatment respectively over a phytoremediation period of 45 days (Figure 17.2).

Effect of *E. crassipes* on Physico-chemical Characteristics of Brass and Electroplating Effluent

Initial physico-chemical parameters of brass and electroplating industry are given in Table 17.1. The effluent in Blackish grey slightly acidic in nature and having high pollution load specially in terms of heavy meal content (Cr, Cd, Ni, Cu etc). The data pertaining to the effect of phytoremediation by *E. crassipes* on various physico-chemical characteristics of industrial effluent are shown in Figure 17.3. The P30 treatment showed the maximum reduction in physico-chemical characters as compared to other treatment. The pH increased from 6.93 to 6.98 after 30 days in P30, whereas EC reduced by 18 per cent in P30 after same period in same treatment. The values of COD, TSS, TDS, TS decreased by 69.6, 41, 63 and 62.4 per cent and the values of Na, K, N and P reduced by 60, 45, 35.8 and 72 per cent in P30 treatment after a phytoremedation period of 30 days as compared to the corresponding values of non phytoremediated effluent.

The heavy metal accumulation by *E. crassipes* found highest and followed an order of Cu (4.230) > Cd (2.108) > Cr (1.920) > Ni (0.492 mg/gdw) at P30 after a phytoremediation period of 30 days, whereas heavy metal accumulation was found minimum at P15 with the order of Cu (2.102) > Cd (0.540) > Cr (0.3000) > N (.278 mg/gdw) after 30 days. The amount and rate of metal accumulation by *E. crassipes* largely depends upon the concentration of industrial effluent. The rate of heavy metal accumulation was found maximum for first 10 days of the phytoremediation study and subsequently decreased for rest of the period.

Discussion

The problem of water pollution aggravated by diverse industrial wastes. This become more complex due to quantitative and qualitative differences of pollutants and nature of industry involved.Although there is growing concern for safe environment and subsequent formulations of government regulations, the industries studied still dispose their wastes untreated or under treated. The values of BOD, COD, TS, TDS, TSS, N,P and heavy metals are found higher than permissible limit of their waste disposal. The

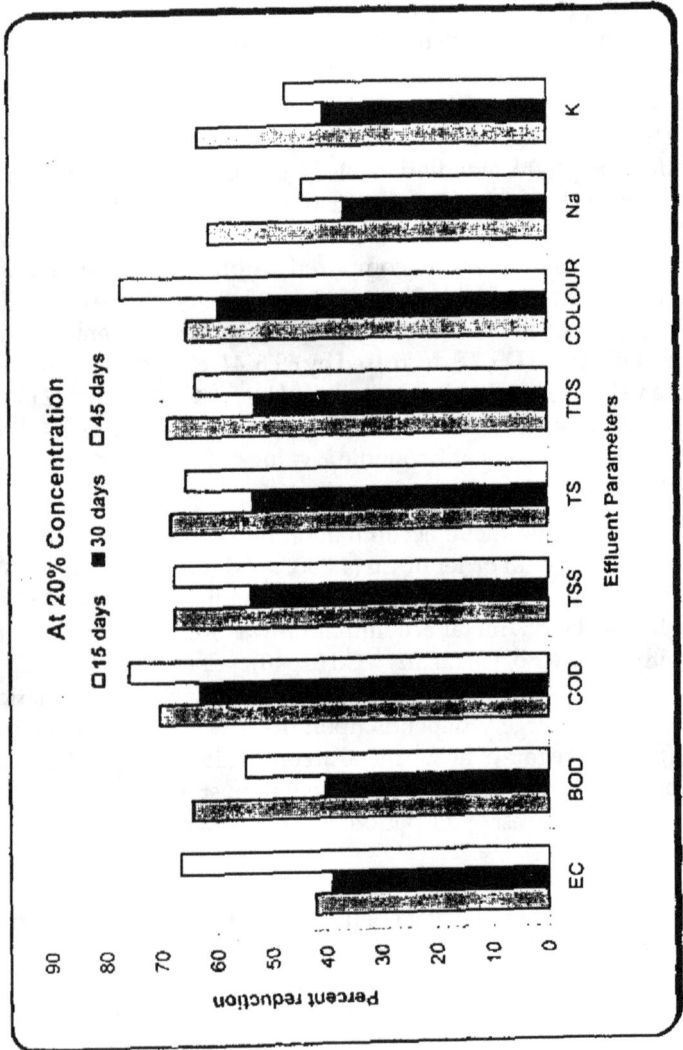

Contd...

Figure 17.2: Per cent Reduction in Physico-chemical Characteristiscs of Distillery Effluent Phytoremediated by *E. crassipes*

At 20% Concentration

☐15 days ■30 days ☐45 days

Percent reduction

Effluent Parameters

EC BOD COD TSS TS TDS COLOUR Na K

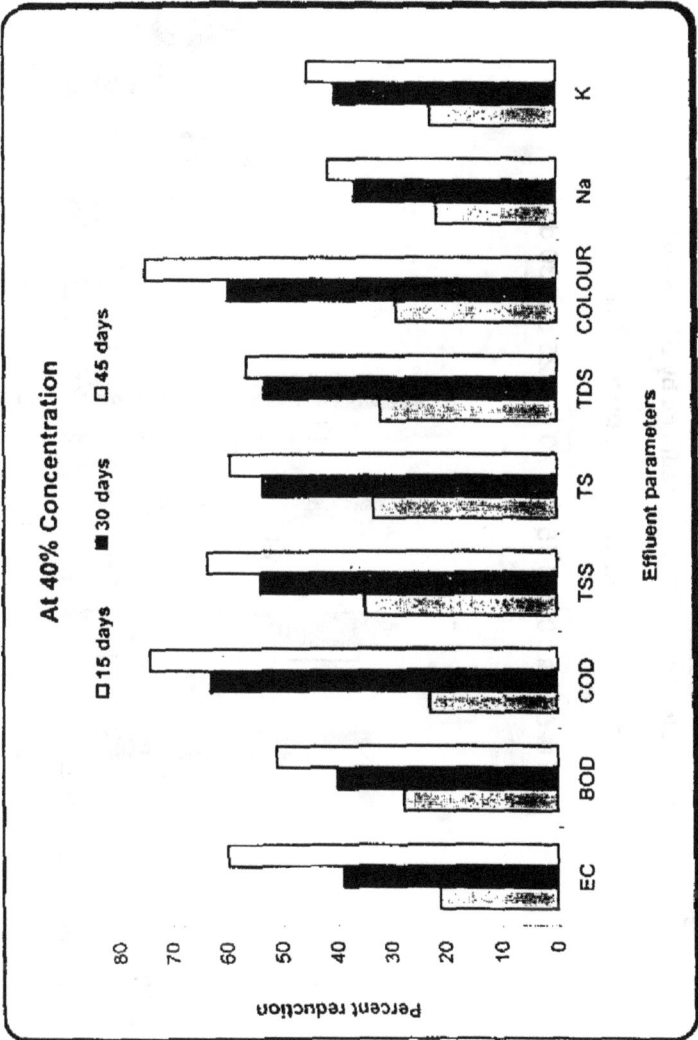

Figure 17.2–Contd...

Figure 17.3: Per cent Reduction in Physico-chemical Characteristics of
Brass Industry Effluent Phytoremediated by *E. crassipes*

At 15% effluent concentration

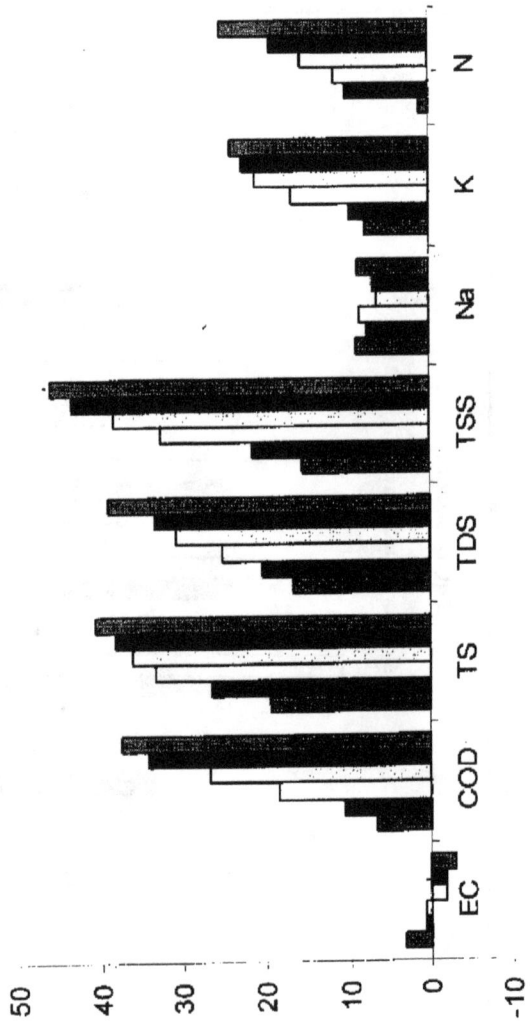

Contd...

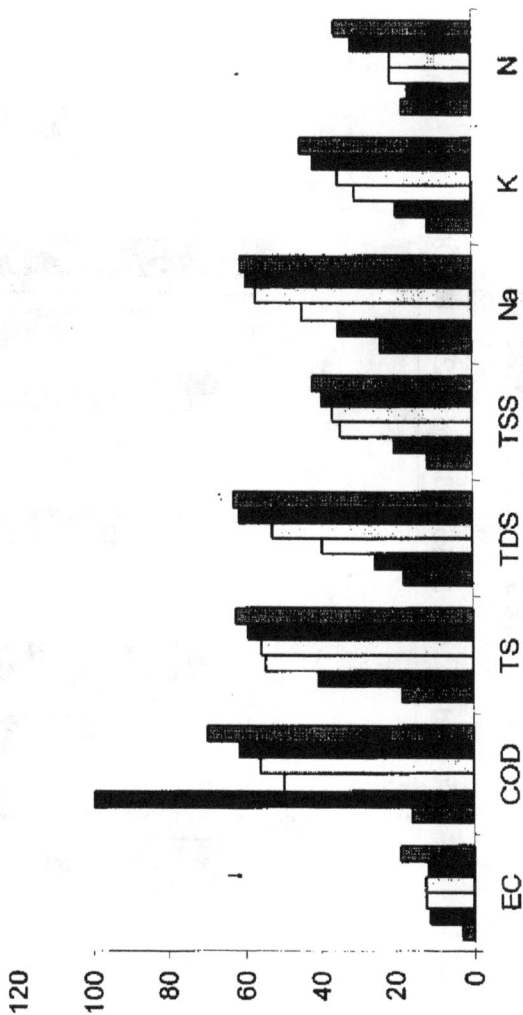

Figure 17.3–Contd...

At 30% effluent concentration

Contd...

Figure 17.3–Contd...

At 45% effluent concentration

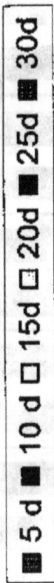

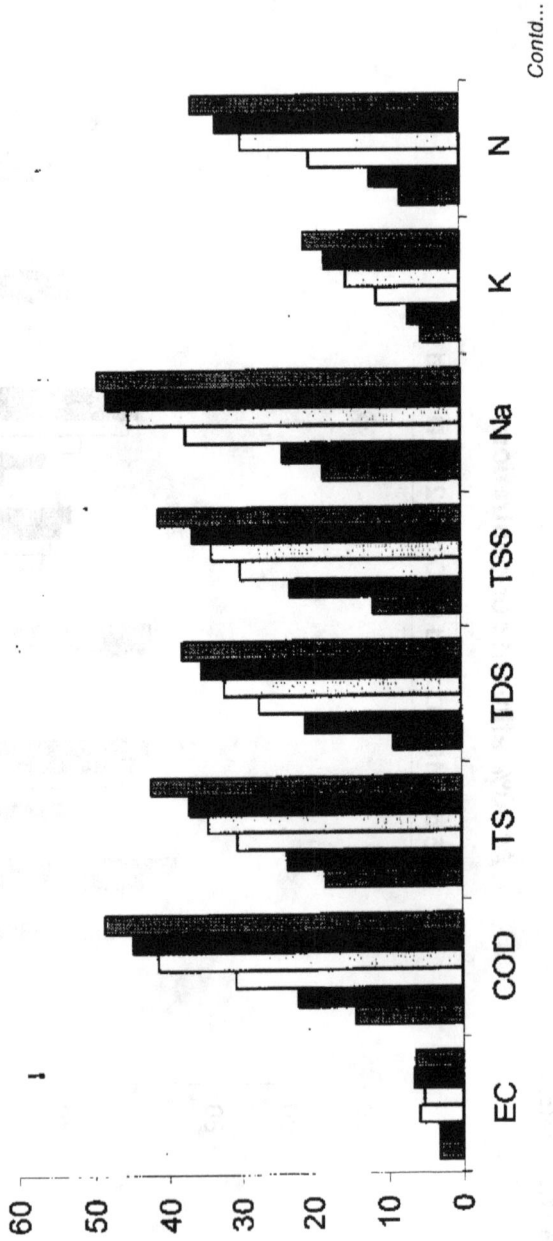

■ 5 d ■ 10 d □ 15d □ 20d ■ 25d ■ 30d

Contd...

Figure 17.3–Contd...

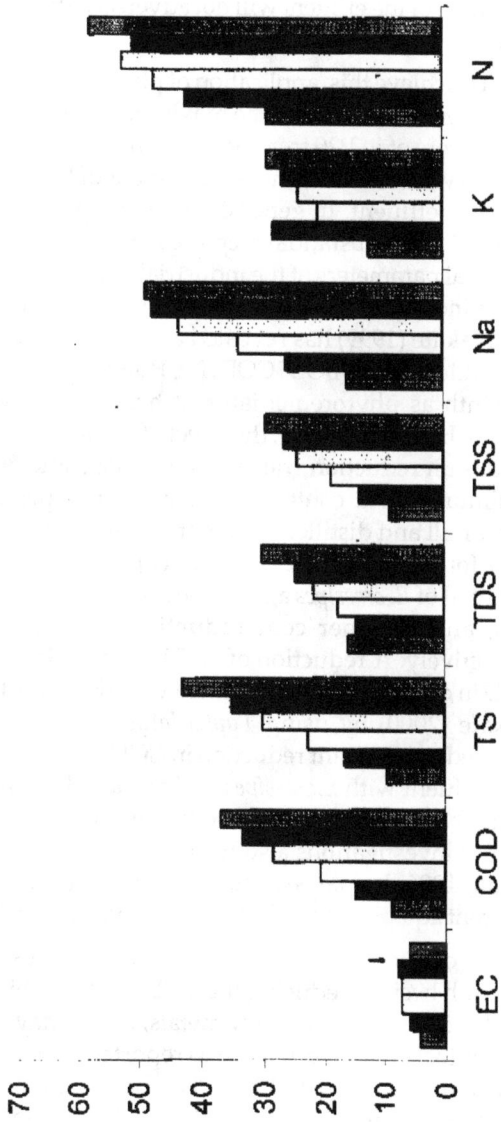

Figure 17.3–Contd...

Contd...

ultimate objective of wastewater treatment is to reduce the concentrations of specific pollutants to the level at which the discharge of the effluent will not adversely affect the environment or pose a health threat.

To achieve this, application of phytoremediation of effluents by water hyacinth revealed the maximum reduction in all the selected parameters of in P20 treatment of Pulp and Paper mill effluent and distillery effluent and in P30 treatment of brass and electroplating industry effluent. In general, increased phytoremediation period resulted in to substantial increase in reduction of selected physico-chemical parameters of the industrial effluents. Aquatic macrophyte based industrial wastewater treatment system studied by Trivedi and Nakate (1999) has revealed a very high degree of reduction in suspended solids, BOD, COD, Ni, P and grease in response to water hyacinth as phytoremediator of hospital and engineering plant wastes. In present study, the effect of phytoremediation depicts that maximum reduction was in colour (80.93 and 80.50 per cent) and minimum in Na content (38.34 and 45.84 per cent) in pulp and paper mill and distillery effluent respectively. Similar results have been found by Saini(1995),who exploited the phytoremediation potential of *E. crassipes* against raw wastewater and recorded 87, 90, 77.6, and 76.6 per cent reduction in BOD,COD,TS and TSS respectively. A reduction of 47.59 per cent TSS and 78.77 per cent COD in diluted distillery effluent was also reported by Trivedi and Nakate (2000) by using *Typha latifolia*. Casabianca *et al.* (1995) reported a significant reduction in COD and BOD of pulp and paper mill effluent with *E.crassipes* and argued that physical setting and plant absorption appear to be the prime cause of such reduction. Similar investigations also have been made by others (Gupta and Sujata, 1996; Vajpayee *et.al*, 1995; Singhal *et al.*, 2003) employing different aquatic macrophytes against a variety of industrial effluent.

In case of brass and electroplating industry effluent water hyacinth besides reducing the COD,TS,TDS,TSS also removed the substantial amount of heavy metals, which may probably be due to the phytochelators formation as reported by Kinnersely (1993) and Howden and Cobbett(1992). The heavy metal accumulation by *E.crassipes* against brass and electroplating industry effluent was recorded as Cu(230 µg/gdw) followed by Cd, Cr and Ni respectively,which finds support from Satyakala and Jamil(1997). The rate of heavy metal accumulation was found higher for initial exposure of 10 days, which might be due to the saturation in accumulation potential of the phytoremediators.

In general, the total phytoremediation observed in present study might be accounted as simultaneous process of phytodegradation, phytoextraction, phytostabilisation and rhizofilteration for significant reduction in COD,BOD, TS, TDS, TSS and heavy metal content of effluent as argued by Cunningham and Ow, 1996; and Salt *et.al*, 1998. However, the presence of microorganism and physical setting resulted in moderate reduction in selected parameters as observed in corresponding non phytoremediated effluent could not be ignored.

Conclusion

Undoubtedly, phytoremediation as a cleanup process is very attractive because it has been shown to be visually unobtrusive, effective, relatively inexpensive and eco-friendly. It is clear from the present study that phytoremediation is not hype but hope as it has potential for improving environmental quality, which otherwise would have been spoiled by the deadliest effluent of Pulp and paper mill, distillery and brass and electroplating industry. However, such studies require in depth upto the stage of practical exploitation, where further research and technical implementation are necessary beyond the pilot stage. The study also emphasizes urgent need to optimize the phytoremediation process; to measure and optimize the underlying economics; and to provide recommendations and convince regulators, policy makers and end-users of this green approach for the treatment of industrial wastewater contaminated by toxic metals and other organic pollutants.

References

Anderson, C.W.N., Brooks, R.R., Stewart, R.B. and Simcock, R. 1998. Harvesting and crop of gold in plants. Nature, 395: 553–554.

APHA, 1998. *Standard Methods for Examination of Water and Wastewater* 19th Edition. American Public Health Association, Americans Water Works Associations, Water Pollution Control Federation, Washington DC, USA.

Banuelos, G.S. and Mayland, H.F., 2000. Absorption and distribution of selenium in animals consuming conola grown for selenium phytoremediation. *Ecotoxicol. Environ. Safety*, 46: 322–328.

Banuelos, G.S., Ajaw, H.A., Mackey, B., Wu, L.L., Cook, C., Akohoue, S. and Zambruzuski, S., 1997a. Evaluation of different plant species used for phytoremediation of high soil selenium. *J. Environ. Qual.*, 26: 639–646.

Barman, S.C. and Lal, M.M., 1994. Accumulation of heavy metals (Zn, Cu, Cd and Pb) in soil and cultivated vegetables and weeds grown in industrially polluted fields. *J. Enviorn. Biol.*, 15: 107.

Barman, S.C. and Ray, M., 1999. Uptake of heavy metals (Cd, Cu, Zn and Ni and comparative study of water plants grown in polluted and unpolluted fields using different varieties of rice D.E.I. *Journal of Science and Engineering Research*, 11(1 and 2): 13–17.

Bizily, S.P., Rugh, C.L., Summers, A.O. and Meagher, R.B., 1999. Phytoremediation of methylamercury pollution: Mer B expression in *Arabidopsis thaliana* confers resistance to organimercurials. *Proc. Natl. Acad. Sci.*, USA, 96: 6808–6813.

Boopathy, R., 2000.Factors limiting bioremediation technologies. *Bioresource Technology*, 74: 63.

Brooks, R.R., 1998. Phytoremediation by volatilization. In: *Plant that Hyperaccumulate Heavy Metals: Their Role in Phytoremediation, Microbiology, Archaeology, Mineral Exploration and Phytomining.* CAB International, Oxon, UK, p. 289–312.

Bunke, G., Gotz, P. and Buchholz, R., 1999. Metal removal by biomass: Physico-chemical elimination methods. In: *Biotechnology, Environmental Processes IVOP*, Wiley VCH, Weinheim, 11a: 431–452.

Bunzl, K., Trautmannsheimer, M., Schramel, P. and Reifenhauser, W., 2001. Availability of arsenic, copper, lead, thallium, and zinc to various vegetables grown in slag-contaminated soils. *J. Environ. Qual.*, 30: 934–939.

Casabianca, C.M.L-de, Coma, C. and De-C.C., M.L., 1991. Treatment of paper industry effluent with *Eichhornia crassipes*. First results (Tartas factory, landes). *Comptes-Rendus-de-l' academic-des-Sciences*, 312(11): 579–585.

Chandra, R., 1999. Distillery effluent treatment by methane production in India. In: *Advances in Industrial Wastewater Treatment*, (Ed.) P.K. Goel. Techno-Science Publications, Jaipur, India, pp. 164–179.

Cunnigham, S.D. and Ow, D.W., 1996. Promises and prospects of phytoremediation. *Plant Physiol.*, 110: 715–719.

Dhamani-Muller, H., Oort, F. and Balabane, M., 2001. Metal extraction by *Arabidopsis halleri* grown on an unpolluted soil amended with various metal-bearing solids, a pot experiment. *Envir. Pollut.*, 114: 77–84.

Dudka, S. and Miller, W.P., 1999. Accumulation of potentially toxic elements in plants and their transfer to human food chain. *J. Envron. Sci. Health*, B344: 681–708.

Eccles, H., 1999. Treatment of metal contaminated wastes: Why select a biological process? *TIBTECH*, 17: 462–465.

Ensley, B.D., 2000. Rationals for the use of phytoremediation In: Phytoremediation of toxic metals. Using plants to cleanup the environment (Eds.) John Wiley and Sons, New York, pp. 1–12.

Felix, H., 1997. Field trials for *in situ* decontamination of heavy metal polluted soil using crops of metal: Accumulating plants. *Zeitschrift Pflanzenernahrung Bodenkunde*, 160: 525.

Fitz, W.J. and Wenzel, W.W., 2002. Aresenic transformations in the soil-rhizosphere-plant system: Fundamentals and potential application to phytoremediation. *J. Biotechnology*, 99: 259–278.

Glass, D.J., 1999. *US and International Markets for Phytoremediation*, 1999–2000 DJ Glass Associates Inc., Needham, MA, USA, pp. 266.

Gupta, A. and Sujatha, P., 1996. Treatment of tannery wastewater by water hyacinth application. *J. Ecot. Environ. Monit.*, 6(3): 209–212.

Heaton, A.C.P., Rugh, C.L., Wang, N. and Meagher, R.B., 1998. Phytoremediation of mercury and methylmercury polluted soils using genetically engineered plants. *J. Soil. Contamination*, 7: 497–510.

Howden, R. and Cobbet, C., 1992.Cadmium sensitive mutants of *Arabidopsis thaliana*. *Plant Physiol.*, 100: 100–107.

Iwamoto, T. and Nasu, M., 2001. Current bioremediation practice and perspective. *J. Biosci. Bioeng.*, 92: 1.

Kinnersly, A.M., 1993. The role of phytochelates in plant growth and productivity. *Plant Growth Regu.*, 12: 207–217.

Kumar, P.B.A.N., Dushenkov, V., Motto, H. and Raskin, I., 1995. Phytoextraction: The use of plants to remove heavy metals from soils. *Environ. Sci. Technol.*, 29: 1232–1238.

Mahabal, B.L., 1993. Principles and design of effluent treatment plant (ETP). In: *Industrial Safety and Pollution Control Handbook*, (Ed.) J. Nagaraj. A Joint Publication of National Safety Council and Associate (DATA) Publishers Pvt. Ltd., Secunderabad, India pp. 257–270.

McGrath, S.P. and Lambi, E., 2001. Plant and rhizosphere processes involved in phytoremediation of metal-contaminated soils. *Plant and Soil*, 232: 207–214.

Mishra, K., 1993. Cytotoxic effect of distillery waste on *Allium cepa* L. *Bull Environ. Contam. Toxicol.*, 50: 199–204.

Parekh, R.C., 1993. Pollution control methods applicable to pulp and paper industry. In: *Industrial Safety and Pollution Control Handbook*, (Ed.) J. Nagaraj. A Joint Publication of National Safety Council and Associate (DATA) Publishers Pvt. Ltd., Secunderabad, India pp. 387–403.

Prasad, M.N.V., 2003. Phytoremediation of metal-polluted ecosystem-Hype for commercialization. *Russ. J. Plant Physiol.*, 50: 686–701.

Saini, R.S., 1995. Evaluating the potential of water hyacinth to treat raw wastewater employing a short determination period of 48 hours. *Poll. Res.*, 14(1): 141–143.

Salt, D.E., Smith, R.D. and Raskin, I., 1998. Phytoremediation. *Ann. Rev. Plant Physiol. Plant Mol. Bio.*, 49: 643–668.

Satyakala, G. and Jamil, K., 1997. Studies on the effect of heavy metal pollution on *Pistia statiotes* L. (water lettuce). *Ind. J. Env. Hlth.*, 39(1): 1–7.

Singhal, V., Kumar, A. and Rai, J.P.N., 2003. Phytoremediation of pulp and paper mill and distillery effluent by channel grass (*Vellisneria spiralis*). *J.S.I.R.*, 62: 319–328.

Sharma, S.D. and Pandey, K.S., 1998. Pollution studies on Ramganga river at Moradabad physico-chemical characteristics and toxic metals. *Poll. Res.*, 17(2): 201–209.

Tandon, H.L.S., 1993. *Methods of Analysis of Soils, Plants, Waters and Fertilizers*. Fertilizer Development and Consultation Organization, New Delhi, 143 p.

Trivedi, R.K. and Nakate, S.S., 1999. Aquatic weed based wastewater treatment in India. R. of Indus. *Pollu. Cont.*, 15(2): 275–279.

Trivedi, R.K. and Nakate, S.S., 2000. Treatment of diluted distillery waste by using constructed wetland. *IJEP*, 20(10): 749–753.

Vajpayee, P., Rai, U.N., Sinha, S., Tripathi, R.D. and Chandra, P., 1995. Bioremediation of tannery effluent by aquatic macrophytes. *Bull. Environ. Contam. Toxicol.*, 55(4): 546–553.

Wenzel, W.W., Bunkowski, M., Puschenreiter, M. and Horak, O., 2003. Rhizosphere characteristics of indigenously growing nickel hyperaccumulators and excluder plants on serpentine soil. *Environ. Pollut.*, 123: 131–138.

Zorpas, A.A., Constantinides, T. Vlyssides, A.G., Aralambous, I. And Loizidou, M., 1999. Heavy metal uptake by natural zeolite and metals partitioning in sewage sludge compost. *Bioresource Technology*, 71: 113–119.

Chapter 18

Potentials of Freshwater Predators as Biocontrol Agents of Mosquito Populations: Constraints and Opportunities

Ram Kumar[1]*, Sarita Kumar[1] & Priyanesh Prasad[2]

[1]Acharya Narendra Dev College (University of Delhi),
Govindpuri, KalkaJi, New Delhi – 110 019
[2]School of Environmental Studies, Delhi University, Delhi – 110 007

ABSTRACT

For years the human population has explored various methods to combat the threat from mosquito borne diseases. However, the harmful effects of chemicals on the non-target populations and the development of resistance to these chemical in mosquitoes along with the recent resurgence of different mosquito-borne diseases, has forced us to explore alternate, simple and sustainable ways of mosquito control. Biological control of mosquito larvae is an environment friendly method and thus needs to be explored. Manipulating or introducing a self-replicative predator in the ecosystem achieves sustained biological control of the pest population. A sound knowledge of prey selectivity pattern, adaptability in the introduced

* Corresponding Author: E-mail: copepod65@rediffmail.com

environment and overall interaction with the indigenous organisms, of the predator to be introduced is warranted. Different aquatic predators of the mosquito larvae include tadpoles, fishes, dragonfly larvae, cyclopoid copepods, planarians, aquatic bugs and mites. This paper presents an overview of potential of various invertebrate and vertebrate predators in fresh water ecosystem to control mosquito larvae.

Keywords: Biocontrol, Mosquito, Predators, Larvivory, Tadpoles, Gambusia, Dragonfly larvae, Copepod.

Introduction

Until recently the role of ecology in environmental management has been more to warn of the damages of pollution than to, propose the sustainable solution of different global and local problems faced by the human society. Very recently some aspects of ecological theory has been applied in the present days bio-manipulation approach for the restoration of eutrophicated water bodies (Gulati, 1990; He *et al.*, 1994). Of the multiple possibilities of applying the ecological theories for human need, is the use of our knowledge on predation and competition in fresh water habitats, whereby manipulating particular trophic levels the desired change could be achieved in the system. It is obvious that different pathogens and disease vectors are part of the ecosystem and, anthropogenically altered conditions provide an edge to some of these pathogens and disease vectors. Mosquito-borne diseases have been a major problem in almost all tropical countries and currently there are no vaccines for these diseases. The most widely used method of control are chemical agents like insecticides which can harm the environment, and also affect the non–target organisms that may be useful for human beings, and overall alter the ecosystem structure. Programs to decimate mosquito populations by trying to kill the adult stage of the mosquitoes frequently fail, because these adults reside alongside with human populations in their households and sometimes these places are not seen as potential refuges for mosquitoes and thereby they escape remedial measures. Mosquitoes live in places where they are difficult to find and kill for instance old tires, trash, and water tanks or any basically anything that can hold water. These are the places where mosquitoes (like *Aedes* spp.) reproduce. Precautions have to be taken so as not to contaminate the drinking water supply and containers with potentially harmful chemicals. And, overall mosquitoes have developed resistance to frequently used pesticides making it even more difficult to control the adult populations.

Hence, long-term and environmentally–friendly methods for relief from mosquito borne diseases are being sought for. A discussion of "vector control" should also mention the potential of genetics for use against mosquito-borne infection. Recent techniques to modify genes of mosquitoes are normally believed to be the best hope for intervening against malaria and dengue. The main purpose is to produce a genetically modified strain of mosquito in laboratory, which should not serve as a carrier of disease and which should be competitively superior in natural habitat such that the wild mosquitoes will be replaced eventually after the release of genetically altered mosquito in nature. However, there are several problems with this approach, and prospects for this technology seem dim.

Therefore, it is essential that we explore more effective and eco-friendly techniques to control mosquito populations. Various living organisms, known as biological control agents found naturally, can be utilized to control the mosquito populations without having to use chemicals and harming the environment in the process. Furthermore, specific biological control agents for specific life stages of the mosquito can be used to ensure a better and a more effective control. For these reasons, it is prudent to use biological control agents that can get into hard to find places, are found naturally, are safe for people and are economical (Rishikesh *et al.*, 1988; Spielman *et al.*, 1993). "Biological control" means the use of different kinds of living things, and/or their derivatives to eliminate pest populations. Many of the biological controlling agents can disperse by themselves enabling them to spread and build up viable populations. In the field of applied ecology there have been many attempts to achieve the biological control of pathogens or vectors by introducing new effective natural enemies to their natural habitats (Arthington and Lloyd, 1989).

Bacterium like *Bacillus thuringiensis israelensis* (BTI) is a proven, environmentally safe mosquito larvicide that is nontoxic for people. It has become commercially available under such names as Teknar, Vectobac and Bactimos. The beauty of this material is that an application destroys larval mosquitoes but spares any predators that may be present. The toxin, however, is destroyed by sunlight.

This imposes a necessity for frequent application. Weekly visits to each breeding site may be required, thereby making it more expensive and rigorous.

Certain algae seem to fill the guts of larval mosquitoes with a mass of material that they are not able to digest, and this is said to hurt growth (Marten, 1986). The promise of such material against the mosquitoes, however, seems limited because these mosquitoes usually breed in sites that are too dark for algae.

Of the numerous fungi that have been described, only *Lagenidium giganteum* seems to have the ability to stop the spread of disease transmission through mosquitos' oospores that can resist droughts can readily be produced in bulk. Oospores survive for many years in soil, but reactivate only about a month after flooding. In practice, the spores are activated by staying in water for 1-2 weeks before being sprayed onto the surface of the site to be treated. About half of the target mosquito population becomes infected by the oospores and larvae continue to die for the next several weeks. The infection may continue for a period of time. This material holds promise for use against vectors breeding in potable water. Although these preparations seem to persist for a long time in water, there are certain additional problems. It can be difficult and costly to apply. The fungus infects and kills only a part of target larval mosquitoes, as well. The level may not be large enough to decrease risk of human dengue infection.

Protozoan parasites provide fascinating prospects for the destruction of mosquitoes. Recent findings show that this might be a possibility (Sweeney and Becnel, 1991) to control mosquito populations. For a biocontrol effort against mosquitoes, microsporidian-infected eggs can be placed into natural breeding sites. Although many of the mosquitoes would escape the lethal effects of this agent, the infection should cause problems for those mosquitoes that escape. Its feasibility in field conditions has been not been tested.

Biological control of mosquito larvae with predators is more convenient and gets rid of the task to frequently repeat applications while using chemicals. Different predators of the mosquito larvae include tadpoles, fishes, dragonfly larvae, cyclopoid copepods, planaria, aquatic bugs and mites. The selection of biological control agent should be based on its self-replicating capacity as the introduced agents, if self–replicating, maintain very close interactions with their prey. They eliminate certain prey and sustain such environments (as and when prey is introduced, it eats the same) for long periods thereafter (Marten, 1994a). However, this will be possible if the predator plays a very important role as a mortality factor in the prey population dynamics. In other words the predators search efficiency in terms of emergence of prey is to be kept extra ordinarily high throughout. It is important to have a sound knowledge of a predator's prey selectivity pattern in general, and mosquito larval selections in the presence of alternate prey commonly present in its habitat in particular (Arthington and Lloyd, 1989; Arthington and Marshall, 1999). In addition, the predator's adaptability in the introduced environment and overall interaction with the indigenous organisms needs to be considered prior to introduction. This paper presents a brief account of different predators of mosquito larvae and possible interactions between its predators and mosquito larval prey in an aquatic ecosystem.

Amphibian Tadpoles

Omnivorous tadpoles are potential predators of the mosquito larvae (Spielman and Sullivan, 1974) and exert a significant impact on the freshwater ecosystem (Blaustein and Kotler, 1993; Blaustein *et al.*, 1996). The larvae of the giant Cuban tree frog (*Hyla septentrionalis*) have been reported to destroy aquatic insects and algae growing in water containers (Morgan and Buttemer, 1996; Webb and Joss, 1997; Goodsell and Kats, 1999; Komak and Crossland, 2000). However, since they are generalist feeders they efficiently remove other predators of mosquito larvae also. Their efficiency to utilize mosquito larvae in the presence of alternate prey has not been elucidated properly. It has been reported that the tadpoles of *Rana tigrina* are more efficient pupal predators (Marian *et al.*, 1983) than other mosquito predators. But being a predator does not mean that one cannot be preyed upon. Selected fishes for example mosquito fishes (Grubb, 1972; Gamradt and Kats, 1996; Carwood, 1997; Komak and Crossland, 2000) and cat–fishes and dragon fly larvae (Woodward, 1983; Travis *et al.*, 1985; Wissinger, 1989) prey on amphibian eggs and tadpoles and contribute to certain anuran declines (Baber and Babbitt, 2003). Of course any introduction of amphibian tadpoles beyond their natural range is difficult and needs utmost reluctance and caution. Tadpoles are rarely found in containers that hold less than several liters of water and seem to have a small impact on mosquito larval population.

Larvivorous Fishes

One of the most widely distributed visually feeding fishes is the western mosquitofish *Gambusia affinis* (Baird and Girard) and the eastern mosquito fish *G. holbrooki*. During the 20th century, several fish species were introduced outside their natural distribution area. Both, the western and eastern mosquito fishes were introduced extensively in >40 countries because of their reputation as mosquito-control agents (Krumholz, 1948; Courtenay and Meffe, 1989). Chinese health authorities have used other species of fish also to exclude *A. aegypti* mosquitoes from breeding in large cisterns or other containers of drinking water (Lu, 1989). Small fishes, such as *Claris fuscus*, *Tilapia nilotica*, and *Macropodus spp.*, have been used in many regions to eliminate the larvae in the domestic water containers with considerable success. The use of catfish appears to be particularly effective also (Neng *et al.* 1987). Larvivorous fishes have been widely used for control of mosquito larvae but it has its own limitations where fishes are expensive and do not survive for long in small places like in containers.

However, little is known about their ecology (Hurlbert *et al.*, 1972; Hurlbert and Mulla, 1981; Meffe and Snelson, 1989). Research on introduced fishes and mosquito fish feeding is particularly scarce although it is crucial to determine the impact of its introduction in the ecosystem. It has invaded the lowest stream sections, most wetlands, shallow lakes and almost all ornamental pools in different parts of world. It has been shown that mosquito fish introduced in 1922 into southern California subsequently reduced populations of native fish throughout the state due to competition, predation, and hybridization. It is now a danger to native fishes that share similar habitats, especially cyprinodontids because of its ecological advantages related to fast growth, early maturity, viviparity (Vargas and Sostoa, 1996, Barrier and Hicks, 1994) and food consumption rates, which can reach 83 per cent of fish weight per day (Wurtsbaugh and Cech, 1983). In addition to competing for resources, *Gambusia* are aggressive and often attack fish more than twice their size in mass (Rowe, 1998) which results in damage to the fins and scales, leaving the fish susceptible to diseases. Researchers have questioned the use of mosquito fish as biological controls, especially when introduced as exotics to supplement native species (Moyle, 1976; McKay, 1984).

Some workers have used this species to test pelagic trophic interactions (Hurlbert *et al.*, 1972; Lancaster and Drenner, 1990), the optimal foraging theory (Bence and Murdoch, 1986) and toxicant bio-accumulation in wetlands. Some dietary studies have focused on its role in control of insects (Walton and Mulla, 1991; Linden and Cech, 1990). In nature, a wide spectrum of different food is available. The utilization of target species by a predator depends upon the prey selectivity patterns of the predator in the presence of alternate natural prey co-occurring in the habitat of target organisms. However, few studies (Crivelli and Boy, 1987) have analysed the dietary patterns of mosquito fishes. Feeding strategy analyses suggested a high individual specialisation, and thus an opportunistic feeding strategy for mosquito fishes both in juvenile and adult stages (Specziár, 2004). Since the start of introductions several studies have pointed out the negative effects of mosquitofishes on small indigenous fish species (Courtenay and Meffe, 1989; Howe *et al.*, 1997; Ivantsoff and Aarn, 1999), amphibians (Morgan and Buttemer, 1996; Webb and Joss, 1997; Goodsell and Kats, 1999; Komak and Crossland, 2000) and invertebrates like zooplankton (Margaritora, 1990), dragonfly larvae (Rowe, 1987), damselflies (Englund, 1999) and fairy shrimps (Leyse *et al.*, 2004).

We have determined the food components of *G. affinis* collected from the local eutrophic pond with noticeable density of mosquito larvae qualitatively and quantitatively on the basis of gut content analysis. We recorded 30–40 per cent of the food component was algae, 15-20 per cent was detritus and the remaining were the animal food (Figure 18.1A). The animal food included 20-30 per cent rotifers, 25-35 per cent cladocerans, 5-8 per cent copepods, 10-15 per cent Ostracods and 10-12 per cent mosquito larvae (Figure 18.1B). Our results are comparable to earlier findings (Gophen *et al.*, 1998; Specziár 2004). Very recently Specziár (2004) recorded 34 per cent algae, 19 per cent detritus and remaining, animal prey. In his study animal food included 11 per cent rotifers, 28 per cent dipterans, 19 per cent ostracods, 19 per cent other insects, 18 per cent copepods and 5 per cent cladoceran. Many factors which affect the dietary preference and width of the food niche are the size of the water body, food availability, population density, the invertebrate community,

Algae 30-40%   Detritus 15-20%
Animal 40-55%

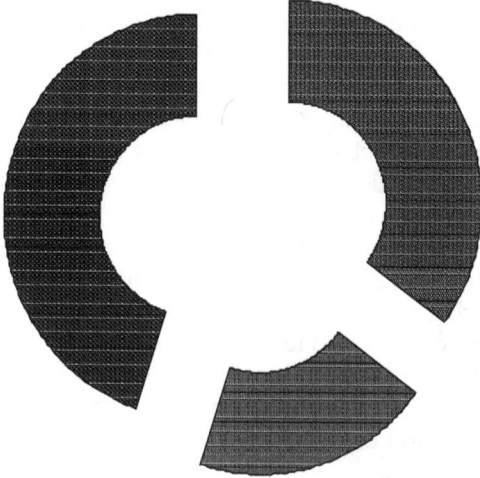

A. Biomass Fraction of Gut- content

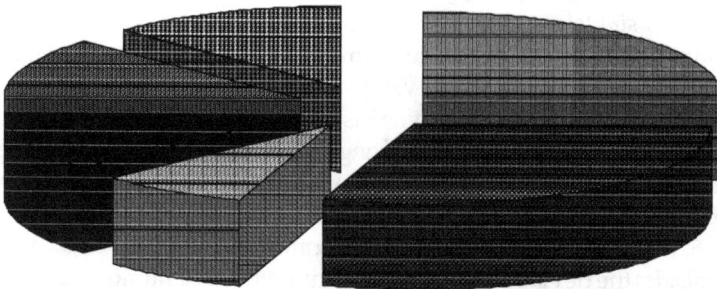

Rotifers 20-30%   Cladocerans 25-35%   Copepods 5-8%
Ostracods 10-15%   Mosquito larvae 10-12%   Others 10-20%

B. Different Fraction of Animal Diet

Figure 18.1: Diet of Mosquito Fish (*G. affinis*) Obtained from Gut Content Analyses. Gut content components are expressed as percentages of estimated biomass, and animal prey as a percentage of numbers.

varieties of fish, primary production of plants and algae, water temperature, oxygen content and the structure of sub-habitats, etc (Mansfield and McArdle, 1998). The diet of eastern mosquito fish consisted of aquatic-and terrestrial invertebrates, filamentous algae and detritus. 1 per cent rate of cannibalism has also been reported in case of eastern *Gambusia*. However, in another study (García-Berthou, 1999) on the eastern mosquitofish G. *holbrooki* of Lake Banyoles, Catalonia, Spain, it was observed that diet of fish was based on littoral cladocerans, particularly *Chydorus sphaericus, Scapholeberis ramneri, Ceriodaphnia reticulata,* and *Pleuroxus laevis,* and nematoceran (basically chironomid) adults. There was a large variety of prey of terrestrial (collembolans, ants) or aquatic neustonic origin (*S. ramneri*, emerging nematoceran adults), showing the microhabitat of mosquito fish closely linked to the water surface. In addition to seasonal and between-site variation, there was an ontogenetic diet shift from microcrustaceans, particularly cladocerans (smallest fish also using diatoms and copepod nauplii) to larger prey, namely nematoceran adults.

On the basis of the results obtained from the field we studied the patterns of mosquito larval selection by the western mosquito fish, G. *affinis* in the presence of alternate cladoceran prey species *Daphnia similoides*, which was abundant in nature. Larval instar related differences in the consumption by the fish, was highly significant (One way ANOVA $p < 0.001$; Figure 19.2A). When a combination of mosquito larvae instar I and *D. similoides* offered as food, elicited in mosquito fish strong preference for mosquito larvae ($a = 0.84$-0.96 $P < 0.001$) regardless of its proportion in the medium (Figure 19.2B), while the fish strongly avoided *D. similoides* ($a = 0.04$–0.16, $P < 0.001$). Given a choice of mosquito larvae Instar IV and *D. similoides* the deviations from the null hypothesis of random feeding were not statistically significant in case of fish ($a = 0.31$, $P > 0.05$ (Our unpublished data). Predation over dipterans was significantly diminished while it was not altered over other insects (trichopteran, terrestrial insects, etc.). Finally, electivity for total zooplankton has not been found to be related to fish density (García-Berthou, 1999).

Significant differences in the extent of carnivory and efficiency of utilizing dipteran larval prey have also been reported between sexes. The sexual dimorphism and size play very important role in feeding behaviour and affect the food web differentially. Females

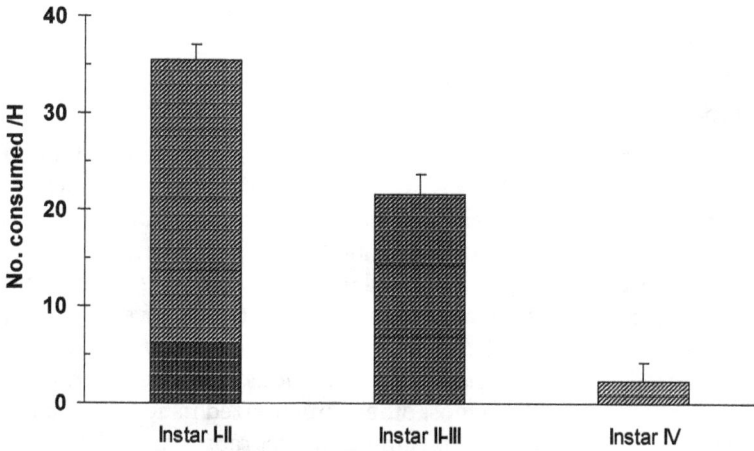

A. Larval Consumption Rates as a Function of Instar Stage

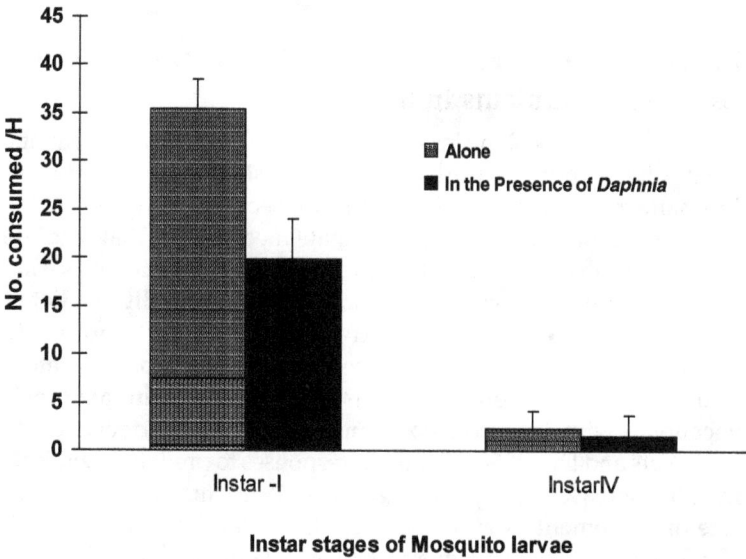

B. Larval Consumption in the Presence of *Daphnia*

**Figure 18.2: Hourly Per Capita Prey Consumption Rates
(Mean ± SE) in *G. affinis* in Relation to the (A) Larval Instar Stage of
the Mosquito Species and (B) in the Presence of *Daphnia similoides***

should predate more intensively over a wide range of items to compensate larger size and reproduction, while males and juveniles should change strategy to avoid competition and exploit resources (Specziár, 2004).

The gut content biomass in females was much higher than in males (One way ANOVA, p < 0.0001). Females are more on zooplankton (cladoceran, copepods, rotifers), and detritus than males. Even if cladocerans were not the most abundant prey in the plankton, they were preferred and removed by females. Rotifer and copepod eggs were frequently found in gut contents of several females. It has been reported that the optimal temperature for *Gambusia* feeding rate was 30-35°C. *Gambusia* have 'voracious' appetites, feeding at rates much higher than most other similar-sized fishes. At optimal temperatures, maximum consumption rates ranged from 0.75 to 1.1 per cent body weight per day for fish ranging from 0.2 to 1.2 g. Mosquito fishes are not well efficient in preying mosquito larvae in winter (<30°C).

Fish Kairomones and Mosquito Oviposition (Predator Escape Mechanisms in Mosquito)

Different ovipositing insects have been reported to avoid aquatic sites where there is high predation risk to their offspring. Mechanisms for predator detection by insects may involve tactile, visual, or chemical cues. The proximate mechanisms that mediate avoidance behaviour are certain chemical exudates released by the predator commonly called as kairomone. Chemically mediated avoidance is an adaptation used by prey to detect and evade predators. Kairomones or semiochemicals emitted from predators (Nordlund, 1981) are normally used by prey to minimize their encounters with dangerous community associates (Kerfoot and Sih, 1987; Kats and Dill, 1998). Common responses to predator chemicals include increased refuge use (Kats et al., 1988), marked changes in rate of movement (Mathis et al., 1993; Ode and Wissinger, 1993; Chivers et al., 1996; Huryn and Chivers, 1999), reduced foraging (Petranka and Hayes, 1998), reduced courtship behaviour, predator avoidance (Flowers and Graves, 1997) and increased growth rates (our unpublished data). Chivers and Smith (1998) and Kats and Dill (1998) reviewed studies showing that fish, amphibians, reptiles, mammals, a bird, and a broad array of invertebrates have evolved chemosensory mechanisms for detecting predators.

Kats and Dill (1998) list 16 studies which involved larvae of aquatic species that provided evidence for chemically mediated detection of predators by insects, and almost all involve the responding to aquatic predators. Where chemical cues are involved, responses to predators are often influenced by the predator's diet. For example, mayfly and damselfly larvae are more likely to respond to fish chemicals if the fish recently consumed conspecific prey (Chivers *et al.*, 1996; Huryn and Chivers, 1999). Prey species are more likely to evolve chemically mediated avoidance of predators when visual detection of predators is limited (Kats and Dill, 1998; Petranka and Hayes, 1998). In addition, prey must possess suitable behaviours and chemosensory mechanisms to detect and respond to predators. Mosquitoes appear to meet all of these requirements since many species are crepuscular or nocturnal and rely heavily on chemosensory mechanisms when selecting oviposition sites. After mating and taking a blood meal, a hemolymph-borne hormone triggers the female to start searching for a suitable oviposition site and to become receptive to chemical signals from the site (Klowden and Blackmer, 1987). Females oviposit on mud or the water's surface (*e.g.*, *Culex* spp.) after sampling the substrate for chemical composition using receptors on the tarsi, antennae, and tip of the proboscis (Davis, 1976; Bentley and Day, 1989). Published evidence now suggests that mosquitoes use chemosensory, information to assess several parameters that reflect habitat quality for offspring, including the availability of nutrients, presence of competitors and predators, and the overall quality and permanence of the water (Davis, 1976; Angerilli, 1980; Chesson, 1984; Bentley and Day, 1989; Petranka and Fakhoury, 1991; Blaustein and Kotler, 1993; Edgerly *et al.*, 1998). It has been demonstrated by Petranka and Fakhoury (1991) that mosquitoes and phantom midges greatly reduced ovipositing rates in experimental pools that contained caged sunfish (*Lepomis*) that were not visible to ovipositing females. Grostal and Dicke (1999) demonstrated similar behaviour in acarine mites, suggesting that the phenomenon may occur across a diverse array of arthropods. Recently it has been convincingly reported that the adult mosquito species has ability to sense the presence of *Gambusia* and mosquitoes reduce egg-laying rates in pools containing mosquito fish odour (Angelon and Petranka, 2002).

Angelon and Petranka (2002) conducted an experiment to determine whether mosquitoes would reduce oviposition rates in pools containing chemicals of the mosquitofish (*G. affinis*). Their experimental treatments consisted of outdoor pools that containing known concentrations of fish chemicals (low, medium, or high) or no fish chemicals (control). The mean number of larvae of the mosquito species *C. pipiens* per pool differed significantly in the experiment and was about three times greater in control pools compared with those receiving medium and high concentrations of fish chemicals. This study has clearly demonstrated that ovipositing female mosquitoes are able to use fish kairomones to make behavioural decisions that affect encounter rates of future offspring with predators.

In nature *G. affinis* inhabits sluggish habitats such as swamps and roadside puddles, and forms schools that may contain hundreds of individuals. Schooling behaviour produces a patchy environment in which predation risk may vary locally depending on the location of schools. Female mosquitoes could avoid *Gambusia* at several spatial scales. For example, females could use the strength of fish chemicals to avoid local sites in ponds with dense schools. Alternatively, females might strongly avoid ponds with *Gambusia* and seek out fishless ponds (Ritchie and Laidlaw-Bell, 1994).

Moreover, recent reviews have not supported the effectiveness of the mosquitofish in controlling mosquito populations and mosquito-borne diseases (Arthington and Lloyd, 1989; Courtenay and Meffe, 1989; Rupp, 1996). There is also evidence that in some cases the mosquito fish may even indirectly increase the survival rate of mosquito larvae by feeding on their cladoceran competitor (Blaustein and Karban, 1990). Despite its reputation as being an effective predator on mosquitoes, Courtenay and Meffe (1989) concluded that *Gambusia* is generally ineffective as a biological control agent. For example, *Gambusia* had a positive impact on controlling mosquitoes in only four of 20 countries where it was introduced for mosquito control or other reasons. One explanation for the ineffectiveness of *Gambusia* in controlling mosquitoes is that ovipositing mosquitoes may seek out fish-free habitats that adjoin large bodies of water containing *Gambusia*. The active avoidance of habitats with high predation risk to offspring may ultimately act to sustain high densities of mosquitoes near areas stocked with *Gambusia*. Ritchie and Laidlaw-Bell (1994) found that ovipositing *Aedes taeniorhynchus* strongly avoided sites with high densities of

G. holbrooki and shifted to adjoining habitats with few or no predatory fish (Stav *et al.*, 1999).

Therefore recent studies suggest that the effectiveness of *Gambusia* in controlling mosquitoes may be compromised if adult mosquitoes respond to fish stocking by shifting to nearby breeding sites that lack fish. It is warranted to reconsider the use of *Gambusia* in biological control programs.

Larvivorous Psorophora

Toxorhynchites, a kind of mosquito with cannibal larvae, have attracted much attention as biological control agents. Certain kinds of toxorhynchites mosquitoes are good for control of *A. aegypti* because they breed in the same kinds of containers. There are problems with the use of toxorhynchites mosquitoes for control of dengue. The technique began to lose luster when it was recognized that it increased prey density. This type of interference effect became evident when the prey was present in excess (Hubbard *et al.* 1988). Certain tests and trials have shown this to be true.

Aquatic Insects

Some of the aquatic insects play an important role in mosquito control (Ellis and Borden, 1970; Pandian *et al.*, 1979). In general almost all aquatic insect predators prey on mosquito larvae and pupae (Ellis and Borden, 1970; Peckarsky, 1984), the aquatic coleoptera (especially notonectids and dysticids) and odonate have been observed to ingest mosquito larvae in natural food assemblage in particular. The notonectid–*Notonecta undulata* has been shown to utilize the second instar of mosquito larvae efficiently (Ellis and Borden, 1970). However, in the presence of alternate prey in the medium it has not been observed to select mosquito larvae efficiently.

The order odonata includes insects like the dragonfly and damselfly. They enjoy a wide distribution and are particularly prominent around any aquatic ecosystem in a tropical country like India. Invariably the adult mates near the water bodies and the female lays her eggs in water soon after. The aquatic larval instars are predators of mosquito larvae. Dragonfly larvae, in particular, may be useful for controlling mosquito larval population. In an experiment in Myanmar (Burma), two dragonfly larvae were introduced in domestic containers having *A. aegypti*. The mosquito larvae virtually disappeared immediately after introduction from each container, and the density of target adults declined about 6 weeks later. The dragonfly and damselfly are real enemies of the mosquito as the larvae of these insects are able to utilize mosquito larvae as food where as their adults are efficient chasers predators of adult mosquito in air (Figure 18.3). However, they not only consume great numbers of mosquitoes, in larval and adult forms, they also feast on different zooplankton, gnats, midges, small moths, little tadpoles (Travis *et al.*, 1985) beetles and other insects (Brendonck *et al.*, 2002). So the quantitative evaluation of consumption of mosquito larvae by the odonate larvae in the presence of other prey is warranted. Breene *et al.* (1990) did not observe any mosquito larvae in the gut of damselfly (*Enallagma civile*) larvae. Their analysis revealed the larvae preyed upon chironimid larvae and in addition they also found Corixid, cladocerans, ostracods and aquatic mites. No remains of mosquito larvae were detected in any of the specimens, even though mosquito larvae (*Aedes, Culex, Culiseta, Mansonia*, and *Psorophora*) were observed in the pond where the damselfly larvae were collected from (Breene *et al.*, 1990).

None of these agents have shown any promise for malaria control, having proven difficult to rear and store, as well as being unstable or inefficient in the field. Their ability to aerially distribute themselves can be an advantage though, as the adults can access any water body and lay their eggs therefore their larvae can be found in small water tubs and tanks also, an ideal place where mosquitoes breed. Probably further in-depth studies need to be done to assess the role of aquatic insects in contributing to the control of mosquitoes.

Larvivorous Organisms in Temporary Ecosystem

Temporary or ephemeral aquatic ecosystems (pools, puddles, floodplains etc.) are natural, endorheic bodies of water, which experience a recurrent dry phase of varying duration. In other words temporary aquatic systems are those in which the entire habitat, for aquatic organisms shifts from being available to become unavailable, for a duration and/or frequency sufficient to influence the entire biota. Generally the hydroperiod in the ephemeral water bodies corresponds to the breeding periods of mosquito and these become the ideal breeding sites for the mosquitoes particularly genus *Aedes* and *Anopheles*. The most important suspected invertebrate predators in temporary pools are turbellarians (Blaustein, 1990; Blaustein and

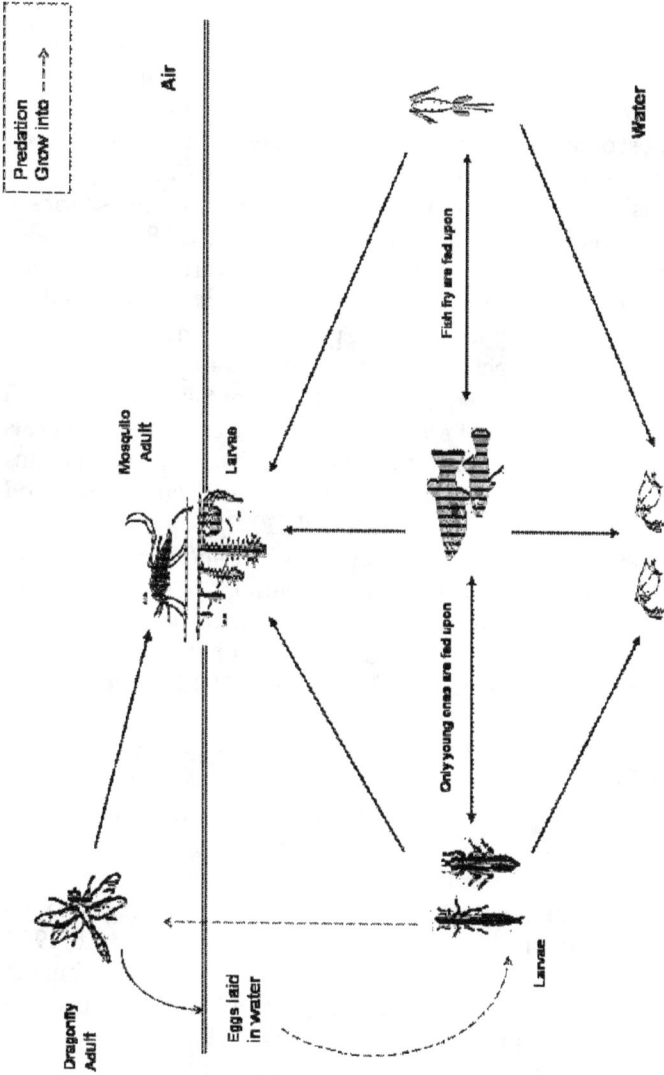

Figure 18.3: Schematic Representation of the Interactions Between Mosquito Larvae and its Predators (diagram not to scale)

Dumont, 1990), notonectids, diving beetles (Coleoptera) and dragonfly larvae (Odonata) (Wiggins *et al.*, 1980; Williams, 1987, 1997; Blaustein *et al.*, 1995; Herwig and Schindler, 1996; Spencer *et al.*, 1999). Turbellarians take a particular position in ephemeral ponds, as they are the only larvivorous organisms able to produce resting eggs to survive dry periods (Blaustein, 1990). They are present and become effective within the first days after the start of rains while most other invertebrate predators only become effective later in the hydroperiod of individual pools or even at a later stage of the rainy season. The most important flatworm predators are species of *Mesostoma* that occur in a wide range of habitats (Blaustein and Dumont, 1990). These species display a wide variety of predation mechanisms. Some species produce a kind of mucus that functions as a toxic web to trap and kill the prey organisms (Dumont and Carels, 1987). Other species just wait to attack the approaching prey ('sit and wait' or 'ambush' predation). Dumont and Carels (1987) furthermore showed that *Mesostoma* cf. *lingua* produces chemicals that are toxic for various prey organisms, *e.g. D. magna*. Turbellarians have been observed to kill and utilize mosquito larvae as a food source (Blaustein, 1990; Brendonck *et al.*, 2002). Some species may also actively search for suitable prey (Schwartz and Hebert, 1982) or reveal prey selectivity (Blaustein and Dumont, 1990). We do not have basic quantitative data related to the mosquito larval selectivity patterns of these flat worms in the presence of other natural prey types (Schwartz and Hebert, 1982; Blaustein, 1990; Brendonck *et al.*, 2002).

Cyclopoid Copepods

Cyclopoid copepods are abundant in eutrophic water bodies and play an important role in the trophic dynamics. Like many predators in the aquatic environment cyclopoid copepods are known to influence very strongly the structural and functional organization of the prey communities on which they feed (Kerfoot and Sih, 1987, Matsumura–Tundisi *et al.*, 1990; Irvine and Waya, 1993; Plabmann *et al.* 1997). Predominantly carnivorous, but with an ability to utilize plant food also (Kumar and Rao, 1999a, b), cyclopids have a wide spectrum of potential food items available in their habitat, including algae, ciliates, rotifers, cladocerans, and copepod nauplii. The patterns of prey selection and feeding rates in this group of invertebrate predators have been the focus of considerable research

(Greene, 1983; Williamson, 1986; Krylov 1988; Wickham, 1995 a,b; Rao and Kumar, 2002).

Some cyclopoids have long been known to have ability of utilizing mosquito larvae as food (Fryer, 1957; Brown *et al.*, 1991 a, b; Marten, 1990; Marten *et al.*, 1994 a, b; Kay, 1996). In the laboratory many species of *Mesocyclops* have been shown to prey upon *A. aegypti* or *Anopheles* larvae (Marten *et al.*, 1994 a,b; Brown *et al.*, 1991a, b; Kay *et al.*, 1992; Mittal *et al.*, 1997). Furthermore, in many ponds and small water bodies a strong negative association has been reported between *Mesocyclops* spp and larvae of *Anopheles albimanus* (Marten *et al.*, 1989, Zoppi De Roa *et al.*, 2002). The ability of certain cyclopoid copepods to destroy larval mosquitoes was noted in 1938 by Hurlbut. These "water fleas" were seen preying on newly hatched larvae. Field experiments in Rongaroa (French Polynesia) later demonstrated that *Mesocyclops* can be used in interventions against *A. aegypti* (Rivière *et al.*, 1987). Field trials have been conducted to determine whether copepods can usefully destroy larval *Stegomyia* mosquitoes. Copepods (*Macrocyclops albidus*) were released into each of about 200 tires arranged in 2 stacks of about 100 discarded tires each located near New Orleans (Marten, 1990). A third stack remained untreated. Larval *A. albopictus* that were numerous in the treated tires at the beginning of the experiment virtually disappeared within 2 months. Adults disappeared about 1 month later and remained scarce for at least another year (Marten, 1990; Marten *et al.*, 1994a, b; Nam *et al.*, 1998). These predators, however, did not reduce the abundance of *Culex salinarius*, another type of mosquito. A series of other field trials are currently being conducted in different parts of the world to determine if this is a good way to prevent transmission of disease. Preliminary results have been encouraging. Cyclopoids now appear to offer high promise as biological control agents for *A. aegypti* mosquitoes (Marten *et al.*, 1994b) and *Anopheles stephensi* (Kumar and Rao, 2003). Malarial mosquito larvae have been observed to be absent from aquatic habitats in Latin America (Marten *et al.*, 1989) that contained natural population of *M. longisetus* (Jennings *et al.*, 1995; Nam *et al.*, 1998: Marten *et al.*, 1994 a, b) and virtually disappeared after *M. longisetus* and other species of *Mesocyclops* introduced to rice fields and small marsh areas in Louisiana (Marten *et al.*, 1994b). The technology is "appropriate" and costs appear to be modest. Biological control using cyclopoid as predator against mosquito larvae did not receive serious consideration in our country.

Cyclopoids are different from other aquatic invertebrate predators that prey on mosquito larvae. If larvae are numerous, they eat a small part of each larva, giving each copepod the capacity to kill 30–40 larvae per day, far more than they actually eat (Kumar and Rao, 2003). Even more important is their large numbers (Humes, 1994). Only about 10 per cent of places with water where mosquitoes might breed have natural population of *M. thermocyclopoides* or other cyclopoids which drastically reduce the survival of mosquito larvae (Brown, *et al.*, 1991a,b; Kumar and Rao, 2003).

The simple life cycle of cyclopoid copepods and their ability to thrive on a wide spectrum of food types makes mass production easy and inexpensive, highly resilient, functioning in open containers of any size or shape. Females are inseminated during adolescence and no further contact with male is required to produce 150–250 eggs weekly for their several months life span (Kumar and Rao, 1999a).

However, not all copepods destroy all mosquitoes. Differential selection by the copepod is related to mosquito larval species and instar stage (Figures 18.4A and B). As cyclopids show distinct prey selectivity behavioral pattern and their prey selection patterns are influenced by many attributes of the prey such as morphology, behaviour and taste (Stemberger, 1985; DeMott, 1995; Wickham, 1995a,b; Rao and Kumar, 2002; Kumar and Rao, 2003), but they are known to differ from species to species (Kumar and Rao, 2001). Some species of cyclopoids usually select smaller prey items from the available prey size spectrum (Williamson, 1980; Stemberger, 1985; Kuno, 1987; Kumar, 2003), while others may actively select the largest prey they can capture (Krylov, 1988, Janicki and DeCosta, 1990). The extent of carnivory in omnivorous cyclopoids appears to be directly related to their body size (Fryer, 1957; Adrian and Frost, 1993). The selectivity for animal prey in those species that include algae in their diet (Adrian, 1991; Santer, 1993; Hopp *et al.*, 1997) is also influenced by the extent of their dependence on algae (Adrian and Frost, 1993; Kumar and Rao, 1999b). Kumar and Rao (2003) studied in the laboratory the prey consumption rates of the most abundant, cyclopoid copepod in local waterbodies, *M. thermocyclopoides* on first and fourth instar larvae of two species of mosquito (*A. stephensi* and *C. quinquefasciatus*) in relation to their density. Since prey vulnerability is a product of prey's encounter

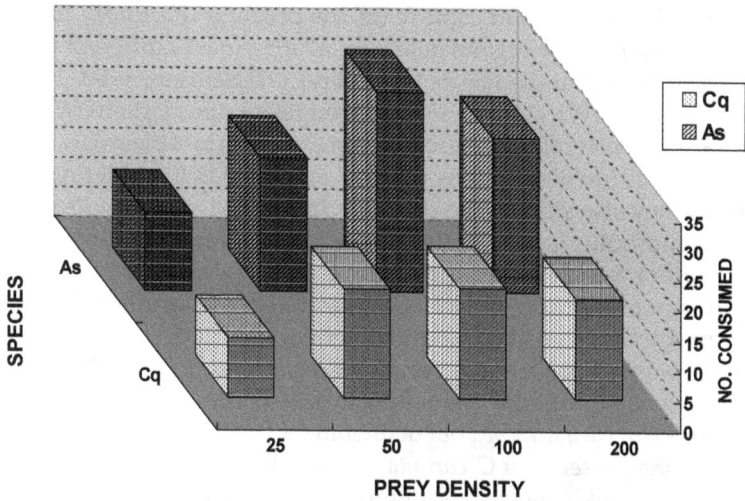

A. Instar I

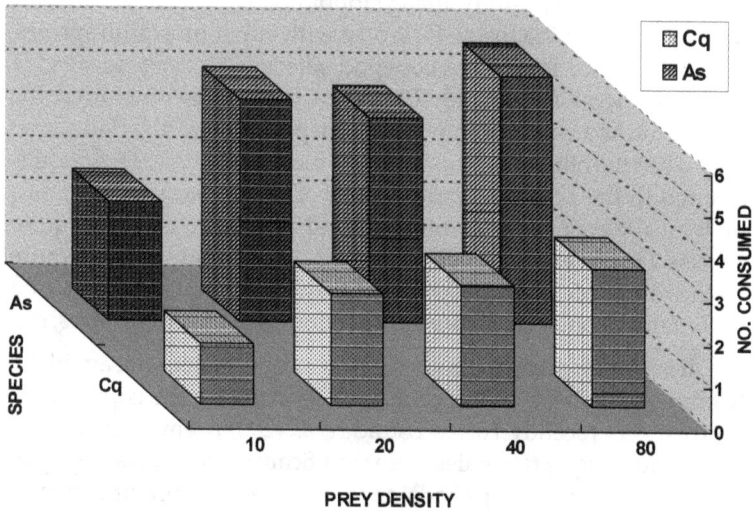

B. Instar IV

Figure 18.4: Daily Per Capita Prey Killing Rates (Mean ± SE) of
M. thermocyclopoides in Relation to the Larval Density of the
Two Mosquito Species *A. stephensi* (*A. s*) and *C. quinquefasciatus*
(*C. q*), (A) Instar I and (B) Instar IV (*Source*: Kumar and Rao, 2003).

rates with, and ease of capture by, the predator (Pastorok, 1981), certain morphological and behavioural attributes of different prey species (which influence capturability) (Williamson, 1980, 1983; DeMott, 1995; Wickham, 1995), and their relative proportions (which affect encounter rates) in the medium, may be able to explain observed differential larval selectivity patterns in copepods. Considering the point they also studied the prey selectivity of *M. thermocyclopoides* with mosquito larvae in the presence of an alternate prey (the cladocerans–either *Moina macrocopa* or *Ceriodaphnia cornuta*) in different proportions. This laboratory study demonstrated that *M. thermocyclopoides* had biocontrol potential against *A. stephensi* even in the presence of alternate prey. However, against *C. quinquefasciatus*, there was significant reduction in larval consumption in the presence of *M. macrocopa* but neither in the presence of *D. similoides* nor in the presence of *C. cornuta*. The predaceous behaviour of *M. thermocyclopoides* against mosquito larvae was comparable with *M. aspericornis* and other cyclopoid species reported earlier (Brown *et al.*, 1991b). In their study the cyclopoid actively selected Instar-I larvae, avoiding the Instar-IV larvae with either mosquito species, and with either instar, the copepod selected *A. stephensi* over *C. quinquefasciatus* (Figure 18.4B). When prey choice included the cladoceran as an alternate prey, the copepod selected the cladoceran only when the other prey was Instar-IV mosquito larvae. Their results pointed to the potential and promise of *M. thermocyclopoides* as a biological agent for controlling larval populations of mosquito species. However, in the light of their study, further studies require on the role of *M. thermocylopoides* against mosquito larvae breeding in containers other small water collection such as coolers, tires, etc.

It is further advantageous that the mosquito does not avoid the habitat on the basis of infochemicals secreted by the copepod. In contrast very recently Torres-Estrada *et al.* (2001) demonstrated that the gravid *A. aegypti* females were significantly more attracted by to ovitraps containing copepods or to ovitraps containing copepod conditioned water. They concluded that the copepod infochemicals may be responsible for attracting gravid *A. aegypti* females and may increase the number of potential prey for the copepod.

M. thermocyclopoides is not a vector of guinea worm disease in India (Joshi, 1996). Although it is dominant in different lakes and ponds only about 10 per cent of places with water where mosquitoes

Table 18.1: Copepod Species Found Promising for Mosquito Control Potential in Different Habitats

Species	Habitat	Mosquito Species	Country	Reference
Mesocyclops longisetus, M.venezolanus,	Water storage containers	Anopheles, Aedes aegypti	Honduras, Luisiana (operational)	Marten et al., 1994a
M.thermocyclopoides, M. albidus	Water storage Urns, Vases and Bromeliads	Aedes aegypti	Brazil	Vasconcelos et al., 1992;
M. thermocyclopoides	200–500 Bowls in Laboratory	Culex quinquefasciatus, Anopheles stephensi and Aedes aegypti	India	Mittal et al., 1997
M. longisetus, M. aspericornis	Laboratory Glass bowls and small artificial pools	Ae. aegypti	Brazil	Santos and Andrade, 1997
Macrocyclops spp Mesocyclops spp.	Domestic containers	Ae. aegypti	Caribbean Islands	Rawlins et al., 1997
Acanthocyclops vernalis, Diacyclops bicuspidatus	Laboratory, Bowls	Aedes canadensis Ae. stimulans		Andrealis and Gere, 1992
Mesocyclops spp.	Natural mosquito breeding habitat	Aedes aegypti	Vietnam (operational)	Nam et al., 1998
Macrocyclops albidus	Residential ditches	Culex quinquefasciatus		Marten et al., 2000
Mesocyclops spp.	Manholes and pits	Aedes aegypti Aedes albopictus	Queensland, Australia	Kay et al. 2002 Dieng et al. 2002
Mesocyclops aspericornis		Aedes aegypti	Thailand	Kosiyachinda et al., 2003
Mesocyclops thermocy-clopoides	Bowls	Culex quinquefasciatus, Anopheles stephensi	India	Kumar and Rao, 2003

might breed have natural population of *M. thermocyclopoides* or other cyclopoids, which drastically reduce the survival of mosquito larvae. Therefore the mass culture of the cyclopoid should be established and active adult cyclopoid should be introduced to different mosquito breeding habitat.

Although highly promising, the practicality of these biological agents in anti-mosquito programs remains to be established India. It has also been recorded that native fishes (Lloyd, 1987; Courtenay and Meffe, 1989) and cyclopoid copepods (Marten, 1994 a,b) including those that the mosquito-fish affects in its natural habitats, are better and more efficient control agents therefore, the use of native fishes and cyclopoid species needs to be promoted.

Conclusion

History has clearly shown that the most effective methods for controlling malaria involve interventions targeted at the vector, either control measures that reduce the population of infectious vectors or measures that reduce vector–human contact rates. Much of the current effort directed at the development of new mosquito control tools has, as Spielman *et al.* (1993) argued "turned deep into the laboratory", as this type of research lack good prospects for managing mosquito borne diseases in the areas of intense transmission. The major requirement of a program that will help to stop transmission of mosquito-borne diseases is the ability to get into obscure bodies of water scattered within and around human settlements where vectorially important mosquitoes mainly breed. A hidden automobile tire presents an obstacle to any effort in a society that respects privacy. Only biological agents carry the potential for overcoming such obstacles, and the most likely agents are those represented by closely related organisms. Toward this end, we require a program of biological research aiming towards an understanding of the factors that limit the number of mosquitoes. Search efficiency of the introduced predator and prey selectivity patterns of larvivorous organisms needs to be explored by offering mosquito larvae in combination with other alternate natural prey. Over all care needs to be taken in case the introduced predator preys or alters the populations of the indigenous flora and fauna of the ecosystem.

A combination of life history, population dynamics, production and eco-ethological traits (*e.g.* fast growth, reduced longevity, viviparity, high productivity, an intermediate position in food chain, plasticity and adaptability in its food use, and no special habitat requirements for reproduction) show that *Gambusia* introduced into different water bodies all over the world, certainly induce an important impact in the structure and functioning of the native biological communities. This question is extremely important to reinforce the recommendation that *Gambusia*, the backbone of biocontrol for one-quarter of a century, not be introduced into new areas. Mosquito larvae are rarely found in permanent waters, the sort of habitat where *Gambusia* flourishes, and not many *Gambusia* are likely to find their way to common mosquito breeding habitats such as tree holes, old tires, tin cans, and undrained swimming pools and boats which can be a suitable habitat for invertebrate predators like cyclopoids and aquatic insects. Specific biological control agents for specific life stages of the mosquito can be used to ensure a better and a more effective control. Therefore, it is prudent to use biological control agents that can get into hard to find places, are found naturally, are safe for people and are economical to go into operational use. Turbellarians, cyclopoid copepods, aquatic insects, and native fishes may prove to be more promising to control mosquito populations.

Acknowledgements

The authors are extremely grateful to Prof. T.R. Rao for his continuous support and guidance during the preparation of this manuscript. The authors also highly appreciate the facilities provided by Acharya Narendra Dev College, University of Delhi.

References

Adrian, R., 1991. The feeding behaviour of *Cyclops kolensis* and *C. vicinus* (Crustacea : Copepoda). *Verh. Internat. Verein. Limnol.*, 24: 2852–2863.

Adrian, R. and Frost, T.M., 1993. Omnivory in cyclopoids copepods: Comparisons of algae and invertebrates as food for three differently sized species. *J. Plankton Res.*, 15: 643–658.

Angelon, K.A. and Petranka, J.W., 2002. Chemicals of predatory mosquito fish (*Gambusia affinis*) influence selection of oviposition site by *Culex* mosquitoes. *J. Chemical Ecol.*, 28: 797–806.

Angerilli, N.P.D., 1980. Influences of extraction of freshwater vegetation on the survival and oviposition by *Aedes aegypti* (Diptera : Culicidae). *Can. Entomol.*, 112: 1249–1252.

Andrealis, T.G. and Gere, M.A., 1992. Laboratory evaluation of *Acanthocyclops vernalis* and *Diacylops bicuspidatus thomasi* (Copepoda : Cyclopoida) as predators of *Aedes canadensis* and *Aedes stimulans* (Diptera : Culicidae). *J. Med. Entomol.*, 29: 974–979.

Arthington, A.H. and Lloyd, L.N., 1989. Introduced Poeciliidae in Australia and New Zealand. In: *Evolution and Ecology of Livebearing Fishes (Poeciliidae)* (Eds.) G.K. Meffe and F.F. Snelson. Prentice-Hall, New York, USA, pp. 333–348.

Arthington, A.H. and Marshall, C.J., 1999. Diet of the exotic mosquito fish *Gambusia holbrooki* in an Australian lake and potential for competition with indigenous fish species. *Asian Fisheries Sci.*, 12: 1–16.

Baber, M.J. and Babbitt, K.J., 2003. The relative impacts of native and introduced predatory fish on a temporary wetland tadpole assemblage. *Oecologia*, 136: 289–295.

Barrier, R.F.G. and Hicks, B.J., 1994. Behavioural interactions between black mudfish (*Neochanna diversus*) and mosquitofish (*Gambusia affinis* Baird and Girard 1854). *Ecol. Freshwat. Fish*, 3: 93–99.

Bence, J.R. and Murdoch, W.W., 1984. Prey size selection by the mosquito fish and its relation to optimal diet theory. *Ecology*, 67: 324–336

Bentley, M.D. and Day, J.F., 1989. Chemical ecology and behavioral aspects of mosquito oviposition. *Ann. Rev. Entomol.*, 34: 401–421.

Blaustein, L., 1990. Evidence for predatory flatworms as organizers of zooplankton and mosquito community structure in rice fields. *Hydrobiologia*, 199: 179–191.

Blaustein, L. and Dumont H.J., 1990. Typhloplanid flatworms (Mesostoma and related genera): Mechanisms of predation and evidence that they structure aquatic invertebrate communities. *Hydrobiologia*, 198: 61–77.

Blaustein, L. and Karban, R., 1990. Indirect effects of the mosquito fish *Gambusia affinis* on the mosquito *Culex tarsalis*. *Limnol. Oceanograph.*, 35: 767–771.

Blaustein, L. and Kotler, B.P., 1993. Oviposition habitat selection by the mosquito, *Culiseta lon-giareolata*: Effects of conspecifics, food, and green toad tadpoles. *Ecol. Entomol.*, 18: 104–108.

Blaustein, L., Kotler B.P. and Ward, D., 1995. Direct and indirect effects of a predatory backswimmer (*Notonecta maculata*)on community structure of desert temporary pools. *Ecol. Entomol.*, 20: 311–318.

Blaustein, L., Friedman, J. and Fahima, T., 1996. Larval Salamandra drive temporary pool community dynamics: Evidence from an artificial pool experiment. *Oikos*, 76: 392–402.

Breene-R.G., Sweet, M.H. and Olson, J.K., 1990. Analysis of the gut contents of naiads of Enallagma civile (Odonata : Coenagrionidae) from a Texas pond. *J. Am. Mosq. Cont. Assoc.*, 6: 547–548.

Brendonck, L., Michels, E., De Meester, L. and Riddoch, B., 2002. Temporary pools are not 'enemy-free'. *Hydrobiologia*, 486: 147–159.

Brown, M.D, Kay, B.H. and Green, J.G., 1991a. The predation efficiency of North eastern Australian *Mesocyclops* (Copepoda: cyclopidae) on mosquito larvae. *Bull. Plankton Soc. Jpn.*, (Suppl.) pp. 329–338.

Brown, M.D., Kay, B.H. and Hendrix, J.H., 1991b. Evaluation of Australian *Mesocyclops* (Copepoda : Cyclopoida) for mosquito control. *J. Med. Entomol.*, 28: 618–623.

Cabral, J.A., Mieiro, C.L. and Marques, J.C., 1998. Environmental and biological factors influence the relationship between a predator fish, *Gambusia holbrooki*, and its main prey in rice fields of the Lower Mondego River Valley (Portugal). *Hydrobiologia*, 382: 41–51.

Carwood, M., 1997. Spawn of an era. *Austr. Geogr.*, 48: 35–51.

Chesson, J., 1984. Effect of notonectids (Hemiptera: Notonectidae) on mosquitoes (Diptera : Culicidae): Predation or selective oviposition? *Environ. Entomol.*, 13: 531–538.

Chivers, D.P., and Smith, R.J.F., 1998. Chemical alarm signaling in aquatic predator/prey interactions: A review and prospectus. *Ecosci.*, 5: 338–352.

Chivers, D.P.,Wisenden, B.D. and Smith, R.J.F., 1996. Damselfly larvae learn to recognize predators from chemical cues in the predator's diet. *Anim. Behav.*, 52: 315–320.

Courtenay, W.R. and Meffe, G.K., 1989. Small fishes in strange places: a review of introduced poeciliids. In: *Ecology and Evolution of Livebearing Fishes (Poeciliidae)*, (Eds.) G.K. Meffe and F.F. Snelson. Prentice Hall, New Jersey, pp. 319–331

Crivelli, A.J. and Boy, V., 1987. The diet of mosquito fish, *Gambusia affinis* (Baird and Girard) (Poeciliidae) in Mediterranien France. *Revue d'Ecologie*, 42: 421–435.

Davis, E.E., 1976. A Receptor sensitive to oviposition site attractants on the antennae of the mosquito, *Aedes aegypti. J. Insect Physiol.*, 22: 1371–1376.

Demott, W.R., 1995. Food selection by calanoid copepods in response to between lake variation in food abundance. *Freshwat. Biol.*, 33: 171–180.

Dieng, H., Mwandawiro, C., Boots, M., Morales, R., Satho, T., Tuno, N., Tsuda, Y. and Takagi, M., 2002. Leaf litter decay process and the growth performance of *Aedes albopictus* larvae (Diptera : Culicidae). *J. Vector Ecol.*, 27: 31–38.

Dumont, H.J. and Carels, I., 1987. Flatworm predator (*Mesostoma* cf. *lingua*) releases a toxin to catch planktonic prey (*Daphnia magna*). *Limnol. Oceanogr.*, 32: 699–702.

Edgerly, J.S., Mcfarland, M., Morgan, P., and Livdahl, T., 1998. A seasonal shift in egg-laying behavior in response to cues of future competition in a tree hole mosquito. *J. Anim. Ecol.*, 67: 805–818.

Ellis, R.A. and Borden. J.H., 1970. Predation by *Notonecta undulata* (Heteroptera : Notonectidae) on larvae of the Yellow-Fever Mosquito. *Ann. Entomol. Soc. America*, 63: 963–973

Englund, R.E., 1999. The impacts of introduced Poeciliid fish and Odonata on the endemic *Megalagrion* (Odonata) damselflies of Oahu Island, Hawaii. *J. Insect Cons.*, 3: 225–243.

Flowers, M.A. and Graves, B.M., 1997. Juvenile toads avoid chemical cues from snake predators. *Anim. Behav.*, 53: 641–646.

Fryer, G., 1957. The food of some fresh water cyclopoid copepods and its ecological significance. *J. Anim. Ecol.*, 26: 263–286.

Gamradt, S.C. and Kats, L.B., 1996. The effects of introduced crayfish and mosquito fish on California newts (*Taricha torosa*). *Cons. Biol.*, 10: 1155–1162.

Garcia-Berthou, E., 1999. Food of introduced mosquito fish: Ontogenetic diet shift and prey selection. *J. Fish Biol.*, 55: 135–147.

Goodsell, J.A. and Kats, L.B., 1999. Effect of introduced mosquito fish on pacific tree frogs and the role of alternative prey. *Cons. Biol.*, 13: 921–924.

Gophen, M., Yehuda, Y. Malinkov A. and Degani, G., 1998. Food composition of the fish community in Lake Agmon. *Hydrobiol.*, 380: 49–57.

Greene, C.H., 1983. Selective predation in fresh water zooplankton communities. *Int. Revue. Ges. Hydrobiol.*, 68: 296–315.

Grostal, P. and Dicke, M., 1999. Direct and indirect cues of predation risk influence behavior and reproduction of prey: A case for acarine interactions. *Behav. Ecol.*, 10: 422–427.

Grubb, J.C., 1972. Differential predation by *Gambusia affinis* on the eggs of seven species of anuran amphibians. *Am. Midl. Nat.*, 88: 102–108.

Gulati, R.D., 1990 Structural and grazing responses of zooplankton community to bio manipulation of some Dutch water bodies. *Hydrobiologia*, 200/201: 99–118.

He, X., Scheurell, M.D., Soranno, P.A. and Wright, R.A., 1994. Recurrent response patterns of a zooplankton community to whole lake fish manipulation. *Freshwat. Biol.*, 32: 61–72.

Herwig. B.R. and Schindler. D.E., 1996. Effects of aquatic insect predators on zooplankton in fishless ponds. *Hydrobiol.*, 324: 141–147.

Hopp, U., Maier, G. and Bleher, R., 1997. Reproduction and adult longevity of five species of planktonic cyclopoid copepods reared on different diets. A comparative study. *Freshwat. Biol.,* 38: 289–300.

Howe, E., Howe, C., Lim R. and Burchett, M., 1997. Impact of the introduced poeciliid *Gambusia holbrooki* (Girard, 1859) on the growth and reproduction of *Pseudomugil signifer* (Kner, 1865) in Australia. *Marine Freshwat. Res.,* 48: 425–434.

Hubbard, S.F., O'Malley, S.L.C. and Russo, R., 1988. The functional response of *Toxorhynchites rutilus* to changes in the population density of its prey, *Aedes aegypti. Med. Vet. Entomol.,* 2: 279–283.

Hurlbert, S.H., Zedler J. and Fairbanks. D., 1972. Ecosystem alteration by Mosquitofish (*Gambusia affinis*) predation. *Science,* 175: 639–641.

Hurlbert, S.H. and Mulla M.S., 1981. Impacts of Mosquitofish (*Gambusia affinis*) predation on plankton communities. *Hydrobiol.,* 83: 125–151.

Hurlbut, H.S., 1938. Copepod observed preying on first instar larva of *Anopheles quadrimaculatus* Say. *J. Parasitol.,* 24: 281.

Humes, A.G., 1994. How many copepods? *Hydrobiologia,* 292/293: 1–7.

Huryn, A.D. and Chivers, D.P., 1999. Contrasting behavioral responses by detritivorous and predatory mayflies to chemicals released by injured conspecifics and their predators. *J. Chem. Ecol.,* 25: 2719–2740.

Irvine, K. and Waya, R., 1993. Predatory behaviour of the cyclopoid copepod *Mesocyclops aequatorialis aequatorialis* in Lake Malawi, a deep tropical Lake. *Verh. Internat. Verein. Limnol.,* 25: 877–881.

Ivantsoff, W. and Aarn, 1999. Detection of predation on Australian native fishes by *Gambusia holbrooki. Marine Freshwat. Res.,* 50: 67–468.

Janicki, A. and Decosta, J., 1990. An analysis of prey selection by *Mesocyclops edax. Hydrobiologia,* 198: 133–139.

Jennings, C.D., Phommasack, B., Sourignadeth, B. and Kay, B.H., 1995. *Aedes aegypti* control in the Lao People's Democratic Republic, with reference to copepods. *Am. J. Trop. Med. Hyg.,* 53: 324–330.

Joshi, G.C., 1996. Guinea worm: A disappearing disease in India. *Proc. Natl. Acad. Sci., India,* 66: 231–244.

Kay, B.H., 1996. The use of predaceous copepods in controlling dengue and other vectors. *Dengue Bull.,* 20: 93–98.

Kats, L.B. and Dill, L.M., 1998. The scent of death: Chemosensory assessment of predation risk by prey animals. *Ecosci.,* 5: 361–394.

Kats, L. B., Petranka, J. P., And Sih, A., 1988. Anti-predator defenses and the persistence of amphibian larvae with fish. *Ecology,* 69: 1865–1870.

Kay, B.H., Cabral, C.P., Sleigh, A.C., Brown, M.D., Ribeiro, Z.M. and Vasconcelos, A.W., 1992. Laboratory evaluation of Brazilian *Mesocyclops* (Copepoda : Cyclopidae) for mosquito control. *J. Med. Entomol.,* 29: 599–602.

Kay, B.H., Lyons, S.A., Holt, J.S., Holynska, M. and Russell, B.M. 2002. Point source inoculation of *Mesocyclops* (Copepoda : Cyclopidae) gives widespread control of *Ochlerotatus* and *Aedes* (Diptera : Culicidae) immatures in service Manholes and pits in North Queensland, Australia. *J. Med. Entomol.,* 39: 469–474.

Kerfoot, C. and Sih, A., 1987. *Predation: Direct and Indirect Impacts on Aquatic Communities.* University Press of New England, Hanover, New Hampshire.

Klowden, M.J. and Blackmer, J.L., 1987. Humoral control of pre-oviposition behavior in the mosquito, *Aedes aegypti. J. Insect Physiol.,* 33: 689–692.

Komak, S. and Crossland, M.R., 2000. An assessment of the introduced mosquito fish (*Gambusia affinis* Holbrooki) as a predator of eggs, hatchlings and tadpoles of native and non-native anurans. *Wildl. Res.,* 27: 185–189.

Kosiyachinda, P., Bhumiratana, A. and Kittayapong, P., 2003. Enhancement of the efficacy of a combination of *Mesocyclops aspericornis* and *Bacillus thuringiensis* var. *israelensis* by community-based products in controlling *Aedes aegypti* larvae in Thailand. *Am. J. Trop. Med. Hyg.,* 69: 206–212.

Krumholz, L.A., 1948. Reproduction in the western mosquito fish *Gambusia affinis* and its use in mosquito control. *Ecol. Monogr.,* 18: 1–43.

Krylov, P.I., 1988. Predation of the fresh water cyclopoid copepod *Megacyclops gigas* on lake zooplankton: Functional response and prey selection. *Arch. Hydrobiol.*, 113: 231–250.

Kumar, R. and Rao, T.R., 1999a. Demographic responses of adult *Mesocyclops thermocyclopoides* (Copepoda, Cyclopoida) to different plant and animal diets. *Freshwater Biol.*, 42: 487–501.

Kumar, R. and Rao, T.R., 1999b. Effect of algal food on animal prey consumption rates in the omnivorous copepod, *Mesocyclops thermocyclopoides*. *Internat. Rev. Hydrobiol.*, 84: 419–426.

Kumar, R., and Rao T.R. 2001. Effect of the cyclopoid copepod *Mesocyclops thermocyclopoides* on the interactions between the predatory rotifer *Asplanchna intermedia* and its prey *Brachionus calyciflorus* and *B. angularis*. *Hydrobiologia*, 453/454: 261–268.

Kumar, R. and Rao, T.R. 2003. Predation on mosquito (*Anopheles stephensi* and *Culex quinquefasciatus*) larvae by *Mesocyclops thermocyclopoides* (Copepoda : Cyclopoida) in the presence of alternate prey. *Internat. Rev. Hydrobiol.*, 88: 570–581.

Kuno, E., 1987. Principles of predator-prey interaction in theoretical, experimental, and natural population systems. *Adv. Ecol. Res.*, 16: 249–337.

Lancaster, H. F and Drenner, R.W., 1990. Expeimental mesocosm study of the separate and interaction effects of the phosphorus and mosquitofish (*Gambusia affinis*) on plankton community structure. *Can. J. Fish. Aquat. Sci.*, 47: 471–479.

Leyse, K.E., Lawler, S.P., Strange, T., 2004. Effects of an alien fish, *Gambusia affinis*, on an endemic California fairy shrimp, *Linderiella occidentalis*: implications for conservation of diversity in fishless waters. *Biol. Cons.*, 118: 57–65.

Li, J.L. and Li, H.W., 1979. Species specific factors affecting predator–prey interactions of the copepod *Acanthocyclops vernalis* with its natural prey. *Limnol. Oceanogr.*, 24: 613–626.

Linden, A.L. and Cech. J.J., 1990. Prey selection by mosquitofish (*Gambusia affinis*) in Carlifornia rice fields: Effect of vegetation and prey species. *J. Am. Mosq. Control. Assoc.*, 6: 115–120.

Lloyd L., 1987. Biological control of insects with fish. *Workshop on Mosquito Vector Control in Australia: Current Status and Future Prospects*. Commonwealth Department of Health, Canberra.

Lu B. Lca, 1989. *Recent Studies on the Dengue Vectors of China*. Working paper, pp. 89–92.

McKay, R.J., 1984. Introduction of exotic fishes in Australia. In: *Distribution, Biology, and Management of Exotic Fishes*, (Eds.) W.R. Courtenay and J.R. Staffer. John Hopkins University Press, Baltimore, USA, 430 pp.

Mansfield, S. and McArdle, B.H., 1998. Dietary composition of *Gambusia affinis* (Family : Poeciliidae) populations in the northern Waikato region of New Zealand. New Zealand. *J. Marine Freshwat. Res.*, 32: 375–383.

Margaritora, F.G., 1990. Influence of *Gambusia affinis* on the features and dynamics of the zooplankton community in the pools of Castel Porziano (Latium). *Rivista di Idrobiol.*, 29: 747–762.

Marian, M. P., Christopher, M.S.M., Selvaraj, A.M. and Pandian. T.J., 1983. Studies on predation of the mosquito *Culex fatigans* by *Rana tigrina* tadpoles. *Hydrobiologia*, 106: 59–63.

Marten, G.G., 1986. Mosquito control by plankton management: The potential of indigestible green algae. *J. Trop. Med. Hyg.*, 89: 213–222.

Marten, G.G., 1989. A survey of cyclopoid copepods for control of *Aedes albopictus* larvae. *Bull. Soc. Vector Ecol.*, p. 232–236.

Marten, G.G., 1990. Evaluation of cyclopoid copepods for *Aedes albopictus* Control in tires. *J. Am. Mosq. Contr. Assoc.*, 6: 681–688.

Marten, G.G., Astaeza, R., Suárez, M.F., Monje, C. and Reid, J.W., 1989. Natural control of larval *Anopheles albimanus* (Diptera : Culicidae) by the predator *Mesocyclops* (Copepoda : Cyclopoida). *J. Med. Entomol.*, 26: 624–627.

Marten, G.G., Bordes, E.S. and Nguyen, M., 1994a. Use of cyclopoid copepod for mosquito control. *Hydrobiologia*, 292/293: 491–496.

Marten, G.G., Borjas, G., Cush, M., Fernandez, E. and Reid, J.W., 1994b. Control of larval *Aedes aegypti* (Diptera : Culicidae) by cyclopoid copepod in peridomestic breeding containers. *J. Med. Entomol.*, 31: 36–44.

Marten G.G., Nguyen, M., Mason, B.J. and Ngo, G., 2000. Natural control of *Culex quinquefasciatus* larvae in residential ditches by the copepod *Macrocyclops albidus*. *J. Vector Ecol.*, 25: 7–15.

Mathis, A., Chivers, D.P. and Smith, R.J.F., 1993. Population differences in responses of fathead minnows (*Pimephales promelas*) to visual and chemical stimuli from predators. *Ethology*, 93: 31–40.

Matsumura-Tundisi, T., Rietzler, A.C., Espindola, E.L.G., Tundisi, J.G. and Rocha, O., 1990. Predation on *Ceriodaphnia cornuta* and *Brachionus calyciflorus* by two *Mesocyclops* species coexisting in Barra Bonita reservoir (SP, Brazil). *Hydrobiologia*, 198: 141–151.

Meffe, G.K. and Snelson, F.F., 1989. An ecological overview of poeciliid fishes. In: *Ecology and Evolution of Livebearing Fishes (Poeciliidae)*, (Eds.) G.K Meffe and F.F. Snelson. Prentice Hall, New Jersey, pp. 13–31.

Mittal, P.K., Dhiman, R.C., Adak T. and Sharma, V.P., 1997. Laboratory evaluation of the biocontrol potential of *Mesocyclops thermocyclopoides* (Copepoda : Cyclopoida) against mosquito larvae. *Southeast Asian J. Trop. Med. Publ. Hlth.*, 28: 857–861.

Morgan, L.A. and Buttemer, W.A., 1996. Predation by the non-native fish *Gambusia holbrooki* on small *Litoria aurea* and *L. dentate* tadpoles. *Austral. Zool.*, 30: 143–149.

Moyle, P.B., 1976. Fish introductions in California: History and impact on native fishes. *Biol.Cons.*, 9: 101–118.

Nam, V.S., Yen, N.T., Kay., B.H., Marten, G.G. and Reid, J.W., 1998. Eradication of *Aedes aegypti* from a village in Vietnam using copepod and community participation. *Am. J. Trop. Med. Hyg.*, 59: 657–660.

Neng, W, Shusen, W., Guangxin, H., Rongman, X., Guangkun, T. and Chen. Q., 1987. Control of *Aedes aegypti* larvae in household water containers by Chinese cat fish. *Bull. Wld. Hlth. Org.*, 65: 503–506.

Nordlund, D.A., 1981. Semiochemicals: A review of the terminology. In: *Semiochemicals: Their Role in Pest Control*, (Eds.) D.A. Nordlund, R.L. Jones and W.J. Lewis. John Wiley and Sons, New York, pp. 319–331.

Ode, P.R., and Wissinger, S.A., 1993. Interaction between chemical and tactile cues in mayfly detection of stoneflies. *Freshwat. Biol.*, 30: 351–357.

Pandian, T.J., Mathavan. S. and Jeyagopal, C.P., 1979. Influence of temperature and body weighy on mosquito predation by the dragonfly nymph *Mesogomphus lineatus. Hydrobiologia*, 62: 99–104.

Pastorok, R.A., 1981. Prey vulnerability and size selection by *Chaoborus* larvae. *Ecology*, 62: 1311–1324.

Peckarsky, B.L., 1984. Predator-prey interactions among aquatic insects. In: *The Ecology of Aquatic Insects*, (Eds.) V.H. Resh and D.M. Rosenberg. Praeger Publishers, New York, USA, pp. 196–254.

Petranka, J.W. and Fakhoury, K., 1991. Evidence of a chemically mediated avoidance response of ovipositing insects to bluegills and green frog tadpoles. *Copeia*, p. 234–239.

Petranka, J.W. and Hayes, L., 1998. Chemically mediated avoidance of a predatory odonate (*Anax junius*) by American toad (*Bufo americanus*) and wood frog (*Rana sylvatica*) tadpoles. *Behav. Ecol. Sociobiol.*, 42: 263–271.

Plabmann, T., Maier, G. and Stich, H.B., 1997. Predation impact of *Cyclops vicinus* on the rotifer community in Lake Constance in spring. *J. Plankton. Res.*, 19: 1069–1079.

Rao, T. R. and Kumar, R., 2002. Patterns of prey selectivity in the cyclopoid copepod *Mesocyclops thermocyclopoides. Aq. Ecol.*, 36: 411–424.

Rawlins S.C., Martinez, R., Wiltshire, S., Clarke, D., Prabhakar, P. and Spinks, M., 1997. Evaluation of Caribbean strains of *Macrocyclops* and *Mesocyclops* (Cyclopoida : Cyclopidae) as biological control tools for the dengue vector *Aedes aegypti. J. Am. Mosq. Contl. Assoc.*, 13:18–23.

Rishikesh, N., Dubitiskij, A.M. and Moreau, C.M., 1988. *Malaria Vector Control: Biological Control*, 118: 1227–1250.

Ritchie, S.A. and Laidlaw-Bell, C., 1994. Do fish repel oviposition by *Aedes taeniorhychus? J. Am. Mosq. Cont. Assoc.*, 10: 380–384.

Rivière F., Kay, B.H. Klein, J.M. and Sechan Y., 1987. *Mesocyclops aspericornis* (Copepoda) and *Bacillus thuringiensis* var. *israelensis* for the biological control of *Aedes* and *Culex* vectors (Diptera : Culicidae) breeding in crab holes, tree holes, and artificial containers. *J. Med. Entomol.*, 24: 425–430.

Rowe, R.J., 1987. *The Dragonflies of New Zealand*. Auckland University Press, Auckland, New Zealand, 260 pp.

Rowe, D., 1998. Management trials to restore dwarf inanga show mosquitofish a threat to native fish. *Wat. Atmos.*, 6: 10–12.

Rupp, H.R., 1996. Adverse assessments of *Gambusia affinis*:an alternative view for mosquito control practitioners. *J. Am. Mosq. Cont. Assoc.*, 12: 155–166.

Santer, B., 1993. Potential importance of algae in the diet of adult *Cyclops vicinus*. *Freshwat. Biol.*, 30: 269–278.

Santos, L.U. and de Andrade, C.F., 1997. Survey of cyclopoids (Crustacea : Copepoda) in Brazil and preliminary screening of their potential as dengue vector predators. *Rev. Saude. Publica*, 31: 221–226.

Schwartz, S.S. and Hebert, P.D.N., 1982. A laboratory study of the feeding behavior of the rhabdocoel *Mesostoma ehrenbergii* on pond Cladocera. *Can. J. Zool.*, 60: 1305–1307.

Speczi´ar, A., 2004. Life history pattern and feeding ecology of the introduced eastern mosquito fish, *Gambusia holbrooki* in a thermal spa under temperate climate of Lake H´ev´iz, Hungary. *Hydrobiol.*, 522: 249–260.

Spencer, M., Blaustein, L. Schwartz, S.S. and Cohen, J.E., 1999. Species richness and the proportion of predatory animal species in temporary freshwater pools: Relationships with habitat size and permanence. *Ecol. Let.*, 2: 157–166.

Spielman A., Kitron U. and Pollack R.J., 1993. Time limitation and the role of research in the worldwide attempt to eradicate malaria. *J. Med. Entomol.*, 30: 6–19.

Stav, G., Blaustein, L. and Margalith, J., 1999. Experimental evidence for predation risk sensitive oviposition by a mosquito, *Culiseta longiareolata*. *Ecol. Entomol.*, 24: 202–207.

Stemberger, R.S., 1985. Prey selection by the copepod *Diacyclops thomasi*. *Oecologia*, 65: 492–497.

Sweeney, A.W. and Becnel, J.J., 1991. Potential of microsporidia for the biological control of mosquitoes. *Parasitol. Today*, 7: 217–220.

Torres-Estrada J.L., Rodriguez, M.H., Cruz-Lopez, L., Arredondo-Jimenez, J.I., 2001. Selective oviposition by *Aedes aegypti* (Diptera : culicidae) in response to *Mesocyclops longisetus* (Copepoda : Cyclopoidea) under laboratory and field conditions. *J. Med. Entomol.*, 38: 188–92.

Travis, J., Keen. W.H. and Juilianna, 1985. The role of relative body size in a predator-prey relatioship between dragonfly naiads and larval anurans. *Oikos*, 45: 59–65.

Vargas, M.J. and de Sostoa, A., 1996. Life history of *Gambusia holbrooki* (Pisces : Poeciliidae) in the Ebro Delta (NE Iberian Peninsula). *Hydrobiologia*, 341: 215–224.

Vasconcelos, A.W., Sleigh, A.C., Kay, B.H., Cabral, C.P., Aranjo, D.B., Ribiero, Z.M., Braga, P.H., and Cavalcante, J.S., 1992. Community use of copepods to control *Aedes aegypti* in Brazil. In: *Dengue: A Worldwide Problem, A Common Strategy, Proc. Internat. Cong. on Dengue and Aedes aegypti Community Based Control,* (Eds.) S.B. Halstead and H. Gomez-Dantes. Mexican Ministry of Health and Rockfeller Foundation, Mexico, pp. 139–144.

Walton, W.E., Tietze, N.S. and Mulla, M.S., 1990. Ecology of *Culex tarsalis* (Diptera : Culicidae): Factors influencing larval abundance in mesocosms in southern California. *J. Med. Entomol.*, 27: 57–67.

Webb, C. and Joss, J., 1997. Does predation by the fish *Gambusia holbrooki* (Atheriniformes : Poeciliidae) contribute to declining frog populations? *Austral. Zool.*, 30: 316–326.

Wickham, S.A., 1995a. *Cyclops* predation on ciliates: Species-specific differences and functional responses. *J. Plankton Res.*, 17: 1633–1646.

Wickham, S.A., 1995b. Trophic relations between cyclopoid copepods and ciliated protists: Complex interactions link the microbial and classic food webs. *Limnol. Oceanogr.*, 40: 1173–1181.

Wiggins, G.B., McKay R.J. and Smith, I.M., 1980. Evolutionary and ecological strategies of animals in annual temporary pools. *Arch. Hydrobiol./Suppl.*, 58: 97–206.

Williams, D.D., 1987. *The Ecology of Temporary Waters.* Croom Helm, London.

Williams, D.D., 1997. Temporary ponds and their invertebrate communities. *Aquat. Conserv. Mar. Freshwat. Ecosyst.,* 7: 105–117.

Williamson, C.E., 1980. The predatory behaviour of *Mesocyclops edx*: Predator preferences, Prey defences and starvation induced changes. *Limnol. Oceanogr.,* 25: 903–909.

Williamson, C.E., 1983. Behavioral interactions between a cyclopoid copepod predator and its prey. *J. Plankton Res.,* 5: 701–711.

Williamson, C.E., 1986. The swimming and feeding behavior of *Mesocyclops. Hydrobiologia,* 134: 11–19.

Wissinger, S.A., 1989. Seasonal variation in the intensity of competition and predation among dragonfly larvae. *Ecology,* 70: 1017–1027.

Woodward, B., 1983. Tadpole size and predation in the Chihuahuan Desert. *South West. Nat.,* 28: 470–471.

Wurtsbaugh, W. and Cech, J., 1983. Growth and activity in *Gambusia affinis* fry: Temperature and ration effects. *Trans. Am. Fish. Soc.,* 112: 653–660.

Zaret, T.M., 1980. *Predation and Freshwater Communities.* Yale Univ. Press, New Haven, 187 pp.

Zoppi De Roa, E., Gordon, E., Montiel, E., Delgado, L., Berti, J. and Ramos, S., 2002. Association of cyclopoid copepods with the habitat of the malaria vector *Anopheles aquasalis* in the Peninsula of Paria, Venezuela. *J. Am. Mosq. Cont. Assoc.,* 18: 47–51.

Chapter 19

Review of Eco-dynamics of the Floodplain Wetland (*Beel*) Ecosystem and Strategy for Enhancing the Fish Production

Utpal Bhaumik & T. Paria

Central Inland Fisheries Research Institute,
Barrackpore, Kolkata – 700 120, West Bengal

ABSTRACT

The new millennium has stepped with lots of hopes, promises and uncertainties. Apart from promoting aquaculture, the country will have to focus her attention to achieve optimum sustainable yield from the open water systems especially from floodplain wetlands, reservoirs etc. In India, floodplain are known as *chaurs and mauns* in Bihar, *beels, baors, jheels, and floodplain wetlands* in West Bengal, Assam and the Northeast Region. The wetlands associated with the floodplain of riverine systems specially those of the Ganga and the Brahamaputra, cover an extensive area of more that 2,00,000 ha. Water bodies with clay bottom are likely to have high turbidity while, rock basin, and those with sand, gravel and humus predominance to have low turbidity. The *beels*, which possess riverine connection, are victimised adversely towards input-output ratio due to continuous water exchange. Water temperature plays an important role in influencing the productivity of water bodies. The *beel* waters receive oxygen mainly from two sources; by absorption from atmosphere at surface level and intense photosynthetic activity of all the chlorophyll bearing lives within the *beels* ecosystems. pH below 4 or 5 severely restricts species diversity and above 10.0 indicates extreme eutrophic environment. Total alkalinity values are the resultant of the entire biological and chemical process taking place in the water body, as such it is also taken as rough index of productivity of water body. Nitrate unlike ammonia, phosphate or metal ions, moves freely through soil along with subsurface water. Ecologically phosphorus is the most critical single nutrient in the maintenance of aquatic productivity. Silicate content in pond water is of immense significance since it is the major nutrient for diatoms, one of the most important component of phytoplankton. In the soil of wetlands organic carbon acts as a source of energy for the microbes responsible for various biochemical processes, besides releasing nutrients including some trace elements. The factors which account for the seasonal variations of plankton are sunshine, water temperature, pH, nitrates and phosphates. Weed infestation problem is more severe in eastern part of the country where 40-70 per cent of water area is covered by weeds. The littoral characteristic was favourable for the growth of periphytonic organisms in the beels. Indo-Gangetic floodplain wetlands available in the are the native for a large variety of fresh water fish fauna which complete their life cycles therein. The fish production in wetland of West Bengal has been reported to be largely dependent on the management procedure adopted thereof. The *beels* inspite of moderate to high productivity capacity lack annual fish yield of desired level.

Introduction

The new millennium has stepped with lots of hopes, promises and uncertainties. The second half of twentieth century had many glorious moments in our country's economic development, especially on the agriculture front. The country is presently self sufficient for food. But the nation is passing through an era of *Population explosion* as evident from the observed population growth rate of 23.56 per cent for the period 1981-91. To keep pace with the situation, the country needs to double the fish production by 2020. The availability of protein rich food has been particularly constrained. Fish is recognized as the most important and easily digestible animal protein and with the available resources, can play a great role in fulfilling the protein requirement in future.

In spite of the phenomenal increase in fish production during last decade, the present per capita availability of fish is still only about 9.85 kg per annum against world average of 12.1 kg per annum and WHO recommendation of providing a minimum of 11.24 kg per capita annually. To fulfill the minimum nutritional requirement (11.24 kg per capita) for the country's projected population, with the assumption that 56 per cent of population would be including fish in their animal protein diet, the total annual fish requirement has been computed to be 12 million tonnes by 2020 A.D. (Gopakumar, 2000a). The above said scenario warrants a concerted effort to increase fish production from inland waters where the resources are vast and diverse with estimated potential of 4.5 million tonnes (Gopakumar,2000b).

Apart from promoting aquaculture, the country will have to focus her attention to achieve optimum sustainable yield from the open water systems especially from floodplain wetlands, reservoirs etc. The floodplain wetlands unlike other open water systems *viz.* Riverine system, Reservoir system, Estuarine system etc, by virtue of their productive potential as well as magnitude, constitute one of the frontline areas which is capable of contributing substantially to country's fish production to the tune of C. 0.02 million tonnes per year. A production rate upto 1,000 kg/ha/yr. is attainable from floodplain lakes when subjected to scientific management against production of 100 kg/ha/yr. under traditional management (Sinha, 1998; 2001). The term wetlands is used for such diverse habitats in different climatic zones of the earth that it is indeed difficult to define in simple terms (Hackson, 1981). Accordingly, wetlands have been defined variously in the past two decades. The definitions range from simple working definitions to highly technical ones.

The Ramsar Convention (1971) defined *Wetlands as areas of marsh, fen, peatland or water, whether natural or artificial, permanent or temporary, with water that is static or flowing, fresh, brackish or salt, including areas of marine water the depth of which at low tide does not exceed six meters.* Cowardin *et al,* (1979), on the other hand, used detailed scientific criteria to define wetlands and according to them: *Wetlands are lands transitional between terrestrial and aquatic systems where the water table is usually at or near the surface or the land is covered by shallow water and wetlands must have one or more the following three attributes (1) at least periodically, the land supports predominantly hydrophytes, (2) the substrate is predominantly hydric soil, and (3) the substrate is non-soil and is saturated with water or covered by shallow water at some time during the growing season of each year.*

In India, there is no single word equivalent to wetland. The marshes are known as *chaurs and mauns* in Bihar, *beels, baors, jheels, and floodplain wetlands* in West Bengal, Assam and the Northeast Region. Large shallow lakes are called *tals* in Uttar Pradesh and Madhya Pradesh while these are called *sars* and *dals* in Jammu and Kashmir. But wetlands have to be recognised and distinguished from other ecosystems by their ecological characteristics alone for their proper management (Sinha, 1997).

The wetlands associated with the floodplain of riverine systems specially those of the Ganga and the Brahamaputra, cover an extensive area of more that 2,00,000 ha. (Table 19.1).

Table 19.1: Distribution of Flood Plain Lakes in India

State	River Basins	Area (ha)
Arunachal Pradesh	Kameng, Subansiri, Siang, Dibang, Lohit, Dihing and Tirap	2,500
Assam	Brahamaputra and Barak	1,00,000
Bihar	Gandak and Kosi (Ganga)	40,000
Manipur	Iral, Imphal and Thoubal	16,500
Meghalaya	Someswari and Jinjiram	213
Tripura	Gomti,Manu and Khowai	500
West Bengal	Hooghly and Matlah (Ganga)	42,500
Total		2,02,213

Source: Sugunan, 1997; Yadav, 1989.

In West Bengal, floodplain wetlands locally known as *beels*, covering an area of 42,000 ha constitute one of the important fishery resources. The origin of the floodplain lakes is as Kaleidoscopic as the often changing tortuous of major rivers and their tributaries. The Brahmaputra and Barak basins in North Eastern region lie in a Zone of acute seismic activity. Frequent earthquakes due to crustal instability have induced local and sudden changes in the basement levels, resulting later in pertinent to water shed changes of the fluvial environment (Jhingran *et al.*, 1976; Das, 1983). This adverse feature coupled with heavy rainfall and cutting action of the stream meanders have resulted in the formation of either typical ox-bow type, lake like or true tectonic depressions in the region. Based on their connection with rivers, the *beels* are classified into two types, (a) Live or Open *beels*–Connected with the river either during monsoon season or throughout the year. The fishery is largely dependent on the rivers through which these are connected via channels. The monsoon floods render the channel active and provide passage for migrating carps which seek the shallower regions of the *beel* for refuge and breeding (b) Dead or Closed *beels*–are the water bodies which have lost connections with the main river generally due to strengthening of river embankment as part of flood control programme. The fishery constituted by miscellaneous fishes, air breathing fishes etc (Yadav, 1987a). The dynamics of water balance as affected by incursion of river water, varying degrees of precipitation and nature of catchment are highly complex and present contrasting pictures of the *beel* area and depth within a year. These water bodies expand and contract. Hydrologically the channel which brings and drains most of the inflow and outflow of *beels* having a riverine connection, principally acts as a *spillway* to limit retention of excess water from storage (Kumar, 1985). Flooding originates from 3 sources (1) Overspill from the river channel (2) Local rainfall and (3) Catchment area due to the flatness of the terrain, increase in volume are achieved by lateral expansion rather than by increase in depth and the water spreads slowly and diffusely outwaters, hampering in its progress by floodplain vegetation. The lakes thus, represent combination of lotic and lentic habitat becoming at times a natural lake ecosystem. However, it never arrives at that relative state because it is always subject to manipulation of inflow and outflow. The *beel* ecosystem is composed of two complementary phases, the aquatic phase and terrestrial phase which alternate seasonally in dominance. During the terrestrial phase, the exposed unmodified plain may be occupied by agriculture or used for grazing both which benefit the fishery by enriching the aquatic environment during flood. The long run may however, lead to cultural eutrophication of the ecosystem (Yadav, 1987b; Welcome, 1979).

In *beels*, where macrophytes are the main producers and the energy fixed by them is not utilised directly by the consumers, the best way to utilise the vast energy resource of detritus is through detritus chain. Studies have shown that the managed *beels* where the main path of energy flow is through detritus chain have resulted in better energy output as fish than the unmanaged *beels* where energy is utilised through grazing chain by unwanted miscellaneous species (Pathak, 1989).

In *beels*, the fishery is mainly dominated by miscellaneous species followed by major carps, catfishes and live fishes. The contribution of detritus feeders is generally poor. Most of the *beels* in West Bengal are culturable, the rights are vested with the Government but are leased out to Co-operative Societies for fishery exploitation. These Co-operatives in order to get remunerative prices, stock the *beels* with Indian Major Carps (IMC) which increases the production significantly. Main species encountered in the *beels* are *Catla catla, Labeo rohita, Cyprinus carpio, Hypophthalmichthys molitrix and Labeo calbasu* among Indian and Exotic carps. Catfishes are represented by *Wallago attu, Mystus sp.* Other groups present are murrels, feather-backs and air-breathing fishes. Usually small dug-out canoe, country boats, catamaran etc. as craft and drag net, gills net, cast and scoop nets as gear are used by the operatives for exploitation of fish from the *beels* (Vass, 1989a).

The resource management implies a basic understanding of the ecology of the *beels*, their productive functions, dynamics of fishery resources and optimum exploitation. Today, the value of wetlands is beginning to be recognised and they are being aimed to be conserved, restored and in some cases created in many countries. The 75 plus countries adhering to the Ramsar convention have agreed to organise their planning systems so as to make wise use of the wetlands. Prior to Rio Conference (1992) with its particular focus on sustainability, the parties to the Ramsar Convention adopted the definition *The wise use of wetlands is their sustainable utilisation for the benefit of humankind in a way compatible with the maintenance of the natural properties of the ecosystem.* The Conference noted that sustainable utilisation is *human use of a wetland so that it may yield the greatest continuous benefits to the present generation whilst maintaining its potential to meet the needs and aspirations of future generations* (Hollis, 1995). The present paper deals the appraisal of the above stated facts in different *beels* of both open and closed system and ecodynamic characterization of the *beels*, basic information to take up developmental programmes for the *beels*.

Biotic and Abiotic Parameters

Maintenance of a healthy aquatic environment and production of sufficient fish food organisms in the water bodies are two factors of primary importance for boosting fish production. Production and growth of fish food organisms in a water body are directly dependent on the availability of nutrient elements in water and soil and also on some physico-chemical parameters relating the them. A comprehensive assessment of the tropic relationship governing the biological productivity of water bodies is difficult without any knowledge of aboitic and boitic factors. Seasonal fluctuations of the different ecological parameters should therefore, be studied closely in order to understand the process of aquatic productivity.

The lakes are the disconnected remnants of tributaries or distributaries of riverine networks. The lakes called *beels* or *baors* in West Bengal are the components of the Gangatic floodplain wetland system and constitute an important resource for inland fisheries of the State.

Water Quality Characteristics

Physical Attributes

Depth

In morphometry most frequently used terms are maximum depth and length of the shore line. The lakes in tropical and arid zones ranges from shallow, very eutrophic nature to the deep, less productive ones (Hackson, 1981). Certain European limnologists reveal that average depth is the factor which determines whether a lake is eutrophic or oligotrophic, computing average depth as the quotient of volume of the lake over the area of the lake. The seasonal water column of the beel ecosystems are shallow water bodies and can be classified as littoral lake with gradual slope and depth not exceeding 6 metre (Welch, 1952). Mukhopadhyay (1997a) indicated that majority of the beels in West Bengal is vulnerable to high water level fluctuations. Low rainfall causes seasons water balance problems for the closed beels. Sugunan *et al.* (2000) reports in shallow *beels*, the whole water body gets heated up rapidly, thereby increasing the spread of chemical and bio-chemical reactions. Dey and Goswami (1982) indicated shallow *beels* also facilitate dense growth of rooted aquatic macrophytes which compete for nutrients and space with phytoplankters and are not part of the autotrophic food chain.

Important physico-chemical parameters of water influencing fish production in fresh water bodies.

Transparency

Water bodies with clay bottoms are likely to have high turbidity while the basins with rock, sand, gravel or humus are likely to have low turbidity. Transparency is an important limiting factor in productivity of a water body. It varies greatly with the nature of the basin, degree of exposure, nature of inflow and sediment etc. Water bodies with clay bottom are likely to have high turbidity while, rock basin, and those with sand, gravel and humus predominance to have low turbidity (Nath, 1997). Other factors affecting transparency are suspended clay and silt, particulate organic matter, dispersion of plankton organisms and the pigments caused by decomposition of organic matter (Kielhorn, 1952; Mc Combic, 1953; Bamforth, 1958; Michael, 1969). Excessive turbidity has a pronounced effect in confining daily heat gains to the surface layer of water (Smith, 1934). Turbidity fluctuations occurring during rainy season by the heavy and deep turbulence is known to prevent development of phytoplankton and also zooplankton with feed on them (Jermolajev, 1958; Michael, 1969). Other factors affecting transparency are suspended clay and silt, particulate organic matter, dispersed planktonic organism and pigments produced by decomposition of organic matter (Kielhorn, 1952; McCombic, 1953; Bamforth, 1958; Michael, 1969). Excessive turbidity has a pronounced effect in confining daily heat gains to the surface layer of water (Smith, 1934). In the present study, the *beels* varied in respect of water transparency mostly influenced by the growth of planktonic organisms and suspended organic matter. The transparency in general was more during monsoon due to the dilution effect of the replenished volume of water. Various workers have also reported similar seasonal fluctuations in lake water transparency (Jermolajev, 1958; Michael, 1969; Kumar, 1985; Nath, 1999). Aquatic organisms vary widely in their relation to different degree of turbidity. Without doubt, many organisms smother in prolonged condition of very high turbidity (Needham and Lloyd, 1930;Nath,2001). Hepher (1962a) found that fertilization of fish ponds with excessive amount of fertilizers increased production in the upper layer of water where favourable light conditions existed but decreased production in the lower layers due to poor transparency.

Flow

The *beels*, which possess riverine connection, are victimised adversely torwards input-output ratio due to continuous water exchange. While this adversely affects the biological productivity of the ecosystem, it helps in delaying eutrophication process. Rapid water renewal also helps in breaking the thermal stratification, if any, in deep *beels*, which is beneficial for nutrient recycling and gasseous exchange. At the same time, continuous water flow does not allow the plankton species to stabilize, resulting in lower plankton density and primary production rates in such beels. In addition, rapid increase in water levels during monsoon months causes death and decay of submerged and marginal macrophytes in all the *beels*, which is beneficial for recycling of nutrients locked up as macrophyte biomass. However, the closed *beels* receive localised flowing condition at the receiving points of catchment run-off during the rainy days (Bhaumik, 2000). Welch (1952) mentioned the works of various workers reporting what appeared to be physiological effects of turbulence, such as suspension of growth in certain algae, acceleration of movements in *Oscillatoria*, alteration in number of mass in *Daphnia* and decrease in locomotion and ingestion in *Paramoecium*.

Temperature

Water temperature plays an important role in influencing the productivity of water bodies. It varies widely depending on the climatic condition of the place where the water bodies are located and it also undergoes wide diuranal fluctuation. Banerjea (1967) stated that of all the physical factors, temperature is most essential for photosynthetic activity which in turn is basic to productivity. In fact, no other single factor has been linked with so many direct effects. Though not much deep, the *beels* irrespective of open or closed were thermally stratified. The variation in surface water temperature was maximum in winter spell when the water was least distributed and the temperature fluctuation within the day was also maximum. Prevalence of such thermoclinic conditions in small tropical lake of some Asian countries (Java, Sumatra and elsewhere) have been reported by Ruttner, (1931). Kumar (1985) recorded stratified temperature in a beel ecosystem of Nadia district in West Bengal. Dwivedi and Chonder (1977) observed moderate stratification of temperature in Keetham-lake Uttar Pradesh. William and Murdococh

(1966) reported that when temperature was high, production was high in Beafort Channel. Bhowmik,(1988) reported that maximum and minimum temperature in *beels* and *baors* of West Bengal varied from 17.5 to 32.0 °C. The more significant effect of higher temperature is the increased rate of biochemical activity of the microbes so that the release of nutrients by decomposition of organic matter at bottom is more at higher temperature with consequent increase in the nutrient status of water (Ganapati and Sreenivasan, 1956). Plecznake and Ozimeck (1976); Dale and Gillespic (1976) and Rai and Dutta Munshi (1979) have also reported that the presence of macrophytes profoundly influence the water temperature. The present study also confirms the same. According to Banerjea (1967); Jhingran and Pathak,(1987) and Dey and Bhattacherjee (1995) in water bodies with high organic contents in bottom mud, large scale mortality takes place in summer months especially after a shower or cold wind. This happens due to overturn of thermally stratified layers, so that bottom layer of anaerobic decomposition zone with reducing gases, distributes itself throughout the volume of water and even the relatively oxygen rich surface layer of water suffer oxygen depletion.

Bhowmik (1968) and Mandal (1972) observed a range of floodplain wetland from 18-33 °C in a water body at Kalyani, from 18-41.5 °C in two water bodies at Burdwan, but Chako and Krishnamoorthy (1954) from 34.2 °C to 42.8 °C in water body in Madras. The highest temperature was recorded during July and August (Dewan, 1973) or in April and May (Ganapati, 1941a, 1941b; Chako and Ganapati, 1949; Darve and Ball, 1961), while the lowest temperature was recorded in the month of January (Dewan, 1973) or during November-December (Ganapati, 1941a, 1941b; Chako and Ganapati, 1949; Darve and Ball, 1961). A diurnal variation of 1.8 °C in the winter month of November and 9.9 °C in the summer month of April was observed in a fish farm in Madras (Ganapati and Sreenivasan, 1956). Production of plankton has been found to be directly related with temperature (Eddy, 1934; Bhowmik, 1968; Mandal, 1972) although other (Imvebore, 1967; Das and Srivastava, 1956; Chako and Krishnamoorthy, 1954) did not find any correlation between the variation in the density of algal population and that in the temperature of water. The higher productivity of tropical waters has been attributed partly to high water temperature (Worthington, 1943). He found an average increase in carp yield of 22 kg/ha for 1 °C rise in temperature, while Backiel and Stagman (1967) observed that in increase of average temperature of 1 °C brought about an increase in the production of carps by about 50 kg/ha. The daily food consumption by carps was also found to be greater at 23-27°C than at 16-18°C.

Dissolved Oxygen

Detailed study of dissolved oxygen along with a knowledge of turbidity and colour of water can provide adequate information on water quality (Hutchinson, 1957). Of all the chemical parameters, dissolved oxygen is considered as prime important factor for regulating metabolic process of plant and animal community as well as indicating the water quality. Hutchinson (1957) has remarked that a series of oxygen determinations along with a knowledge of turbidity and colour of water could provide more information about the nature of water than any other chemical data. The *beel* waters receive oxygen mainly from two sources; by absorption from atmosphere at surface level and intense photosynthetic activity of all the chlorophyll bearing lives within the *beels* ecosystems. The addition of atmospheric oxygen to the beels depends on the number of factors such as temperature, water movement, wind velocity etc. Phenomenally, the *beel* waters were more oxygenated during the photoperiod in compared to the non-photoperiod of the day. Similar diurnal variation in oxygen content of lake water have been reported by various workers (Hutchinson, 1957; Bhowmik, 1968; Jana, 1973) and the ill effects of complete deoxygenation on fish survival have also been recorded by (Moore, 1942). Bhaumik (2000) stated that leaving aside the diurnal pulse, largely the reflection of temperature and photosynthetic-respiration relationship and dissolved oxygen showed stratified distribution in water column of the beel ecosystems. The difference in oxygen concentration was also related to the seasonal variation; showing the maximum in monsoon (1.1 to 3.0 ppm) followed winter (0.4 to 2.7 ppm) and summer (0.2 to 2.6 ppm). Reason for the maximum stratification of oxygen in monsoon may be attributed to the high rate of surface mixing of atmospheric oxygen due to showering of rain drop (Banerjea, 1967). Oxygen content observed to be poor during the period of the high temperature such as summer season (Pearsall, 1921; 1931;Bhowmik, 1968;1988; Chakrabarti, 1980and Sugunan et al., 2000). The D O values in the bottom layer of the *beels* indicate discrepancy and depletion of oxygen characteristic of eutrophic waters; reading as low 0.2 ppm were recorded during early morning of summer. Kumar (1985) also reported similar observation. Dehadrai (1972), reported oxygen depletion in shallow eutrophic waters as a common feature in swamps of North Bihar. There is considerable loss of oxygen at the soil water inter phase due to accumulation of organic matter. Gas bubbles arising from the sediments remove oxygen from the water. Mainly caused by the concurrence of calm hot weather with the massive organic aggregate, the dissolved oxygen have been reported to be alarmingly low level in eutrophic water (Dehadrai, 1972). The short and long term variations in dissolved oxygen in lakes give a good measure of their trophic state (Goswami, 1985). Oligotrophic waters show variation from saturation, while eutrophic ones may range from virtual to 250 per cent saturation. Many investigators studied the changes in dissolved oxygen content of water of the ponds during different seasons at different places and during different hours of the day and observed wide fluctuations ranging from 3.2 to 12.5 ppm in the water bodies at Kalyani, Nadia, West Bengal (Bhowmik, 1968), from 4.0 to 12.0 ppm at Bakreswar, Birbhum, West Bengal (Jana, 1973), and from 6.5 to 18.1 ppm at Burdwan, West Bengal (Chakraborti, 1980). In very small bodies of water, the range of variation is found to be very high. Dissolved oxygen content in a water bodies is mainly dependent on temperature and rate of photosynthesis. Oxygen content is poor during periods of high temperature such as summer season (Pearsall, 1921; Bhowmik, 1968 and Chakraborti, 1980), while it is high in winter due to low water temperature (Reid, 1961; Moitra and Bhattacharya, 1965 and Chakraborti, 1980). Minimum dissolved oxygen is noted during the early morning and maximum in the afternoon. Such increase of dissolved oxygen is attributed to photosynthetic activity (Bamforth, 1962 and Michael, 1969). Attempts were made to correlate the oxygen content with the volume of plankton of the water body, and it was observed that the phytoplankton peak corresponds to the high oxygen values (Das and Srivastava, 1956; Moitra and Bhattacharya, 1965 and Mandal, 1972), while the zooplankton peaks are associated with low oxygen values (Das and Srivastava, 1956). Studies on the optimum range of dissolved oxygen for the survival and growth of fish reveal that dissolved oxygen below 3.0 ppm indicates the possibility of asphyxia due to oxygen deficiency and a minimum of 5.0 ppm of oxygen is required for a productive fish pond (Banerjea, 1967). On the other hand, very high concentration of dissolved oxygen may be lethal to fish fry during the rearing of spawn in nursery ponds (Alikunhi, 1952).

Water pH

The pH negative log of hydrogen–ion concentration controls the chemical state of nutrients in the water bodies. Changes in pH influence important plant nutrients such as phosphate, ammonia, iron and trace metals. pH below 4 or 5 severely restricts species diversity and above 10.0 indicates extreme eutrophic environment. In lakes pH is regarded important as (*i*) limiting and (*ii*) as an index of general environment condition. In the present investigation the hydrogen-ion concentration in surface water of the beels was 8.0 and above excepting in few occasions. However, pH at the bottom of the water column was marginally different with decreased concentration of hydrogen ion. The pH values of water in the afternoon are almost always higher than those in the morning due to photosynthesis and respiratory activities of various organisms in the water (Dewan, 1973; Michael, 1969). It generally tends to maintain a nearly constant value throughout the year and the seasonal fluctuation is narrow (Kato, 1941; Heron, 1961; Michael, 1969). Several workers (Chakravorty *et al.*, 1959; David *et al.*, 1969; Miller and Rabe, 1969; Dewan, 1973) have, however, reported highest pH values during summer and lowest during monsoon (Chakravorty *et. al.*, 1959; Ray *et al.*, 1966; David *et al.*, 1969; Dewan, 1973). A slightly alkaline water pH was optimum not only for the fish, but also fish food organisms. Thus, Michael (1969) observed that when pH ranged between 7.3 and 8.4 the water provided optimum conditions for the growth of plankton. It is important to note that there is some evidence of different species of a taxonomic group having an individual pH range. The present study bears the agreement of alkaline pH with the study of Bhowmik (1988) where the pH value of the *beels* and *baors* of West Bengal was recorded between 6.8 and 9.1.

A number of investigators (Nees, 1946, Hutchinson, 1957) studied the pH of the water body in relation to fish growth and observed that largest yields were obtained from water which was just on the alkaline side of neutrality between pH 7.0 and 8.0. According to Banerjea (1967), water with an almost neutral reaction with pH 6.5 to 7.5 is best suited for fish production and an average production is expected in the pH range of 7.5 to 8.5, but Sahai and

Sinha (1969) found no relationship of pH of water with pond or lake productivity. The pH values of water in the afternoon are almost always higher than those in the morning due to photosynthesis and respiratory activities of various organisms in the water (Dewan, 1973 and Michael, 1969). It generally tends to maintain a nearly constant value throughout the year and the seaonsal fluctuation is quite narrow (Heron, 1961 and Michael, 1969). Several workers (Chakravorty *et al.*, 1959; David *et al.*, 1969; Miller and Rabe, 1969 and Dewan, 1973) have, however, reported highest values during summer and lowest during monsoon (Chakravorty *et al.*, 1959; Ray *et al.*, 1966; David *et al.*, 1969 and Dewan, 1973). Water pH in a water bodies might be dependent on its plankton content. Thus, many investigators (Das and Srivastava, 1956; Moitra and Bhattacharya, 1965; Bhowmik, 1968 and Mandal, 1972) reported that high pH was related to heavy blooms of phytoplankton, while low pH indicated a rise in zooplankton. A slightly alkaline water pH was optimum not only for the fish, but also for fish food organisms. Thus, Michael (1969) observed that when pH ranged between 7.3 and 8.4, the water provided optimum conditions for the growth of plankton.

Total Alkalinity

Alkalinity or acid combining capacity of natural freshwater systems is mainly caused by carbonates and bicarbonates of calcium and magnesium, calcium being the major constituent. These, along with dissolved carbon dioxide in water form an equilibrium system, which is of primary importance in the ecology of the environment. In natural water body the total alkalinity value is inversely related to the water level (Mathew, 1969), and a perennial pond having nearly same water level may be alkalitropic as was found in Cuttack (Saha *et al.*, 1971). Since total alkalinity values are the resultant of the entire biological and chemical process taking place in the water body, as such it is also taken as rough index of productivity of water body (Laal, 1981). Alkalinity or acid combining capacity of natural freshwater bodies is mainly caused by carbonate and bicarbonate of calcium and magnesium, calcium being the major constituent. These along with dissolved carbon dioxide in water form an equilibrium system, which is of primary importance in ecology of environment. In natural water body the total alkalinity values is inversely related to the water level (Mathew, 1969). The high alkalinity of the *beel* water was because of the present of salts generated through the death and decay of the macrophytes, benthic organisms and also the plankton. Obviously the high alkalinity value was recorded in *beels* infested with high density of macrophyte-associated fauna and benthic biomass. Many workers observed that decrease in water level of a pond due to evaporation caused steady increase in alkalinity (Hazelwood and Parker, 1961 and Michael, 1969), while rainfall used a lowering of total alkalinity of a water body by increasing the water level (Bamforth, 1958). Mandal (1972) and Sahai and Sinha (1969) studied alkalinity in the fish ponds at Burdwan and in the Ramgarh lake in Gorakhpur, U.P. and observed a range from 34.0 to 205.0 ppm and from 38.0 to 125.0 ppm respectively. They also observed a direct relationship between alkalinity and pH. Bhowmik (1968) observed that the methyl orange alkalinity followed a unimodal pattern of distribution in the Kalyani fish farm. Different workers at different places have observed maximum productivity for fish at different alkalinities, such as more than 50.0 ppm by Ohle (1938), more than 90.0 ppm by Moyle (1946), more than 100.0 ppm by Alikunhi (1957), from 200.0 to 500.0 ppm by Schaperclaus (1933), from 80.0 to 150.0 ppm by Michael (1964), and more than 50.0 ppm by Banerjea (1967). A waterbody is usually unproductive or low productive when the total alkalinity becomes less than 10.0 ppm (Ohle, 1938, Banerjea, 1967). According to Moyle (1946) a pond is low to medium productive when the total alkalinity ranges from 20.0–40.0 ppm and medium to high productive when it ranges from 40.0 to 90.0 ppm. Carlander (1955) found a linear relationship between alkalinity and logarithm of fish production.

Free CO_2

Many have studied the fluctuations of free CO_2 of water in ponds and lakes at different places and during different seasons in a year as well as at different hours in a day and observed a range from 17 to 39 ppm in lakes at Lucknow (Das and Srivastava, 1956), from nil to 10 ppm in water bodies at Kalyani (Bhowmik, 1968), and from 0.4 to 15 ppm in water bodies at Burdwan (Mandal, 1972). Free CO_2 plays a very important role in influencing the productivity of the water bodies. Many workers have studied the fluctuations of free CO_2 of water in ponds and lakes at different places and during different seasons in a year as well as at different hours in a day and observed a range 17 to 39 ppm in lakes at Lucknow (Das and Srivastava, 1956;1965), from nil to 10 ppm water body at Kalyani (Bhowmik,

1968), and from 0.4 to 15 ppm in water body at Burdwan (Mandal, 1972). CO_2 is absolutely necessary for the photosynthesis, free CO_2 content may sharply go up, and may even disappear. Thus, free CO_2 was absent in Amaravathy reservoir, Madras (Sreenivasan, 1964), and in Ramgarh lake, Gorakhpur, U.P. (Sahai and Sinha, 1969) during certain periods. Such absence of free CO_2 in water was found to be related with presence of heavy phytoplankton populations (Michael, 1969). In some cases the absence of free CO_2 in water may also be due to reaction with carbonate present in large quantity in the water (Sahai and Sinha, 1969). In the present study, free CO_2 content in surface water of the beels was always below detection levels and the gas was examined to be present at the bottom layers in moderate to low concentration. The pronounce absence of the free CO_2 at the subsurface level of the water was in confirmation of the observation made by Reid (1961), who reported that at pH 8 and above the free CO_2 is usually absent. The available free CO_2 in the column water was indicative of photosynthetic activities and bio-respiration in the environment.

CO_2 is absolutely necessary for the photosynthesis by phytoplankton. During the periods of intense photosynthesis, free CO_2 content may fall sharply, and may even disappear. Thus, free CO_2 was absent in Amaravathy reservoir, Madras (Sreenivasan, 1964), and in Ramgarh lake, Gorakhpur, U.P. (Sahai and Sinha, 1969) during certain periods. Such absence of free CO_2 in water was found to be related with the presence of heavy phytoplankton populations (Michael, 1969). In some cases the absence of free CO_2 in water may also be due to reaction with carbonate present in large quantity in the water (Sahai and Sinha, 1969). Some workers (Moyle, 1946 and Michael, 1969) did not find any correlation with productivity while others (Wolny, 1967) noted that CO_2 was important and might limit phytoplankton development. Wolny (1967) observed that free CO_2 plays an important role in fish ponds by producing calcium bicarbonate from calcium carbonate and by maintaining the pH of the water nearly constant through the buffer system of CO_2–$CaHCO_3$-$CaCO_3$. Although presence of free CO_2 in fish water body is desirable, but its presence at a level higher than 15 ppm for long duration is detrimental to fishes (Swingle, 1967). High CO_2 content appears to be more toxic in presence of low oxygen content (Welch, 1952). During monsoon months the content of free CO_2 was found to be high due to poor photosynthesis because of cloudy weather and also due to the release fo CO_2 as a result of decomposition of organic matter and minimum value was reported during April, May or September presumably due to high temperature (Dewan, 1973).

Specific Conductivity

Total contribution of dissolved solids is considered as a useful parameter in describing the chemical density as a fitness factor and as a general increase of edaphic relationship that contribute to the productivity of water (Jhingran, 1991). Electrical conductivity, which gives the amount of ionised material, is an important measure of total dissolved solid. The conductivity of an electrolyte in solution is directly proportional to the ionic strength of that solution and the total conductivity is the sum of the several conductivity resulting from various ionizable salts present. According to Welch,(1948) the other things being equal, the richer a body of electrolytes the greater its biological productivity. Sreenivasan (1967) has reported an optimum range as 250–400 μmhos/cm^2 and opined that specific conductivity above 400 μmhos/cm^2 do not limit or favour productivity.

Water Nitrogen

As constituents of protein, nitrogen occupies a highly important place in aquatic ecosystem. Nitrate unlike ammonia, phosphate or metal ions, moves freely through soil along with subsurface water. It is most highly oxidised form of nitrogen and is usally the most abundant form of combined inorganic nitrogen in lakes and streams. Large number of investigators have monitored the fluctuations of nitrate content of water in ponds and lakes ad different places and during different seasons in a year and observed a direct relationship between it any phytoplankton (Bhowmik, 1968, Saha et al., 1971, Mandal, 1972, Chakravorty et al., 1959). Thus, high concentration of nitrate and phosphate in a fish pond might be one of the causes of permanent bloom of phytoplankton such as *Microcystis sp.* (George, 1962). Many investigators have monitored the fluctuation of nitrate content in water bodies at different places and during different seasons in a year and observed a direct relationship between nitrate and phytoplankton population (Chakrabarti et al., 1959; Bhowmik, 1968; Saha et al., 1971; Mandal, 1972). By and large, the systems were comparatively richer in nitrate value during summer and winter seasons, while the monsoon concentration of the nutrient was comparatively lower because of the dilution effect. The fluctuation trend in nitrate level indicated mesotrophic to eutrophic condition of the beels (Goldman and Horne, 1983).

Several investigators have studied the reductions of nitrate to ammonia or free nitrogen. Nitrates are reduced to ammonia when the redox potential attains a value of +0.35 V but they are reduced to free nitrogen by denitrifying bacteria when the dissolved oxygen concentration in the water falls to a very low value (Hepher, 1952). Chu (1943) observed that for optimal growth of plankton, nitrogen should range between 0.3 to 1.3 ppm, while Banerjea (1967) opined that for a productive pond it should be above 0.2 ppm.

Water Phosphate

Ecologically phosphorus is the most critical single nutrient in the maintenance of aquatic productivity. The main supply of phosphorus in water body comes from leaching of soils of the catchment area by rains (Swingle, 1947; Woodwansee, 1958; Heron, 1961). Primary production of water body has been found to be a function of water soluble inorganic phosphorus when the concentration of other essential nutrients are in their optimum range (Das and Dehadrai, 1986). The summer concentration of this nutrient was always higher in compared to other seasons. The minimum level of phosphate was recorded in monsoon season in most of the *beel* ecosystems (Bhaumik, 2000). The phosphate cycle of the *beels* was in correlation with the dissolved oxygen and known to play important role in controlling the rate of phosphorus release from the sediment to the photic zone (Munwar, 1970; Goldman, 1972). Hiclking (1962) stated that it was unlikely to have a pond where application of phosphate would not beneficial. Both phosphorus and nitrogen occur in natural waters in very small quantities which are far below the upper limit for optimal growth of plankton, and often do not reach lower optimal concentration (Chu, 1943). Therefore, maximum phosphate concentration in pond water is observed during the rainy season (Michael, 1964). When phosphatic fertilisers are added to a water body, the water is richer in phosphate, but within a short period, the phosphate concentration is reduced almost to the low initial value, presumably due to its immediate uptake by phytoplankton (Smith, 1945; Hutchinson and Bowen, 1950). Riglar (1956), using radioactive phosphorus (p^{32}) showed that when phosphates are added at low doses to pond water, over 95 per cent of it is taken up by plankton, but when applied in large quantities a portion of it is absorbed by phytoplankton whereas the excess quantity may be fixed as iron and aluminium phosphate or absorbed on the ferric hydroxide gel of the bottom soil if it is acidic (Ohle, 1937). But when the total alkalinity or calcium in the pond water is moderately high, the phosphate may be precipitated as calcium phosphate (Matida, 1956, Hepher, 1958). Thus, fixation of phosphate is minimum if the soil is of neutral reaction (Banerjea and Mandal, 1965). Most of the investigators (Chakraborty *et al.*, 1959, Das and Srivastava, 1956; Bhowmik, 1968; Moitra and Mandal, 1970 and Mandal, 1972) observed a direct relationship between phosphate concentration plankton production. Moyle (1946) studied a large number of lakes and ponds and observed the phosphorus concentration range (a) 0-0.02 ppm P-as low productive, (b) 0.02-0.05 ppm P-as fairly productive, (c) 0.05-0.10 ppm P-as good, and (d) 0.10-0.20 ppm P-as very good productive whereas Banerjea (1967) observed the phosphorus concentration range (a) 0-0.05 ppm P_2O_5-as poor, (b) 0.05-0.20 ppm P_2O_5-as medium to high, and (c) above 0.2 ppm P_2O_5 as highly productive.

Silicate

Silicate content in pond water is of immense significance since it is the major nutrient for diatoms, one of the most important component of phytoplankton. Although silica is a common mineral largely occurring in the earth's surface, the silicate concentration in freshwaters is not generally high. The reactive silicate is probably the only form available for the planktonic growth. Silica cycle is very different form the cycle of nitrogen, phosphorus and other nutrients. Although silica is common mineral largely in the earth's surface, the silicate concentration in freshwater is not generally high; it ranges between nil and 7.5 ppm (Hutchinson, 1957). Many workers (Bhowmik, 1968; Mathew, 1969; Chakrabarti, 1980; Das and Srivastava, 1956) have observed direct relationship of between silicate content and diatom population in the water body. However, the silicate level much above the limiting concentration of 0.5 ppm is favourable for planktonic production (Bhowmik, 1968; Jana, 1973; Mandal, 1972 and Chakrobarti, 1960; Goldman, 1972).It ranges between nil to 7.5 ppm (Hutchinson, 1957). The concentration of silicate in pond water is believed to be controlled by redox potential through reduction of ferric silicate in the surface layer of bottom mud (Jhingran, 1977). Many (Bhowmik, 1968; Mathew, 1969; Chakrabarti, 1980 and Das and Srivastava, 1956) have observed a direct relationship between silicate content and diatom population in the water body. Many investigators have also studied the fluctuations of silicate of the water body in ponds and lakes at different places and during different seasons in a year and observed a range from 3.1 to 7.8 ppm in the pond at Kalyani (Bhowmik, 1968), from 6.0 to 16.0 ppm in the pond at Bakreswar, Birbhum (Jana, 1973), from 4.2 to 15.5 ppm in the ponds at Burdwan (Mandal, 1972), from 5.0 to 10.2 ppm in the pond at Burdwan (Chakrabarti, 1980).

Important Characteristics of Bottom Soils Influencing Fish Production in Freshwater Bodies

Soil Texture

The bottom soil composition of a *beel* is the resultant contribution of various factors and for this reason great variation exists. Even within a restricted area, two beels may differ significantly in bottom type and bottom associated feature also. Eventually, a large number of beels have been studied by many workers in respect of their development and evaluation of the basin together with a succession of biologically different association of organisms. The modification of shore water line and depth is inherent in beels as a result of dynamics operating within the wetland and processes outside the wetland.

Chemical Attributes

Soil pH

The pH of bottom soil influences the chemical composition of the water and hence, the productivity of the aquatic system. The soil pH was always acidic to neutral. This is indicative of productive nature of the system. Banerjea (1967) was of the opinion that the soil reaction in the neutral range (pH 6.5 to 7.5), was most favourable, whereas, that in the moderately acid (pH 5.5 to 6.5) or in moderately alkaline (pH 7.5 to 8.5) average in terms of productivity. Schaperclaus (1933) observed that with a markedly acid soil, water is deprived of adequate amount of calcium and magnesium resulting in short supply of nutritional salts. Lakhsmanan *et al.* (1967) observed low fish growth in Assam with soil pH ranging between 4.8 and 5.7.

The pH of bottom soil influences the chemical composition of the water *vis a vis* and the productivity of the water body. Schaeperclaus (1933) observed that with a markedly acid soil the water would be deprived of adequate amount of calcium and magnesium and the nutritional salts would be in short supply. Lakshmanan *et al.* (1967) observed poor fish growth in some ponds in Assam with soil pHranging between 4.8 and 5.7. Banerjea (1967) was of the opinion that the soil reaction in the neutral range (pH 6.5 to 7.5) was most favourable, whereas that in the moderately acid (pH 5.5 to 6.5) or in the moderately alkaline (pH7.5 to 8.5) range is average in terms of productivity. The pH of the bottom mud may fluctuate during the different times in a year and also undergo changes due to manuring and fertilization. Mandal (1972) observed a variation in soil pH from 6.1 to 7.3 in a fish pond at Burdwan during the different seasons in a year. Saha *et al.* (1971) noted a decrease both in soil as well as water pH due to fertilization of ponds with heavy doses of ammonium sulphate.

Available Nitrogen

Nitrogen in aquatic system is present in mostly inorganic form, the fraction present as amino acids, peptides and easily decomposable proteins is called available nitrogen. Apart from combined nitrogen and phosphorus, and for diatoms silica, there is little likelihood of other nutrients becoming limiting for the phytoplankton production in ponds or small lakes (Hutchinson, 1944). Phosphorus tends to be the limiting factor in winter, while nitrate tends to be so in summer (Thomas, 1962). Addition of nitrogen and phosphorus to shallow ponds increased the abundance of plankton and bottom fauna (Mc Intire and Bond, 1962). Many investigators have monitored the fluctuations of nitrate content of bottom soil in ponds at different places and during different seasons in the year and observed a unimodal pattern with a peak during late monsoon (Bhowmik, 1968) or during autumn (Chakravorti, 1980). Attempts have been made to correlate fish production in a pond with soil nitrogen. In small culture ponds, fish production is found to be closely correlated with fractions of both nitrogen and phosphorus in the pond soil rather than in the pond water. Banerjea (1967) classified the fresh water bodies under five categories on the basis of available nitrogen content of the soil. He observed that fish production was generally poor in water bodies having available nitrogen content below 25 m/100 g soil, water bodies having available nitrogen content in the range 50-75 mg/100 g soil appeared to be more favourable for fish production.

Available Phosphorus

Available phosphate content in bottom mud is considered to be more important than the total phosphate (Jhingran, 1977a) because of the fact that PO_4^{-3} ion in soil forms insoluble compounds with iron and aluminum under acidic and with calcium under alkaline condition, rendering the phosphate ion unavailable to water phase. Banerjea (1967) classified the Indian fish ponds into three categories of productivity on the basis of available phosphate content (expressed as P_2O_5) in the bottom soil, namely poor productive (>30 ppm), average productive (30-60 ppm) and productive (>60 ppm). According to Sreenivasan (1967), pond soils having available phosphorus less than 0.5 ppm were poorly productive (<500 kg fish/ha): those recording values between 0.5 to 1.0 ppm were moderately productive (500-1000 kg of fish/ha), and those having values higher than 100 ppm were highly productive (>1000 kg of fish/ha. Many investigators have studied the effect of phosphate fertilization if fish ponds at different places and observed the striking effects of it on freshwater pond fish yield (Dendy, 1963). The available phosphorus in the bottom is considered to be most critical element in productivity and to be more important than total phosphate (Chang and Kaclspm, 1957;Jhingran, 1977). Because of the fact PO_4^{-3} ion in soil form soluble compound with iron and aluminium and with calcium under alkaline condition, renders the phosphate ion unavailable to water phase. According to Sreenivasan (1967a; 1967b) soil having available phosphorus less than 0.5 ppm is poorly productive; between 0.5 and 1.0 ppm moderately productive and higher than 100 ppm highly productive. Sugunan *et al.* (2000) reported that in contrast to other nutrient parameters, available phosphorus values were lowest in closed and weed choked *beels* (traces to 3.18 mg/100 g of soil), higher in closed but moderately weed infested *beels* (traces to 7.6 mg/100 g of soil), and highest in open one (traces to 10.08 mg/100 g of soil). This observation is based on the study made on large number of *beels* of West Bengal.

The experiments carried out at the Tropical Fish Culture Research Institute, Malacca showed a linear relationship between phosphate fertilization and fish production (Hickling, 1962;Prowse,

1968). Super phosphate was found to be the most useful fertilizer for freshwater ponds in Taiwan, where one kg of P_2O_5 produced 10 kg of fish (Lin and Chen, 1967). Much work has already been done on the release and fixation of phosphates under different conditions in fish ponds. The mud in very productive fish ponds remains mainly in a reduced state owing to very rapid mineralization of organic matter together with a high uptake of oxygen. In such cases, the phosphorus added to the fish ponds is very likely to be made unavailable due to the formation of insoluble Fe_2PO_4 (Golterman, 1967) or absorption by Fe $(OH)_3$. The phosphorus bound to $Fe(OH)_3$ can be dissolved by H_2S or H_2CO_3 and Ca $(HCO_3)_2$ more easily than the phosphorus of $FePO_4$ (Ohle, 1937). The addition of too much $CaCO_3$ to a pond is disadvantageous due to precipitation of phosphate as calcium phosphate (Golterman, 1967). Banerjea and Mandal (1965) studied the distribution of different inorganic forms of phosphorus in fish pond soils and found iron phosphate to be highest in acidic soil, followed by the neutral and the alkaline soil in the above order. They also found no apparent correlation of the quantity of aluminum bound phosphate with the pH of the soil. Banerjea and Ghosh (1970), in a subsequent study in this regard, observed that as the p^H of the soil increased, relative proportion of iron phosphate fraction tended to decrease, while that of calcium phosphate fraction tended to increase. Bhowmik (1968) studied the seasonal distribution of phosphate in a impounded freshwater at Kalyani, and observed a unimodal pattern with the peak occurring in late monsoon or autumn, whereas Mandal (1972) found a bimodal pattern of fluctuation in two fish ponds at Burdwan with the peak occurring during summer and autumn. Not only the absolute quantities of nitrogen and phosphorus present in water bodies souls in available for, but their ratios also play an important role in controlling the productivity of water bodies. The most favourable N: P ration as suggested by many (Swingle and Smith, 1938; Wolny, 1967) is 4: 1.

Organic Carbon and C : N Ratio

In the soil of wetlands organic carbon acts as a source of energy for the microbes responsible for various biochemical processes, besides releasing nutrients including some trace elements. Inorganic salts of nitrogen and phosphorus added to the water body were not effective alone in increasing the size of the zooplankton and benthos population unless sufficient organic matter is present in or added to the soil (Dobie, 1967). Organic carbon acts as a source of energy for the microbes responsible for various biochemical processes besides releasing nutrients including some trace element in aquatic system. The organic carbon is an integral part of certain chemical compound. High level of organic carbon is recorded during summer season. The fluctuation of organic carbon in aquatic system with the change of places and during different seasons have been reported by various workers (Bhowmik, 1968; Mandal, 1972). Kumar (1985) reported that organic carbon values ranged from 3.8 to 4.8 per cent in a *beel* located at Kalyani.

Satomi (1967) studied the physiological significance of carbon sources in fertilized fish ponds of Japan and observed that phytoplankton absorbed nitrogen and phosphorus excessively when carbon sources became a limiting factor in the growth of phytoplankton. In water body, bacterial activity depends not only on the carbon content but also on the C: N ratio of the substrate. The activity was found to be poor when the ratio was less than 10 and high when the same was 20: 1 or above. Golterman (1967) opined that in very productive water body, there was a very rapid mineralization of organic matter together with a high rate of uptake of oxygen which suggest that the pond mud was predominantly in the reduced state. In aquatic systems, bacterial activity depends not only on carbon content but also C/N ratio of the substrate. Carbon and nitrogen ratio is an indication of rate of nitrogen utilisation in the system, in other words, the mineralisation efficiency in the system. Singh (1960) observed a significant correlation of C/N ratio with available nitrogen content. Banerjea (1967) opined the C/N ratio less than 5.0 as very poor productive; in the range 5.0 to 10.0 as productive and between 10.0 and 15.0 ideal for aquatic systems. Banerjea (1967) attempted to correlate the productivity of fish ponds with organic carbon content of the bottom mud, and classified the ponds as follows: Org. C<0.5 per cent -poor, 0.5-1.5 per cent -average, and 1.5-2.5 per cent as optimal. Organic carbon above 2.5 per cent was not conducive for fish production. According to him, C: N ratio less than 5 was indicative of very poor production, the range 5-10 indicated poor production, while the range 10-15 represented the ideal condition for production of fish. Many workers have studied the fluctuations of organic carbon in ponds at different places and during different seasons in the year. A bimodal patterns of fluctuation of organic carbon was observed by Bhowmik (1968) and Mandal (1972) in some beels.

Plankton

The factors which account for the seasonal variations of plankton are sunshine, water temperature, pH, nitrates and phosphates (Philipose *et al.*, 1976). Utilization of nitrate by phytoplankton involves its conversion ultimately to ammonia before assimilation in the cell material (Boney, 1976). Relationship between available nitrogen and phytoplankton was found to be direct (Welch, 1952). Release of phosphate and nitrogen from soil colloids decomposition of organic matter and conversion of insoluble salts contribute to the sudden rise of nutrients enriching the pond water during monsoon season. (Hutchinson, 1944). Year to year fluctuation in plankton quality is a general phenomenon in freshwater impoundment (Welch, 1952) and the factors attributed to it are variation in rainfall, depth of water body, silting and other chemical factors. Pearsall (1930) and Strom (1927) established that the water favouring green algae are chemically distinct from those harbouring diatoms and blue green algae. Similar views have been expressed by Gonzelves and Joshi (1946); Rao (1953); Philipose (1967) and Zafar (1964). Hutchinson (1967) recorded two major groups of planktonic green algae, the *Volvocales* and *Chlorococoales* on the one hand and the *Desmids* on the other, which have different physiological requirement and so different ecological preferences.

Sewage doped swamps harbour large percentage of *Volvocales* due to abundance of nitrogen and phosphates in these bodies. (Munnawar, 1970). Philipose (1967) too, recorded that water rich in nitrates and phosphates support a rich bloom of *Volvocales*. As reported by Griffiths (1923) and Zafar (1964) *Chlorococcales* prefer habitat rich in oxygen and nitrogenous organic matter. Gonzelves and Joshi (1946) indicated higher pH values of water for favourable growth of *Chlorococcales*. According the Patrick (1948) it is the nitrate form of nitrogen which is most utilized by diatoms.

The importance of phosphorus in the distribution of diatoms has been emphasized by many workers (Gunther 1936). Ruttner (1964) observed that the diatoms are capable of absorbing phosphate in much larger quantity than their immediate requirement. The excess is said to be stored in their bodies and utilized later. The ecological distribution of Myxophyceae has been discussed by Pearsall (1932) and Philipose (1967). These workers have identified sunshine, temperature, dissolved oxygen, free carbon dioxide, alkalinity, nitrate, phosphate, and organic matter as the probable factors which in unison or in combination govern the propagation of Myxophyceae population. Perarsall (1932) and Philipose (1967) concluded that blue green algae increased in population when nitrate and phosphate values are low. Pearsall (1932) also attached great importance to higher concentration of organic matter in the development of Myxophyceae population. Singh (1960) also made similar observation and inferred that in tropical waters, changes in the concentration of oxidizable organic matter do not influence the development of blue-green algae as its concentration never falls down to a level to become limiting factor for their growth. The cosmopolitan nature of the zooplankton of freshwater is well recognized (Byars, 1960). Kumar (1985) the protozoan population has generally been found to be low. The density and period of occurrence varied widely among the species. Azeello and Difflugia occurred in good number while other organisms showed sporadic appearances. Pennak (1953) opined that protozoan thrive well in water temperature ranging from 16-25 °C. The rotifers have versatile capacity to survive in different environment as some of them are primary consumers feeding on various phytoplankton some feed on detrital element and bacteria and some have been described as raptorial predators. Kumar (1985) reported the loricate rotifers (*Brachionus* sp., *Keratella* sp., and *Polyarthro* sp.) were dominant over illoricates (Asplanchna sp.) possibly due to higher alkaline and temperature values. Michael (1969) reported the influence of alkalinity on the growth of loricate forms. Kumar (1985) observed the quantitative dominance of Copepods among the zooplankton which is a striking feature in the Kalyani beel. The larval copepods were present during the entire period of investigation. Definite peaks were observed during April-July indicating their intense breeding during this period.

Sugunan *et al.* (2000) observed population of phyto and zoo plankton in flood plain wetlands of West Bengal at lower level during the south-west monsoon which increased thereafter when the environment become stable and the plankton population established uitilisizing inorganic nutrients and organic matter brought in by the incoming flood or run-off water. Thus, they recorded plankton population during winter (25 to 4,658 μ/l) whereas in summer it was many fold higher 281 to 40,836 μ/l. Bhowmik (1988) also recorded maximum plankton population during summer season predominated by phytoplankton whereas during winter predominated by zooplankton. Jha (1997) recorded higher plankton population in the closed type of flood plain lakes in Bihar.

Species Diversity

The diversity indices are based on the concept that the structure of normal communities may be changed by environmental perturbations and extent of change in plankton community structure may be used to assess the intensity of environmental stress. Stability of the ecosystem can be studied by comparing the species diversity of different community (Sugunan, 1989). The winter season with favourable temperature, dissolved oxygen and other physico-chemical parameters besides the solar penetration augmented the species diversity in all the system irrespectively of open or closed systems Beaver *et al.* (1998). Sugunan *et al.* (2000) reported that diversity indices in respect of zooplankters in different beels of South and North Bengal have shown many variations.

Similarity Coefficient

The similarity coefficient drawn following Bray and Curtis (1957) among the systems and between the seasons, indicates the trend in species similarity of plankton in the water bodies. Correlation between the plankton and sediment nutrient levels indicated soil nitrate to bear close impact on planktonic growth, and phosphorus. Banik *et al.* (1994) also made similar attempt to correlate rotiferans with the limnological parameters and observed temperature, dissolved oxygen, bicarbonate and phosphate phosphorus showing significant correlation with the occurrence of total rotifers.

Macro Zoobenthos

The low number of species and density of benthic life is attributed to low bottom oxygen values. Similar conclusion was drawn by Moore (1942) on his study of a eutrophic lake. Roback (1974) in his study in Kam lake demonstrated that out of the 83 *Dipteran* identified in the lake, only five could tolerate bottom environment. Further, high level of organic material in sediment restrict tube development of insect species (McLachlan and Contrell, 1976). Since, the zooconosis include both stenobenthosis and eurybimoitic forms, these are generally used in estimating the degree of eutrophication (William, 1967; Chutter, 1972). Most of them are based on bio-indicator of different morphological groups of *Chironomids* (Curry, 1962) mathematical interpretation of indicators or correlation between components of zoobenthos and environmental factors. Srivastava (1955) recorded the minimum in January and maximum in August in the tropical climate of north India.

Benthic community in beels reveals the level of energy being transferred through this phase and indicates trophic status of the lake. Organically rich bottom coupled with a conducive physicochemical environment encourages fast colonisation by the benthic communnity (Sugunan, 1989). Benthic organisms have been studied at large with the macroscopic organisms and change of a rhythmic spurt has been reported to occur in benthic fauna with the succession of seasons in all lakes. A mixed population of diversified fauna constituted the benthic population of the investigated ecosystems. Fluctuation in population density and species composition were conspicuous as revealed from the observations made during the investigated period. Monsoon density of the macrobenthic fauna was always lower in compared to other seasons of the years. Kumar (1995) reported seasonal influence on the diversity and richness of macrobenthic fauna and also found bimodal peaks in Zygopteran and trimodal peaks in Anisopteran larvae from the ponds of Santhal Parganas. However, Rowson (1930) drew correlation between the bottom quality and benthic population and recorded maximum number per unit area in sandy bottom with high vegetative growth. Bhaumik (2000) also observed high density of benthic fauna in highly macrophyte-infested system. On the other hand Moon (1955) found that the littoral fauna of the lake Windermere is a continuous state of movement and the fauna to be sensitive to the damages in the surface level. This also supports the seasonal impact of benthic fauna of the *beel*. Sugunan *et al.* (2000) opined that the beels of West Bengal support rich growth of benthos, the average density ranging from 90 to 13,238 nos./m^2. Parameswaran and Vass (1995) stated benthos of the *beels* of West Bengal is generally dominated by molluscs, insect larvae, nymphs and *Oligochaetes* which is in agreement with the present investigation.

Similarity Coefficient

The stagnation at the bottom bringing about a condition which some of the species could not tolerate (Eggleton, 1931). Laal (1981) observed the peak and tough of macro-invertebrates life in the swamps of North Bihar in the months of September and June. Rai and Datta Munshi (1979) found largest number of zooconosis in Novermber, Septmeber and December in three different swamps in Bihar. Laal (1981) revealed that *Oligochaets* and *Chironomids* were present throughout the period of their observations. The pre monsoon rise in *Oligochaets* density followed by a rise in temperature in summer. Stressed on the combination of factors like temperature, dissolved oxygen, nature or substratum, wage action, grazing and predation etc. as major factors.

Macrophyte

In India, weed infestation problem is more severe in eastern part of the country where 40-70 per cent of water area is covered by weeds (Philipose *et al.*, 1976). According to Chakraborty (1961) 50-60 per cent of cultivable waters in West Bengal, Assam, Bihar, Orissa are unfit for fish cultivation solely due to weed invasion. Realising the nuisance of weed infestation, considerable work has been done in India and abroad on the extent and impact of infestation (Westlake, 1965; Varshney and Singh, 1976 biology and ecology (Westlake, 1963; Boyd, 1968; 1969; 1970 (a); 1970 (b); Mitchell, 1969; Sahai and Sinha, 1970; Mitchell, 1976 and Brij Gopal, 1976), Control measures and utilization (Swingle, 1957; Mitra, 1997; Robson, 1976; Ramchandran, 1968; Philipose, 1940). Light quality and quantity (Westlake, 1965) temperature and alkalinity have profound influence on infestation and regeneration. The vegetation of tropical lakes, particularly in small lakes in which there may be considerable changes in water level during the year, the basin may be occupied by large number of species of maximum growth and flowering.

The macrophytes are known to be sheltering a large population of mixed groups of organisms for various purposes mainly for nourishment and life process completion. The tropical beels being very good ecosystem for the balanced growth of vegetation and animal components are also suitable for harbouring a large number of macrophytic associated fauna. The littoral nature and high macrophyte infestation were favourable situation for the growth of a mixed population of molluscs, insects, annelids and crustacean population in support of the foliage and other body structures of the plants. Peter (1968) reported that in the weed-choked shallow basined eutrophic waters, the benthic fauna usually leave the bottom due to hypoxic or anoxic conditions prevailing therein and mingles with the macrophytes. Maitland (1978) stated that weed-choked shallow studying waters period variety of niches for nymphs and larvae of insects adopted from boring into stems and leaves of plants. The invertebrate community in the *beel* can be conveniently divided into two groups *"loosely attached"* molluscans and insects (Kumar, 1985). Sozeka (1975) has recorded the quantitative differences of *Annelids* and *Dipteran* in different zones of freshwater lake. Laal (1981) while working in North Bihar swamps reported dominance of the gastropods in macrophytes.

Considering the area of coverage, the winter was recorded as most favourable season for the growth of the macrophytes. The monsoon changes in ecological condition particularly the water depth and flow affected the macrophytic infestation in all the beel ecosystems. Rooted submerged group of macrophytes like *Ceratophyllum demersum, Hydrilla verticillata, Najas graminae, Potamogeton sp, Vallisneria spiralis, Ottelia sp.* etc were the major contributors all throughout the year in all the beels. Sharma (1995) also recorded dominance by submerged and emergent vegetation particularly *Phragmites kanko, Hydrilla verticillata* and *Ceratophyllum demersum* in Kawar lake of Bihar. Disappearance and reappearance of macrophytic species were often recorded in the *beel* systems particularly of marginal and floating varieties. However, such a condition favouring rapid growth rate and spread of macrophyte infestation particularly by *Eichhornia crassipes* and *Salvinia sp* have been felt to be of great concern for the existence of the wetlands by Choudhuri *et al.* (1994). According to Kaul and Zutshi (1966) light intensity attenuates rapidly with depth depending on water quality and density of Macrophytes. Swingle (1957) has shown that aquatic plant growth can be enhanced or inhibited by modifying the water in which the plants are grown. According to Soulthrope (1967) the principal chemical factors which effect weed growth are temperature, sunlight, size, shape of the basin, water quality including transparency, dissolved oxygen and nutrient status. Misra (1959) has formulated a scheme of succession operating with alternation of hydrosore and xerosore and inevitable consequences of extreme seasonal changes in shallow temporary or fluctuating water bodies. Among floating weeds, next in degree of abundance comes, *Lemnids* and *Azolla*. They propagate by multiplication of fronds, however in

S. polyrhiza turaions are formed which also contribute to the regeneration potential. The increase in fronds after 36 days in laboratory conditions in *N. indcus* and *S. polyrrhiza* has been reported to be 10.8 per cent. The percentage increase in buds in *S. polyrrhiza* is 56 per cent and *L. minor* 86 per cent (Kaul and Bakaya, 1976).

Periphyton

The term *periphyton* was introduced by Behning in 1924 for the plant growth on buoys, ships and mooring in the Volga. Hunt (1952) defined periphyton as an assemblage of algae and minute animalcule covering submerged objects with slimy coatings. Neel's definition of periphyton as an assemblage of microscopic organisms that form or live in coating upon rocks and other submerged objects has been regarded by Gumtow (1955) to correspond closely to Ruttner's (1964) *Aufwuch*. In subsequent years, mostly in Europe, a number of studies on the seasonal succession of periphyton of inland waters were made. The investigation were conducted in different types of water like lakes (New Combe, 1950), reservoirs (Sladeckova, 1962; Sladeck and Sladeckova, 1963) streams and rivers (Neel, 1953) and Gumtow 1955) ponds (Hammenn, 1953) and on slow sand filter beds (Brook, 1955).

Studying the role of attached algae in lacustrine waters Wetzel (1964) revised the confusing terminology and classified four kinds of *Aufwuch viz. epizoic, epiphytic, epilithic* and *epipelic*. In India, Philipose (1940) made observation on the periphytic forms of marginal grasses, stones and shells of molluscs in the pond and later durin 1976 published detailed work on the distribution in time and space of periphyton of a perennial pond of Cuttack. Misra and Singh (1963) and Vass (1978) also reported the periphytic fauna of freshwater and brackishwater impoundment respectively. Kumar (1985) opined though many of the periphytic forms recorded in the Kalyani *beels* are the same as reported by Philipose *et al.* (1976) but there were obvious differences with regard to quantum and seasonal succession on account of different habitat.

The littoral characteristic was favourable for the growth of periphytonic organisms in the beels (Bhaumik, 2001). A good growth of submerged vegetation supported periphytic growth consisting of various groups of phyto and zooplanktonic organisms providing adequate substrate required for the purpose (Chakraborti, 1997). The biomass production of periphyton varies greatly in different ecological conditions (Odum, 1957). The solar penetration, more transparency and available nutrients influenced periphytic growth in the beel ecosystems and as a result, variability in diversity and density of the organisms was observed among the beels. During monsoon growth of periphyton was always less might be due to the change in physico-chemical environment in the ecosystem and also due to the effect of water flow, turbulence and increased in depth of water. The diversity in algal periphyton has been studied by Datta Munshi and Singh (1995) in the wetlands of North Bihar. As reported by them 51 species of the periphyton algae were identified in the wetland ecosystem.

Primary Productivity

Primary productivity is an important parameter to assess the productivity of a water body. It varies widely depending on the climate conditions, intensity of solar radiation, transparency of water and also on the nutrient concentration in water. The primary production of ponds varied widely from 10-2500 mg C/M^2/hr. The nutrient rich ponds were found to show an average production greater than 10,000 mg C/m^2/day late in summer (O'Brien and Noyelles Jr., 1974). According to Hepher (1962), production in the unfertilized ponds may also be fairly high and reach values of 138-190 mgC/m^2/hr in summer, but production in the fertilized water bodies is usually 4 to 5 times higher than that in the unfertilized ponds. In intensely manured fish ponds, photosynthesis is maximum at the surface, which shows a steep decreasing trend with depth, the compensation depth range between 40 cm to 70 cm (Hepher, 1962, Noriegacutis, 1979). Primary productivity of a water body depends on its nutrient status. When the nutrient status is poor, the primary productivity and the fish production is also poor (Singh and Desai, 1980) indicating that the primary productivity is positively correlated with nutrient concentration. The primary productivity which pertains to the organic substance synthesized in a prescribe space and per unit of time largely depend on the interaction of light, temperature and phytoplankton production. The workers like Croome and Tyler (1975) and Khan and Zutshi (1980) have reported high production during higher light intensity and vice-versa. Yadav *et al.* (1987) and Pathak (1997; 2001) reported moderate primary productivity with little variation in values from tropical waters. The *beels* investigated, by virtue of shallowness and high nutrient content, showed moderate to high productivity throughout the period. The net production values when considered for productivity evaluation of the beel indicated variations in the values with the change of season and also of the systems. In general the closed system as per the net production values has been observed to be more productive in compared to the open ones. Further, the seasonal influence on primary production in the beels was pronounced. The net production was maximum during winter in all the beel systems. While correlating the net production values with the water and sediment nutrients, dissolved oxygen and plankton, on the same significant influence of were recorded in all the beels.On the contrary, Wong (1969) found no indication of any direct relation between inorganic phosphate absorption rates and photosynthetic rates. Many investigators have monitored the primary productivity of water bodies at different places and during different seasons of the year and observed only one peak of primary production during summer or early summer in the reservoirs (Singh and Desai, 1980). The knowledge of primary production in tropics is still limited except the works by Sreenivasan (1964 and 1968); Ganapati and Sreenivasans (1970); Kaliamurthi (1978) and Natarajan and Pathak (1980). Information on biological productivity of swamps, ox-bow lakes and other wetlands is almost scanty (Laal'1981 and Yadav, 1988).

Fish Production Potentiality

Odum (1962) felt that a harvest of 1.2 per cent of primary production of fish is ideal. Natarajan and Pathak (1983) stated that the energy source for all living organisms is sun, a vast incandescent sphere of gas, which releases energy by nuclear transmission of hydrogen to helium in the form of electromagnetic waves. The biotic communities (producers, consumers and decomposers) in an ecosystem are linked with one another with energy change. Sugunan et al.,(2000) estimated yield potential in eighteen beels of West Bengal on the basis of energy fixation by plankton and macrophytes. Bhaumik (2001) reported that the fish fauna of the West Bengal beels in totality consisted of 41 species under 29 genera and 17 families.

Fish Fauna

The strictly limnetic fishes rapidly reduced with decreased depth and surface area. But the Indo-Gangetic floodplain wetlands available in the from of lakes, depression and so on, are the native for a large variety of fresh water fish fauna which complete their life cycles therein. Apart from this, innumerous fish species do migrate to the wetlands (*beels*) either for breeding and nourishing or growing purposes. Sharma *et al.* (1994) identified 51 species of fishes belonging to 15 families and 10 orders in Konar wetland in Bihar. Sugunan *et al.* (2000) while studying the open and closed beels of West Bengal have reported fish fauna of 21 families consisting 51 species under 36 genera. The riverine population of fishes contributed significantly to the ichthyofauna of the system.

Fish Production

The fish production in wetland of West Bengal has been reported by various workers (Bhowmik, 1988; Parameswaran and Vass, 1995; Mukhopadhyaya, 1997b; Sinha, 1999; Sugunan *et al.* 2000; Sugunan and Bhattacharya, 2000) and mentioned to be largely dependent on the management procedure adopted thereof. However, in general partial stocking with fast growing Indian and Exotic carps greatly influenced the production turn up. Obviously, it affected the total production. While correlating the contributing parameters like plankton, benthos, dissolved oxygen and nutrient of water and soil with the fish production, a clear relationship with benthic and planktonic organisms emerged in a closed system and open systems. The relationship with the soil nutrient levels with fish production was conspicuous in open system and the fisheries therein, is largely dependent on planktonic food chain. Such situation of under-exploitation of the potentiality was due to uncontrollable flood and resultant loss of the fast growing stocked species of Indian major carps (IMC) and in lieu of infiltration of cat fishes and miscellaneous groups of riverine small fishes contributing major fishery in the *beel*. Sugunan *et al.* (2000) reported the production potentiality of number of weed choked closed *beels* in the range of 84.4 to 264.6 kg/ha/yr in the West Bengal which differ from the observations made in the closed *beels* of the present investigation.

Bhaumik (2001) reported fish production potential closed beel 1,218.75 kg/ha/yr to 1, 134.37 kg/ha/yr and open beel 1,687.5 kg/ha/yr to 1,901.25 kg/ha/yr in West Bengal. Based on the diurnal dissolved oxygen values of individual beel ecosystem, the production potential were estimated.

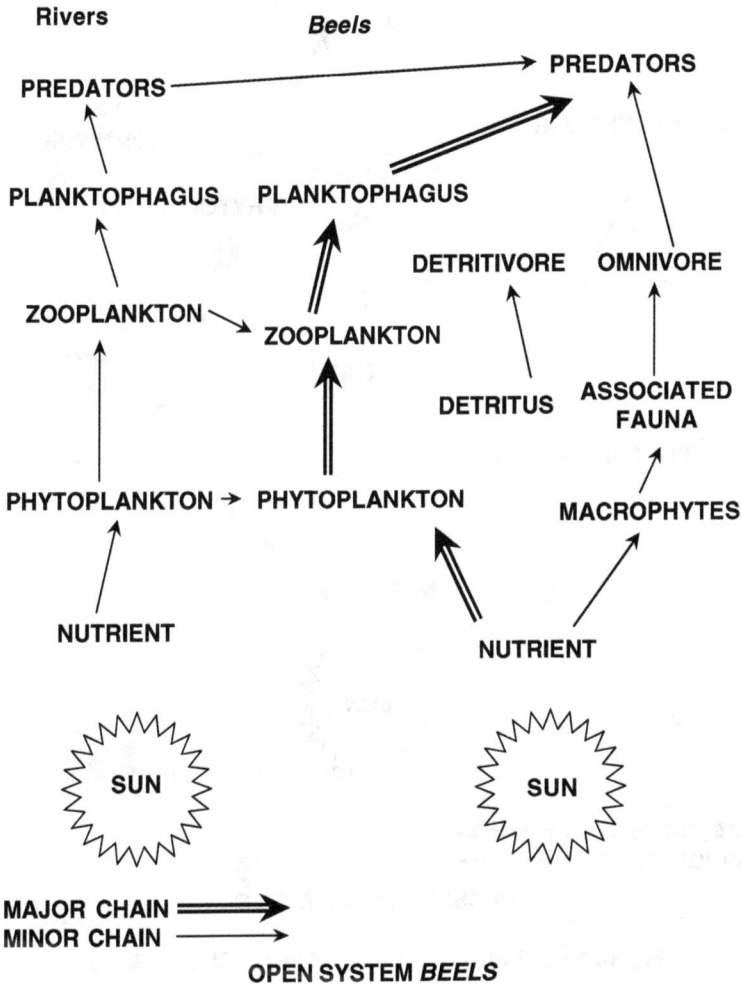

Figure 19.1: Pathway of Energy-Flow in Open *Beels*

Conclusion

Beels are highly productive inland fisheries resource of eastern stated of the country. Over the years, inspite of being managed by the Fishermen Co-operative Societies, the yields from these water bodies have been showing declining trend barring few where stocking is followed with fast growing Indian and exotic carp

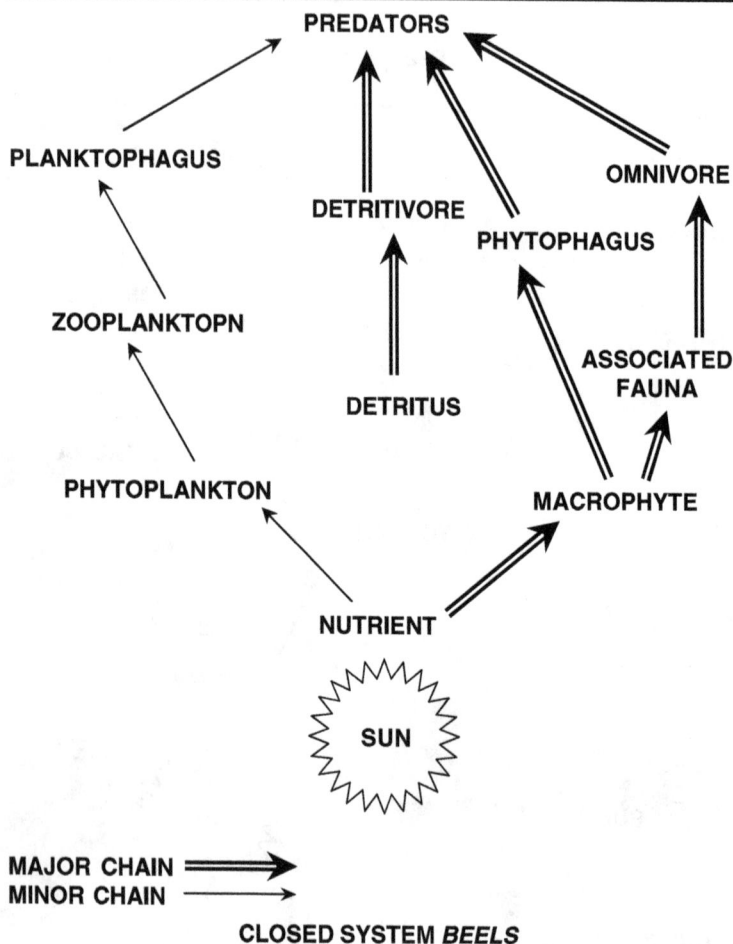

Figure 19.2: Pathway of Energy-Flow in Closed *Beels*

combinations. The *beels* inspite of moderate to high productivity capacity lack annual fish yield of desired level. As revealed from the species diversity and group wise composition of fish there is enough scope for production enhancement through rational exploitation of the available food spectrum, manipulation of fish species composition, density and stocking ratio. If necessary transplantation of the lost population of endemic and economically important species need be indulged.

References

Alikunhi, K.H., 1952. On the food of young carp fry. *J. Zool. Soc. India*, 4: 77–84.

Alikunhi, K.H., 1957. Fish culture in India. *Fin. Bull. Indian Coun. Agri. Res.*, 20, New Delhi.

Backiel, T. and Stagman, K., 1967. Temperature and yield in carp ponds. *Proc. of World Symp. on Warm water pond fish culture FAO Fish. Rep.*, 4(44): 334–342.

Bamforth, S.S.S., 1958. Ecological studies on planktonic protozoa of a small artificial pond. *Limnol. Oceanogr.*, 3: 398–412.

Bamforth, S.S.S., 1962. Diurnal changes in shallow aquatic habitats. *Limnol. Oceanogr.*, 7: 348–353.

Banerjea, S. M and Mandal, L. N., 1965. Inorganic transformation of water soluble phosphate added in fish ponds as influenced by the nature of pond soils. *J. Indian Soc. Soil Sci.*, 13: 167–173.

Banerjea, S.M., 1967. Water quality and soil condition of fish ponds in some States of India in relation to fish production. *Indian J. Fish.*, 14(1 and 2): 115–144.

Banerjea, S.M. and Ghosh, S. R., 1970. Studies on the correlation between soil reaction and different forms of bound phosphorus in pond soils. *J. Inland Fish. Soc. India*. 2: 113–120.

Banik, S., Debnath, R., Debbarman, S. and Kar, S., 1994. Occurrence of rotifers in a seasonal wetland in relation to some limnological conditions. *J. Feshwat. Biol.*, 6 (3): 221–224.

Beaver, J.R., Miller-Lanke, A.M. and Acton, J.K., 1998. Mid–Summer zooplankton assemblages in four types of wetlands in the upper midwest, USA, *Hydrobiologia*, 380: 209–220.

Bhaumik, Utpal., 1998. Strategy of extension towards development of fisheries in wetlands, *Meenbarta*, Dept. of Fisheries, West Bengal: 61–65.

Bhaumik, Utpal, 2000. Community–based co-operative approach for management of fisheries in small reservoirs, *In*. Management of Fisheries in small reservoirs. *Ed*. Bhaumik,Utpal, *CIFRI bull*. 106: 127–140.

Bhaumik, Utpal, 2001. Eco-dynamics and Fisheries management of some beels and its impact on the Socio–economic status of the operatives in West Bengal. Kurukshetra University, Haryana, p 309.

Bhowmik, M.L., 1968. Environmental factors affecting fish food in freshwater fisheries, Kalyani,West Bengal, India, *Unpublished Ph.D. Thesis, University of Kalyani*: 238 p.

Bhowmik, M.L., 1988. Limnology and productivity of some *beels* and *baors* of West Bengal with reference to recent development, *Environ. and Ecol.*, 6(1): 42–47.

Boney, A. D., 1976. Phytoplankton, *The camelot Press Ltd, Southampton*: 116p

Boyd, C. F., 1968. Freshwater plants–a potential source of protein. *Econ. Bot.*, 22: 359–368.

Boyd, C. F., 1969. Production mineral assimilation, absorption and biochemical assimilation of *Juscia americana* and *Alternanthera philoxeroides*. *Arch. Hydrobiol.*, 66: 139–160.

Boyd, C. F., 1970a. Chemical analysis of some vascular plants. *Arch, Hydrobiol.*, 67: 78–85.

Boyd, C. F., 1970b. Amino–acid, protein and calorie content of vascular aquatic macrophytes. *Ecology*. 51: 902–906.

Bray. J. R. and Curtis, C. T., 1957. An ordination of the upland forest communities of southern Wisconsin. *Ecol. Monogr*. 27: 325–490.

Brij Gopal., 1976. A note on the growth and ecology of *Salinia molestna*. *In. S. E. Asia. Proc. of Reg. Sem. on Noxious. Aq. Vet*. held at New Delhi, December 12–17, 1973: 177–182.

Brooke, A. J., 1955. The attached algal flora of slow sand filter beds. *Hydrobiologia*, 7 (1 and 2): 103–117.

Byars, J. A., 1960. A freshwater pond in New Zealand. *Aust. J. Mar. Fresh. Wat. Res., II*: 222–240.

Carlander, K. D., 1955. The standing crop of fish in lakes. *J. Fish. Res. Bd. Can.*, 12: 543–570.

Chacko, P.I., and Ganapati, S.V., 1949. Some observations of the Adayar river with special reference to its hydrographical conditions, *Indian Geogr. J.*, 24: 1–15.

Chacko, P.I., and Krishnamoorthy, J.B., 1954. On the plankton of three freshwater fishponds in Madras City. *Indian Symp. Mar. Freshwater Plant. Indo. Pacific Fish coun.*: 103–107.

Chakrabarti, P., 1980. Studies on the hydrobiology of some freshwater fisheries. *Unpublished Ph.D. Thesis, Burdwan University, India*: 251 p.

Chakraborti, P.K., 1997. Role of plankton, benthos and periphyton in the production–cycle in the floodplain wetland ecosystem. *In.* Fisheries enhancement of small reservoirs and floodplain lakes in India. *Eds.* V. V. Sugunan and M. Sinha, *CIFRI Bull.*, 75: 95–102.

Chakraborti, S. C., 1961. Weed control in India, *In. Sem. on Weed Cont*, held in Bombay during October 3–4, 1961.

Chakravorty, R.D., Roy, P. and Singh, S.P., 1959. A quantitative study of the plankton and the physico-chemical conditions of the river Yamuna at Allahabad in 1954–55. *Indian J. Fish.*, 6(1): 166–203.

Chang, S. S. and Kaclspm. M. L., 1957. Fractionation of soil phosphorus. *Soil. Sci.*, 84: 133–144.

Chowdhuri, H, Ramaprabhu, T and Ramachandran N., 1994. *Ipomoea carnea* Jacq. And new aquatic seed problem in India. *J. Aq. Plant mangement*, 32: 37–38.

Chu, S. P., 1943. The influence of mineral composition of the medium on the growth of planktonic algae. Part II. The influence of the concentration of inorganic nitrogen and phosphate phosphorus. *J. Ecol.*, 31: 109–148.

Chutter, F. M., 1972. An emperical biotic index of the quality of water in South African streams and rivers. *Wat. Res.*, 6: 19–30.

Cowardin, L.M., Carter, V., Golet, F.C. and LaRoe, E.T., 1979. Classification of wetlands and deepwater habitats of the United States, *U.S. Fish and Wildlife Service, Washington D.C., USA*, FWS/OBS–79/31: 103p.

Curry, L. L., 1962. A study of the ecology and taxonomy of freshwater midges (Diptera; chironomid) of Michigan with special reference to their role in turn over to radio–active substances in the hydrosol. *A.E.C. Prog. Rept.*.149p.

Dale, H.M. and Gillespic, T.J., 1976. The influence of floating vascular plants on the diurnal fluctuations of temperature near the water surface in early springs, *Ecology.*, 26: 399–418.

Darve, V.S. and Ball, D.V., 1961. Hydrology of the Kelwa backwater and adjoining sea. *J. Univer. Bombay,* 29(3–5): 39–48.

Das, P., 1983. Studies on management and limnobiology in relation to fish production in some fresh water ponds. *Unpublished Thesis, Burdwan University.*267p.

Das, R. K. and Dehadrai, P. V., 1986. Soil, water intraction and nutrient turn over in a weed infested swamp. *J. Inland Fish. Soc.* India, 18 (2): 13–19.

Das, S.M. and Srivastava, V.K., 1956. Some new observations on plankton from freshwater ponds and tanks of Lucknow, India, *Sci. and Cult.*, 21: 446–467.

David, A., Ray, P., Govind, B.V., Rajagopal, K.V. and Banerjee, R.K., 1969. Limnology and fisheries of Tungabhadra Reservoir, *CIFRI Bull.* 13: 70–83.

Dehadrai, P.V., 1972. Progress of work on the techniques of culture of air–breathing fishes in swamps in Bihar, Assam and Mysore. *In.* 2nd Workshop on Co-ordinated Project on Air–breathing fishes held at Patna.

Dendy, J. S., 1963), Farm ponds. *In.* Linmology in North America, *Ed.* D. G. Frey,, Madison University of Wisconsin Press. 595–620 pp.

Dewan. S., 1973. Investigations into the ecology of fishes of a Mymensing Lake. *Unpublished Ph.D. Thesis, Bangladesh Agricultural University, Mymensingh, Bangladesh.*

Dey, S. C. and Goswami, M. M., 1982. Studies on the hyrobiological conditions of some commercially important lakes of Kamrup district of Assam and their bearing on fish production, *Final Technical Report,* NEC (GOI. 177 p.

Dey, S.C. and Bhattacharjee, P.C., 1995. Conservation and sustainable development of floodplain wetland in Assam. Howes, J.R. (*Ed.*), Asian Wetland Bureau, Kuala Lumpur: 103–107.

Dobie, Jhon., 1967. Experiments in the fertilization of Minnesota fish rearing ponds. *FAO Fish. Rep.* 44(3): 274–284.

Downie, N.M. and Health, R.W., 1970. Basic statistical methods, Harper and Row Publishers, New York: 356 p.

Dwivedi, S. N. and Chondar, S. L., 1977. Hydrology and fishery of Keetham lake (Agra), *CIFE Newsletter;* 32 p.

Eddy, S., 1934. A study of freshwater plankton communities. III *Biol. Monogr.;* 1–93.

Eggleton, F. M., 1931. A limnological study of the profoundal bottom fauna of certain freshwater lakes, *Ecol. Monogr.* I: 361–388.

Ganapati, S. V. and Sreenivasan, A., 1970. Energy flow in natural aquatic ecosystem in India. *Arch. Hyrobiol.,* 66 (4): 458–498.

Ganapati, S.V., 1941a. Studies on the chemistry and biology of ponds in Madras City–seasonal changes in the physical and chemical conditions of a garden pond containing aquatic vegetation. *J. Madras Univ.,* 13(1): 55–69.

Ganapati, S.V., 1941b. An ecological study of a garden pond containing abundant zooplankton, *Proc. Ind. Acad. Sci.,* B., 17(2): 41–50.

Ganapati, S.V. and Sreenivasan, R., 1956. Thermal studies of the Chetput fish farm, Madras, *Fish Stat. Rep. Madras:* 303–313.

George, M. G., 1962. Diurnal variations in two shallow ponds in Delhi, *Curr. Sci.,* 30: 268–269.

Goldman, C.R., 1972. The role of minor nutrients in limiting the productivity of aquatic ecosystem. *Symp. on nutrients and eutrophication,* I: 21–23.

Goldman, R.C. and Horne, J.A., 1983. Limnology, Mc Graw Hill, International Book Company, Tokyo: 464 p.

Golterman, H. L., 1967. Influcence of soil on the chemistry of water in relation to productivity. *FAO Fish. Rep.,* 44 (3): 27–42.

Gonzalves, E. A. and Joshi, D. B., 1946. Freshwater algae near Bombay. *J. Bombay Nat. Hist. Soc.* 46 (1): 154–176.

Gopakumar, K., 2000a. Indian fisheries: an overview. The fifth Indian Fisheries Forum, *Souvenir:* 1–6.

Gopakumar, K., 2000b. Fisheries of India, problems and prospects, *Abstract, First Indian Fisheries Science Congress, Chandigarh:* 1.

Goswami, M.M., 1985. Limnological investigations of a tectonic lake of Assam, India and their bearing on fish production. *Unpublished Ph.D. Thesis, Guwahati University*: 395 p.

Griffiths, B. M., 1923. The phytoplanktons of freshwater bodies and the factors determining its occurrence and composition. *J. Ecol.*, 2: 184–213

Gumtow, R. B., 1955. An investigation of the periphyton in a riffle on the West Gallatin River, Montano. *Trans. Amer. Mier. Soc.* 74 (3): 278–292.

Gunther, E. R., 1936. A report on the Oceanographical investigations in the Peru coastal current. *Discovery Report.*, 13: 107–276.

Hackson, L., 1981. A manual of lake morphometry. Springer–Verlag, New York: 78 p.

Hammenn, I., 1953. Okologische and Biologishe untersuchungon an susswasser peritrichen. *Arch. F. Hyrobiocl.*, 47: 177–228.

Hazelwood, H.D. and Parker, R.A., 1961. Population dynamics of some freshwater zooplankton. *Ecology*, 42: 266–274.

Hepher, B., 1952. The fertilization of fish ponds, 2. Nitrogen, *Bamidgeh*, 4(10–12): 220–223.

Hepher, B., 1962. Primary production in fish ponds and its application to fertilization experiments. *Limnol. Oceanogr.*, 7(2): 131–36.

Heron, J., 1961. The seasonal variation of phosphate, silicate and nitrate in waters of the English lake District. *Limnol. Oceanogr.*, 6: 338–346.

Hickling, C.F., 1962. Fish culture, Faber and Faber, London., 295 p.

Hollis, G.E., 1995. The functions of floodplain wetlands within integrated river basin management: international perspectives– *Ibid*; 1–8.

Hunt, B.P., 1952. Food relationship between Florida spotted gar and other organisms in Ismiami canal Dade country Florida, *Trans. Amer. Fish. Soc.*, 82: 13–33.

Hutchinson, G.E., 1944. Limnological studies on Connecticut, VII. A critical examination of the supposed relationship between phytoplankton periodicity and chemical changes in lake water. *Ecology*, 25: 3–26.

Hutchinson, G.E., 1957. A treatise on limnology, Vol. I.*John Wiley and Sons Inc.* N.Y., 1015 p.

Hutchinson, G.E. and Bowen, V.T., 1950. Limnological studies in Connecticut. IX. A quantitative radiochemical study of the phosphorus cycle in Linsley Pond. *Ecology*, 31: 194–203.

Imvebore, A.M.A., 1967. Hydrology and plankton of Eliyele Reservoir, Ibadan, Nigeria. *Hydrobiol.*, 30(1): 154–176.

Jana, B.B., 1973. Seasonal productivity of plankton in a freshwater pond in West Bengal, India. Int. Revue. *Ges. Hydrobiol.*, 58(1): 127–143.

Jermolajev, E.G., 1958. Zooplankton of the inner bay of Fundy, *J. Fish. Res. Bd. Canada*, 15: 1219–1228.

Jha, B.C., 1997. Salient ecological features of *mauns* and *charus* of Bihar. *In Fisheries enhancement of small reservoirs and floodplain lakes in India.Eds.* V. V.Sugunan and M. Sinha. *CIFRI Bull.* 75: 167–174.

Jhingran, A.G. and Pathak, V., 1987. Ecology and management of *beels* in Assam–a case study of Dhir *beel. Workshop on development of beel fishery in Assam, Assam Agricultural University*, April 21–22, 1987: 16–28.

Jhingran, A.G., Thakur, V.C. and Tandon, S.K., 1976. Structure and tectonics of the Himalayas. *In. Him. Geol. Sem.*, Pt. I: 1–39.

Jhingran, V. G., 1977a. A note on progress of work under coordinated project on brackishwater fish farming. *CIFRI;* Mimeo: 9 p.

Jhingran, V. G., 1977b. Fish and Fisheries of India. *Hindustan Publishing Corporation, Delhi*, 954 p.

Jhingran, V.G., 1991. Fish and Fisheries of India, *Hindustan Publishing Corporation*, Delhi: 727 p.

Jhingran, V.G., Natarajan, A.V., Banerjea, S.M. and David, A., 1988. Reprinted. Methodology on reservoir fisheries investigations in India. *CIFRI Bull.* 12: 102 p.

Kaliamurthy, M., 1978. Organic production in relation to environmental features nutrients and fish yield of lake Pulicat. *J. Inland Fish. Soc. India*, 10: 68–75.

Kato, G., 1941. Studies on the freshwater regions in the compound of the Palau Tropical Biological Station. (2) Temperature, O_2-content and pH of the water *Kagaku Nanyo* (Sci. of the South Sea. 3: 29–36.

Kaul, V. and Bakaya, U., 1976. The noxious floating, Lemnids–*Salvinia* aquatic weed complex in Kashmir. Aquatic weeds in *S. E. Asia Reg. Sem.* on *noxious Aqa. Veg.*, New Delhi, December 12–17, 1973: 183–192.

Kaul, V. and Zutshi, D. P., 1966. Some ecological considerations of floating islands in Srinagar lakes. *Proc. Nat. Acad. Sci.* India, 36 B III: 273–381.

Kielhorn, V.W., 1952. The biology of the surface zone zooplankton of a Boreo–arctic Atlantic Ocean area. *Jour. Fish. Res. Bd. Canada,* 9: 223–264.

Kimball, H.H., 1935. Intensity of solar radiation at the surface of the earth and its variation with the altitude, season and time of the day. *Mon. Weather Rev.,* 63: 393–398.

Kumar, A., 1995. Population dynamics and species diversity of odonata larvae in fish farming wetland of Santhal Pargana (Bihar), India, *Proc. Nat. Acad. Sci.* 65 (4): 401–410.

Kumar, Kuldip (1985. Hydrobiological investigations of a freshwater beel with special reference to its fish production potentialities.*Unpublished Ph.D. Thesis, Bhagalpur University,Bihar, India.*

Laal, A.K., 1981. Studies on the ecology and productivity of swamps in North Bihar in relation to production of fishes and other agricultural commodities. *Unpublished Ph.D. Thesis, Bhagalpur University, Bihar, India.*

Lakhsmanan, M. A. V., Murthy, D. S., Pillai, K. K. and Banerjee, S. C., 1967. On a new artificial feed for fry, *FAO Fish. Rep.* 44 (3): 373–387.

Lin, S. Y. and Chen, T. P., 1967. Increase of production in freshwater fish ponds by the use of inorganic fertilizers, *FAO Fish. Rep.* 44(3): 210–225.

Maitland, P. S., 1978. *Biology of freshwaters.* Blackwell Book Company, London:272p

Mandal, B.K.(1972. Limnological investigation on freshwater fisheries of Burdwan. *Unpublished Ph.D. Thesis, Burdwan University, West Bengal, India.*

Mathew, P. M.(1969. Limnonogical investigations on the plankton of Govindgarh lake and its correlation with physico–chemical factors. *In. Proc. Sem. Ecol. Freshwat. Resevoirs, Barrackpore:* 46–55.

Matida, Y., 1956. Study of farm fish culture, 3 fates of fertilized elements and the relationship between the efficiency of fertilizer and biochemical environment in the pond. *Bull. Freshw. Fish. Res. Lab., Tokyo.*, 6(1):27–39.

McCombic, A.M., 1953. Factors influencing the growth of phytoplankston. *J. Fish. Res. Bd., Can.*, 10(5): 253–282.

McIntire, C. D. and Bond, C. E., 1962. Effects of artificial fertilization on plankton and benthos abundance in four experimental ponds. *Trans. Amer. Fish. Soc.*, 91 (3): 303–312.

McLanchan, A. J. and Contrell, M. A., 1976. Sediment development and its influence on the distribution and tube structure of *Chironomus pulmosis (Chironomidae; Diptera)* in a new impoundment, *Freshwater Biol.*, 6: 437–443.

Michael, R. G., 1964. Limnological investigations on pond plankton, macro fauna and chemical constituents of water and their bearing on fish production. *Unpublished Thesis. University of Calcutta:* 225p

Michael, R.G., 1969. Seasonal trends in physico–chemical factors and plankton of freshwater fishpond and their role in fish culture. *Hydrobiol.*, 33(1): 145–160.

Miller, G.W. and Rabe, F.W., 1969).A limnological comparison of two small Idaho reservoirs.*Hydrobiol.*, 33(3 and 4): 523–546.

Misra, J. N. and Singh, C. S., 1963. A preliminary study on periphyton growth in a temporary pond, *Proc. Symp. Recent. Advances in Tropical Ecology*, Part I: 311–315.

Misra, R., 1959. The study of tropical vegetation in Madhya Pradesh and Gangetic Valley. *Study of Trop. Veg.* 1: 74–83.

Mitchel, D. S., 1969. The ecology of vascular hydrophytes in lake Kasiba. *Hydrobiologia*, 34: 448–464.

Mitchel, D. S., 1976. The growth and management of *Eichhornia crassipes* and *Salvinia sp.* in their native environment. *In. S.E. Asia. Proc. Of Reg. Sem. on noxious. Aq. Veg.* held at New Delhi; December 12–17, 1973: 167–176.

Moitra, S.K. and Bhattacharya, B.K., 1965. Some hydrological factors affecting plankton production in a fish pond at Kalyani, West Bengal. *Ichthyologica*, 4(1–2): 8–12.

Moon, H.P., 1955. Flood movements of the littoral fauna of Windermere. *Tour, Anim. Ecol.*, 4: 216–228.

Moore, W.G., 1942. Field studies on oxygen requirements of freshwater fishes. *Ecology*, 23: 317–324.

Moyle, J.B., 1946. Some indices of lake productivity. *Trans. Amer. Fish. Soc.*, 76: 322–334.

Mukhopadhyaya, M.K., 1997a. Ecology of the beels in West Bengal.*In* Fisheries enhancement of small reservoirs and floodplain lakes in India. *Eds.* V.V. Sugunan and M. Sinha. *CIFRI Bull.* 75: 187–90.

Mukhopadhyaya, M.K., 1997b. Present status of fish production in beels of West Bengal with suggestions for fish yield optimization. *Ibid:* 193–198.

Munwar, M., 1970. Limnological studies on freshwater ponds of Hyderabad, India. I. The biotope. *Hydrobiologia*, 25(1): 127–162.

Natarajan, A. V. and Pathak, V., 1980. Bioenergetic approach to the productivity of man made lakes. *J. Inland. Fish. Soc. India*, 12 (1): 1–14.

Natarajan, A.V. and Pathak, V., 1983. Pattern of energy flow in freshwater tropical and subtropical impoudments. *CIFRI Bull.* 36: 27 p.

Nath, D., 1997. Methods of evaluating primary productivity in small water bodies. *CIFRI Bull.* 75: 65–74.

Nath, D., 1999), Role of biotic factors in management of fisheries in open water system, *In:* open water fisheries technologies and extension methods, *CIFRI. Bull.* 96: 86–90.

Nath, D., 2001. Methods of evaluating primary productivity in small reservoirs. In. Management of fisheries in small reservoirs. *Ed.* Utpal Bhaumik, *CIFRI Bull* 106: 73–80.

Needham, J. G. and Lloyd, J. T., 1930. The life of inland waters, 2nd ed. Storing field: 438 p.

Neel, J. K., 1953. Certain limnological features of a polluted irrigation system. *Trans. Amer. Fish. Soc.*, 72: 119–135.

Nees, J., 1946. Development and status of pond fertilisation in Central Europe. *Trans. Amer. Fish. Soc.*, 76: 335–358.

New Combe, C. L., 1950. A quantitative study of attachment material in Soelon Lake, Michigan. *Ecology*, 31: 203–215.

Odum, E.P., 1962. Relationship between structure and function in the ecosystem. *Jap. J. Ecol.*, 12: 108–118.

Odum, H.T., 1957. Trophic structure and productivity of silver spring. Florida. *Ecol. Monogr.*, 27: 55–112.

Ohle, W., 1937. Kolloidgele ala Nahrstoffregulatoren dergewasser. *Naturwissenschaften*, 25 (29): 471–474.

Parameswaran, S. and Vass, K. K., 1995. In overview of floodplain wetlands in West Bengal, *In*. Conservation and sustainable use of floodplain wetlands. *Ed.* R. J.Howes. *Asian Wetland Bureau*, Kuala Lumpur: 52–66.

Patrick, R., 1948. Factors affecting the distribution of diatoms. *Bot. Rev.*, 14 (8): 473–524.

Pearsall, W.H., 1921. A theory of diatoms periodicity. *J. Ecol.*, 1(2): 165–183.

Pearsall, W.H., 1930. Phyplankton in English lake, I, The proportion in the water of some dissolved substances of biological importance. *J. Ecol.*, 18: 306–320.

Pearsall, W.H., 1932. Phytoplankton in English lake, II The composition of phytoplankton in relation to dissolved substances. *J. Ecol.*, 20: 241–262.

Pennak, R. W., 1953. Freshwater invertebrates of United States, Ronald, New York: 7–69.

Peter, T., 1968).Population changes in aquatic invertebrates living on two water plants in a tropical man made lake, *Hydrobiol.*, 32: 449–485.

Philipose, M. T., 1940. The ecology and seasonal succession of algae in a permanent poll in Madras. *M. Sc. Thesis* Madras University; 220 p.

Philipose, M.T., 1967. Cholorococcales, *ICAR Publication*, New Delhi.

Philipose, M.T., Nandy, A.C., Chakraborty and Ramakrishna, K.V., 1976. Studies on the distribution in time and space of the periphyton of a perennial pond at Cuttack, India. *CIFRI Bull.* 21: 16 p.

Plecznake, E. and Ozimeck, T., 1976. Ecological significance of lake of macrophytes. *Int. J. Ecol. Environ. Sci.*, 2: 115–128.

Prowse, G. A., 1968. The latin square fertilizer trials. *Trop. Fish. Cult. Res. Inst. Wkg. Pap.*, 1: 28 p.

Prowse, G. A., 1969. Energy flow in relation to productivity in fresh and brackish waters. *In: International Congress of the Pacific Science* Association held at university of Malaya, May, 1969.

Rai, D.N. and Datta Munshi, J.S., 1979. The influence of thick floating vegetation (waterhyacinth) on the physico–chemical environment of a freshwater wetland. *Hydrobiologia*, 62: 65–69.

Ramchandran, V., 1968. The method and technique of using anhydrous ammonia for aquatic weed control. *Proc. Indo. Pacific Fish. Conf.*, 10 (2): 146–153.

Ramsar Convention Bureau., 1971. Convention on wetlands of International importance. Asian wetland Burean, Kuala Lumpur.

Rao, C. B., 1953. On the distribution of algae in a group of six small ponds. *J. Ecol.*, 41: 62–71.

Raunkiar, C., 1934. The life form of plants and statistical plant. *Geography. Claredon Press, Oxford.* England: 125 p.

Ray, P., Singh, S.B. and Sehgal, K.L., 1966. A study of some aspects of the river Ganga and Yamuna at Allahabad (U.P.) in 1958–59. *Proc. Nat. Acad. Sci. India*, 36B(3): 235–272.

Reid, G.K., 1961. Ecology of inland waters and estuaries, Reinhold Publishing Corporation, New York; 375 p.

Riglar, F. H., 1956), A tracer study of the phosphorus cycle in lake water. *Ecology*, 37:550–562.

Roback, S. S., 1974. Insects (Arthropoda: Insects), *In.* Pollution ecology of freshwater invertibrates, *Eds.* C. W. Hart and H.Fullar, *Academic Press*, New York: 313–376.

Robson, T. O., 1976. A review of distribution of aquatic weeds in tropics and subtropics. *In. S. E. Asia. Proc. of Reg. Sem. on noxious. Aq. Veg.* held at New Delhi, December 12–17, 1973; 25–30.

Rowson, D. S., 1930), The bottom fauna of lake Simcos and its role in the ecology of lake. *Univ. Toronto. Stud. Pun. Ont. Fish. Res. Lab.,* 40: 1–83.

Ruttner, F., 1931. Die schichtung in tropischen Sean, *Vet. d. Int. Ver. f. theor. U. angew, Limn.* 5: 44–67.

Ruttner, F., 1964. Fundamentals of Limnology, 3rd ed. Toronto University Press; 295p.

Saha, G. N., Sehgal, K. L., Mitra, E. and Nandy, A. C., 1971. Studies on the seasonal and diurnal variations in physico-chemical and biological conditions of a perennial freshwater pond. Cuttack, Orissa, *J. Inland Fish. Soc. India.,* 3: 80–102.

Sahai, R. and Sinha, A.B., 1969. Investigation on bioecology of inland waters of Gorakhpur (U.P., India) Limnology of Ramgarh lake. *Hydrobiol.,* 34(3 and 4): 433–477.

Satomi, Y., 1967. Physiological significance of carbon sources in fertilized fish ponds. *FAO fish. Rep.* 44(3): 257–264.

Schaeperclaus, W., 1933. Text book of pond culture, *U.S. Fish and wildl. Serv. Fish. Leaf,* 311: 261 p.

Sharma, U. P., 1995. Role of macrophytes in the ecosystem of Karwar Lake weltnad (Begusarai), *J. Freshwat. Biol.* 7 (2): 123–128.

Sharma, U. P., Pandey, K. N. and Prakash, V., 1994. Ecology of fishes of Karwar Lake wetland (Begusarai), *J. Freshwat. Biol.* 6 (2): 123–128.

Singh, R. K. and Desai, V. R., 1980. Limnological observations on Rahand reservoir. III–Primary productivity., *J. Inland Fish. Soc. India.,* 12 (2): 63–68.

Singh, V. P.(1960. Phytoplankton ecology of the Inland waters of Uttar Pradesh, *Proc. Sym. Algal., ICAR,* New Delhi: 243–271.

Sinha, M., 1997. Floodplain wetlands an important inland fishery resources of India, *Aquac.* Asia, 2 (4): 8–10.

Sinha, M., 1998. Wetlands and their implications on fisheries. *Meenbarta,* Dept. of Fisheries, West Bengal: 11–14.

Sinha, M., 1999. Inland fisheries development–achievements and destinations for twenty–first century. *In. National Seminar of Ecofriendly management of resources for doubling fish production–strategies for 21st century,* Souvenir: 25–35p.

Sinha, M., 2001. Reservoir–and important fishery resources of India and its utilization for increasing fish production in the country. *In.* Management of fisheries in small reservoirs. *Ed.* Utpal Bhaumik, *CIFRI. Bull.* 106: 1–15.

Sladeckova, A., 1962. Limnological investigation methods for the periphyton community. *Bot. Rev.,* 28: 286–350.

Sledeck, V. and Sladeckova, A., 1963. Limnological study of the reservoir Sedlice near Zeliv, XIII, Periphyton production. *Sci. Pap. Inst. Chem. Tech. Prague, fac. Tech. Fuel. and wat.,* 7(2): 77–133.

Smith, M.W., 1934. Physical and biological conditions in heavily fertilized water. *J. Biol. Bd. Canada,* 1(1): 67–93.

Smith, M.W., 1945. Preliminary observations upon the fertilization of Crecy lake, New Brunswick. *Trans. Amer. Fish. Soc.,* 75: 165–174.

Soulthrope, C. D., 1967. The biology of acustic vascular plants, *Edward and Herold (Publisher) Ltd.,* London.

Sozeka, G.J., 1975. Ecological relations between invertebrates and submerged macrophytes in the lake littoral. *Ecol. Pol.,* 23(3): 393–415.

Sreenivasan, A., 1964. The limnology, primary production and fish production in tropical pond. *Limnol. Oceanogr,* 9(3): 391–396.

Sreenivasan, A., 1967a. Application of limnological and primary productivity studies in fish culture. *FAO. Fish. Rep.* 44 (3): 101–113.

Sreenivasan, A., 1967b. Fish production in some rural demonstration ponds in Madras (India) with an account of the chemistry of water and soil. *FAO Fish. Rep.* 44: 179–197.

Sreenivasan, A., 1968. The limnology and fish production in two ponds of Chingleput (Madras), *Hydrobiologia,* 32 (1–2): 131–137.

Srivastava, V.K., 1955. Benthic organisms of a freshwater fish tank. *Curr. Sci.*, 250: 158–159.

Strom, K. M.(1927. Recent advances in Limnology. *Linn. Soc. London, 140 th Session*: 96–110.

Sugunan, V.V., 1989. Limnological features of beels–biotic factors. *In.*Training in management of beel (oxbow lake) fisheries. *CIFRI Bull.* 63: 128–135.

Sugunan, V.V., 1997. Floodplain wetlands, small water bodies, culture based fisheries and enhancement–conceptual frame work and definitions. *In* Fisheries enhancement of small reservoirs and floodplain lakes in India. *Eds.* V. V. Sugunan and M. Sinha, *CIFRI Bull.* 75: 13–21.

Sugunan, V.V. and Bhattacharya, B.K., 2000. Ecology and fisheries of beels in Assam, *CIFRI Bull.* 104: 65 p.

Sugunan, V.V., Vinci, G.K., Bhattacharya, B.K. and Hassan, A., 2000. Ecology and fisheries of beels in West Bengal. *CIFRI Bull.* 103: 53 p.

Swingle, H. S., 1957. Control of pond weeds by herbivorous fishes, *Proc. 5th Weed Sci.*, 17: 56–59.

Swingle, H. S. and Smith, E. V., 1938. Increasing fish production in ponds, 4th. *North Amer. Wildlife Conf. Trans.*: 332–338.

Swingle, H.S., 1967. Standardizations of chemical analysis for waters and pond muds, *FAO Fish Rep.*, 4: 397–421.

Thomas, J. D., 1962. The food and growth of brown trout (*Sulmo trutta*) and its feeding relationships with Salmon parr (*Salmo salar*) and eel (*Anguilla anguilla*) in river Teify, West Wales, *J. Anim. Ecol.*, 31: 175–205.

Varshney, C. K. and Singh, K.P., 1976. A survey of aquatic weed problem in India. *Proc. Reg. Sem. Noxious Aqua. Veg.,*: 31–42.

Vass K. K., Bhanot, K. K. and Ghosh, A. N., 1978. Studies on periphyton production in two ponds of brackish water fauna at Kakdwip, West Bengal. *J. Inland Fish. Soc. India* 10: 32–38.

Vass, K.K., 1989. Beel fisheries resources in West Bengal. *In.* Training in management of beel (oxbow lake) fisheries. *CIFRI Bull.*63: 29–35.

Welch, P.S., 1948. Limnological methods, Blakiston, Philadelphia, USA. 381 p.

Welch, P.S., 1952. Limnology (2nd ed.) Mc Graw–Hill, New York. 538 p.

Welcome, R.I., 1979. Fisheries ecology of floodplain rivers, Longman Ltd.. 317 p.

Westlake, D. F., 1963. Comparison of plant productivity, *Biol. Rev.*, 38: 385–425.

Westlake, D. F., 1965. Some basic data for investigation of the productivity of aquatic macrophytes, *In*. Primary production in aquatic environments *Ed*. R. C. Goldman University of Califonia Press. Berkeley.

Wetzel, R. G., 1964. A comparative study of the primary productivity of higher aquatic plants, periphyton and phytoplankton in a large shallow lake. *Int. Revue. Ges. Hydro. Biol.*, 49: 1–61.

William, J.L., 1967. Comparison of some diversity indices applied to population of benthic macroinvertebrates in a stream receiving organic wastes. *J. Wat. Pollut. Con. Fed.*, 39: 1673–83.

William, R.B. and Murdococh, M.B., 1966. Phytoplankton production in the Beaufor Channel, North Carolina. *Limnol. Oceanogr.*, 1(1): 73–82.

Wolny, Powel., 1967. Fertilization of warm water fish ponds in Europe, *FAO Fish. Rep.*, 44 (3): 64–81.

Wong, S. L., 1969. The measurement of primary production and its relation to inorganic phosphate absorption in a freshwater lake. *Hydrobiol*. 34 (3–4): 378–391.

Woodwansee, R. A., 1958. The seasonal distribution of the zooplankton of chicken key in Biscaync Bay, Florida. *Ecology.*, 39: 247–262.

Worthington, E.B., 1943. Freshwater fisheries in British Colonial Empire. *Nature*, 141: 353–355.

Yadav, Y.S., 1987. Studies on the ecology of an oxbow lake in context to the development of beel fishery in Assam. *Assam Agril. University Guwahati, April* 21–22, 1987: 70–74.

Yadav, Y.S., 1988. Riverine flood plain fishery resources. *Ed.* A. G.Jhingran and V. V. Sugunan. *CIFRI Bull.* 57: 134–142.

Yadav, Y.S., 1989. Beel fisheries resources in North–East India. *CIFRI Bull. No.* 63: 8–14.

Zafar, A. R., 1964. On the ecology of algae in certain fishponds of Hyderabad, India. II Distribution of unicellular and colonial forms. *Hydrobiologia*, 24 (4): 556–566.

Chapter 20

Disposal of Various Concentrations of Different Effluents from J.K. Paper Mill to the River Nagabali, Orissa

Lakshman Nayak, Suchitra Mangal,
Manoj Kumar Samantara & Gourab Kumar Sethi

P.G. Department of Marine Sciences, Berhampur University
Berhampur – 760 007, Orissa

ABSTRACT

The effluent of the J.K. Paper Mill drains into the river Nagabali. The effluents were analysed for physico-chemical parameters and trace metals. The physico-chemical parameters like temperature, pH, total suspended solids, dissolved oxygen, total alkalinity, total hardness, total chloride, biological oxygen demand and chemical oxygen demand from two stations. The study was undertaken to assess the concentration of effluents of the paper mill discharged into the river. The maximum temperature of 30°C and high pH value of 8.0 were observed. The high value of total suspended solids and dissolved oxygen were 341 mg/l and 5.9 mg/l respectively. Total alkalinity and total hardness varied from 115 mg/l to 159 mg/l and 81 mg/l to 154 mg/l respectively. The highest total chloride value was 14.9 mg/l. The biological oxygen demand and the chemical oxygen demand values varied from 09 mg/l to 49 mg/l and 47 mg/l to 181 mg/l respectively. Among the four trace metals the maximum concentration being 0.680 mg/l was observed in iron. It is observed that the Station 2 is very much polluted than the Station 1.

Keywords: Disposal, Concentration, Effluents, Nagabali River.

Introduction

Water covers about 71 per cent of earth's surface and provides the most extensive medium for aquatic life because of the several unique properties from ecological point of view. The life on earth first started from aqueous medium and evolved up to human beings. So, from every corner of existence of life on earth, the most important factor is water. The water mass of earth mainly inflow to the ocean but the ocean water cannot be utilized by human beings for their consumption directly and the fresh water can be readily used by the human beings for their consumption. The fresh water mainly includes river systems, lakes, ponds and under groundwater systems etc. The total annual water supply by rivers to the ocean amounts to about $40.10^3 km^3$, where already 1.1 per cent of this amount is discharged by the river Ganga. River provides the daily water supply for agriculture, industries and humans.

The fresh water goes on changing its physico-chemical properties day by day, which is harmful for living beings. In due course of time it can be included under pollution. Of late, various effluents are getting down to the medium from different sources.

The pollution is occurred due to the pollutants and these pollutants come from many sources like industries, sewage, agriculture etc. Mainly the industrial effluent comes from the industries like paper and pulp textiles, fertilizers, chemicals etc.

The paper industries are the major consumer of water and most of the water is discharged as a waste. A quantity of 90,000 gallons of water is required to produce one tonne of paper and the discharge effluents is up to 80,000 gallons. The effluents has intense brownish colour, alkaline, rich in inorganic and organic sodium compounds, dissolved inorganic salts, chlorinated lignins, fatty acids, resin acid, biological oxygen demand (BOD) and chemical oxygen demand

(COD) (Gillespie, 1985*)*. When the pulp and paper wastewater, treated by aeration and pH correction was used for irrigation of trees, the height and collar diameter of *Eucalyptus camadulensis, Leucaena leucocephala* and *Dendrocalamus strictus* were increased (Neely and Dhondizal, 1985*)*. Paper factory effluent was found to reduce the germination, root and shoot, growth, chlorophyll and protein content, free amino acids and soluble sugar of plant (Rout, 1990). There was inhibition in the germination of rice seeds with high concentration of cardboard factory effluent (Dixit *et al.,* 1986).

A survey of industrial cities in India was made few years back showed that industrial wastewaters contributed by volume between 8-16 per cent of the total wastewater generated. The remaining 84-92 per cent came from domestic sector or non point sources. The figure now at the beginning of 21st century will be 38 per cent and 67 per cent respectively due to rapid rate of industrialization. The biological oxygen demand (BOD) is an important parameter for assessment of water quality; the pollution load is about 50 per cent each for industrial and domestic wastewater.

During past 20-25 year there has been an increasing awareness of potential hazards that exit from the pollution environment by the toxic substances which are often associated with mining industrial wastes and their discharges. The pulp and paper industries are one of the older industries in our country. But there has been a tremendous expansion of paper industries near the rivers during last two decades. Thus the paper industries are one of the largest major industries, which contributes a lot towards the pollution of water environment. (Zingde *et al.,* 1981). Much of the work has not been carried out in the effect of effluent of the industries discharged to the rivers from India in general and J.K. Paper Mill of Orissa in particular. Therefore, this is an attempt to investigate the concentration of different physico-chemical parameters along with trace metals disposed to the river Nagabali from the J.K. Paper mill.

Materials and Methods

J.K. Paper mill is situated at Jaykaypur of Rayagada district, which is about 220km away from Berhampur town. Rayagada is considered as an industrial town of Orissa and Jeykaypur is located on the slopes of eastern ghat plateau in the southern part of the Orissa bordering the state of Andhra Pradesh. Its geographical position is 83°-25' east longitude and 19°-10' north latitude. Its average height above the mean sea level is 758ft. The main source of water supply is from river Nagabali, which is a perennial river. The total land area is 635acres, out of which the factory occupies 393.98 acres and the colony occupies 242 acres.

Nagabali river occupied a unique position in the tribal district of Orissa. It originates from the interior region of Kalahandi district (Orissa). The effluents from this industry are discharged to river Nagabali itself. Its drainage area is 9,410 km^2 from which 3,746km^2 is in Orissa. The length of the main drainage channel is about 217km.It's peak discharge is 6,800mm^3 and annual flow is 2430mm^3.The main tributaries are Jhunjavati, Subermukhi and Vagavati. River Nagabali, to which the effluents of paper mill are discharged, has strong current due to its topography. Many villages like J.K.Pur village, Chandilivillage, Chekaguda etc., are found situated at the riverbank of Nagabali. Two stations were chosen for sample collection. Station 1 is located away from the discharge point known as upstream station. The station 2 is located down the discharge point known as downstream station.

The effluents discharged from the J.K. Paper mill were collected during January 2002 to December 2002 from two stations of the river Nagabali. The water samples were collected on each month to study the physico–chemical parameters such as temperature, pH, total suspended solids, dissolved oxygen, alkalinity, total hardness, total chloride, biological oxygen demand (BOD) and chemical oxygen demand (COD). The water temperature and pH were recorded immediately after collection using a standard centigrade thermometer of 0.01°C accuracy and a field pH meter respectively. Winkler's method was adopted for determination of dissolved oxygen (Strickland and Parsons, 1972). Biological oxygen demand was determined using the dilution technology as described in Strickland and Parsons (1972). Nutrients contents were estimated using filtered sample water adopting in the spectrophotometric methods (APHA, 1998). The average monthly data were taken into account. The water samples were also collected from these two stations for detection of heavy metals like nickel, copper, iron and mercury. The samples were dried using the standard procedure and digested with the addition of nitric acid (AR). Blanks were run with each set. The solutions were made 100ml. each in pre acid wash measuring flask with metal free double distilled water. The concentration of nickel, copper, iron and mercury in mg/l. wet weight were determined by using atomic absorption spectrophotometry.

Results

River ecology is of a great importance in the study of environment because the surface pollutants are coming through this channel. Therefore, the water pollution of Orissa is mainly considered due to river pollution. The industries contribute more waste to the sewage, streams and river etc. The main source of water pollution is discharge of toxic chemical substance to the river by the industries and nature's response to pollution by aerobic digestion. The pollutant causes the decrease in the oxygen concentration in water, which is detrimental to aquatic life and ultimately lead to eutrophication of water body, if the oxygen depletion is not maintained.

Primitive method of disposal of sewage and wastes has also adversely affected the environment, besides that killing the wild life. Deforestation and overgrazing has disturbed the ecosystem. Thus disturbing the ecological balance, which resulted shallowness, high temperature, nitrogen, phosphorus, carbonate compounds increased to the surrounding medium. However, high increase of these factors promotes the abundant growth of algae in this environment.

The temperature, pH, total suspended solids, dissolved oxygen, conductivity, total dissolved solids, total chloride, biological oxygen demand (BOD), chemical oxygen demand (COD) and nutrients were analysed from different drainage systems of the J.K. Paper mill and their average value were presented in Table 20.1. The temperature from different drainage system varied from 35°C to 65°C. The highest pH value from different drainage system varied from 1.2 to 10.5. The highest pH value of 10.5 was observed from the alkali drain of pulp mill. The total suspended solid (TSS) value was highest being 1200 mg/l. from the paper machine and water reclamation plant. Dissolved oxygen value was negligible at all drainage pipes. Conductivity varied from 750 μmho/cm to 3800 μmho/cm from all the drainage systems. Total dissolved solid was highest being 1900 mg/l. from the alkali drain of the pulp mill. The total chloride was negligible from all drainage pipes except the combine effluent treatment plant

Table 20.1: Average Value of Different Parameters from Different Drainage System of J.K. Paper Mill, Rayagada During January 2002 to December 2002

Sample	Temp (0°C)	pH	T.S.S. (mg/l)	D.O (mg/l)	Conductivity (µmho/cm)	T.D.S. (mg/l)	Total Chloride (mg/l)	B.O.D. (mg/l)	C.O.D. (mg/l)	Phosphate (mg/l)	Nitrogen (mg/l)
Acid Drain of Pulp Mill	55 to 60	1.2 to 2.0	100 to 150	Neg	1100 to 1200	552 to 600	Nil	60 to 85	200 to 300	1.73 to 2.33	3.0 to 3.7
Alkali drain of pulp mill	50 to 55	9.5 to 10.5	200 to 250	Neg	3500 to 3800	1750 to 190001	—	280 to 320	1480 to 1850	5.6 to 7.5	9.9 to 12.1
Paper machine and water reclamation plant down	35 to 40	6.0 to 7.5	1000 to 1200	Neg	1200 to 1350	600 to 780	—	22 to 50	450 to 500	1.29 to 1.74	2.2 to 2.7
Power block and soda recovery drain	38 to 48	7.5 to 9.0	800 to 1000	Neg	750 to 800	350 to 400	—	50 to 70	350 to 500	1.73 to 2.33	3.0 to 3.7
Pulp mill and other combine drain	50 to 55	7.5 to 8.5	540 to 780	Neg	2100 to 2300	1050 to 1200	—	160 to 180	750 to 900	9.2 to 13.4	17.6 to 21.4
Combine effluent to E.T.P	40 to 45	7.5 to 9.0	650 to 850	Neg	1700 to 2000	900 to 1000	15 to 20	140 to 170	600 to 800	13.0 to 17.5	23.0 to 28.0

Contd...

Table 20.1–Contd...

Sample	Temp (0°C)	pH	T.S.S. (mg/l)	D.O (mg/l)	Conductivity (μmho/cm)	T.D.S. (mg/l)	Total Chloride (mg/l)	B.O.D. (mg/l)	C.O.D. (mg/l)	Phosphate (mg/l)	Nitrogen (mg/l)
Final discharge after E.T.P.	35 to 38	7.0 to 8.0	35 to 40	1.0 to 2.0	1700 to 2000	850 to 1000	8 to 10	22 to 25	230 to 245	0.5 to 1.5	2.0 to 5.0
General standards for discharge of effluents (marine coastal waters)	Shall not exceed 5°C above the receiving water temperature	5.5 to 9.0	100	—	—	700 to 2000	1.0	100	250	5.0	20

(ETP). Biological oxygen demand (BOD) varied from 22 mg/l. to 320 mg/l. in all the drainage pipes. Similarly the chemical oxygen demand (COD) varied from 200 mg/l. to 1850 mg/l. from all the drainage pipes. The highest value of phosphate and nitrogen was 17.5 mg/l. and 28.0 mg/l. from combined effluent discharge pipe respectively. The seasonal variation of different parameters like temperature, pH, total suspended solids (TSS), dissolved oxygen (DO), total alkalinity, total hardness, total chloride, biological oxygen demand (BOD), total chemical oxygen demand (COD) values were represented in Table 20.2. Trace metal concentration of nickel, copper, iron and mercury for two stations were also depicted in Table 20.3.

Discussion

Temperature is basically the most important factors, which affects certain chemical and biological reaction especially on the aquatic organisms. The average temperature ranged from 22°C to 30°C. The upstream river temperature was less than the downstream temperature during the study period, which is in general agreement with the findings of Upadhyay (1988).

The hydrogen ion concentration (pH) affects the taste of water. Low concentration of pH favours corrosion control. The high concentration of pH helps in effective chlorination, chemical coagulation, disinfection and water softening. The pH ranged from 7.1 to 8.0 during the present study. The pH value of 6.5 to 8.5 is recommended for drinking purposes. The present study is in agreement with the result of Sexana *et al.* (1986). pH of water has given the idea of the intensity of pollution (Verma and Shukla, 1970).

The highest value of total suspended solids (TSS–285 mg/l) has been observed in the present study. This high values acts as the indicator of polluted water as confirmed by Surabhi (1994).

The dissolved oxygen (DO) value varied from 3.7 ml/l to 5.9 ml/l, which is generally low compared to any healthy environment. This low level of dissolved oxygen suggests that the water is deteriorating. Slightly high value of dissolved oxygen (DO) at station 1, which may be due to influence of freshwater influx and not contaminated with the discharged water from the paper mill. The result of dissolved oxygen can be comparable with the results obtained from Chilka lagoon by Nayak and Behera (2004).

Table 20.2: Seasonal Variation of Different Parameter from Two Stations of River Nagabali During January 2002 to December 2002

Month	Site	Temp (°C)	pH	T.S.S. (mg/l)	D.O. (mg/l)	Alkalinity (mg/l)	Total Hardness (mg/l)	Total Chloride (mg/l)	BOD (mg/l)	COD (mg/l)
January	Sta–1	23	7.2	95	5.9	118	85	12.8	10	55
	Sta–2	25	7.5	248	4.7	148	137	14.5	45	168
February	Sta–1	26	7.3	89	5.8	124	89	11.9	09	61
	Sta–2	27	7.5	235	4.6	158	147	13.8	49	174
March	Sta–1	26	7.1	92	5.8	115	82	13.1	10	52
	Sta–2	28	7.6	341	4.2	145	149	14.8	42	181
April	Sta–1	28	7.2	84	5.6	129	87	12.1	10	49
	Sta–2	29	7.5	275	4.2	151	154	14.7	40	178
May	Sta–1	29	7.3	78	5.8	115	84	13.1	09	54
	Sta–2	29	7.8	244	4.2	149	149	14.9	38	175
June	Sta–1	28	7.5	79	5.7	121	82	12.9	09	51
	Sta–2	30	7.8	262	4.0	152	147	14.9	42	174
July	Sta–1	29	7.2	84	4.9	115	81	12.5	10	52
	Sta–2	30	7.4	278	3.8	143	135	14.7	42	172

Contd...

Table 20.2–Contd...

Month	Site	Temp (°C)	pH	T.S.S. (mg/l)	D.O. (mg/l)	Alkalinity (mg/l)	Total Hardness (mg/l)	Total Chloride (mg/l)	BOD (mg/l)	COD (mg/l)
August	Sta-1	28	7.3	87	4.7	118	81	12.1	09	51
	Sta-2	30	7.5	268	3.7	149	138	14.5	36	174
September	Sta-1	27	7.3	92	4.6	115	82	12.3	09	47
	Sta-2	28	7.5	285	3.7	145	142	14.3	45	178
October	Sta-1	26	7.4	89	5.1	123	84	12.1	10	47
	Sta-2	27	7.9	276	3.9	152	149	13.9	39	174
November	Sta-1	23	7.4	90	5.6	121	85	11.9	09	49
	Sta-2	24	8.0	275	4.3	159	145	14.7	42	172
December	Sta-1	22	7.3	93	5.8	118	87	11.7	10	53
	Sta-2	23	7.9	252	4.6	147	139	14.7	42	169

Sta–1: Upstream of the river; Sta–2: Downstream of the river.

Table 20.3: Trace Metal Concentration (mg/l) of Ni, Cu, Fe and Hg from Two Stations of River Nagabali During January 2002 to December 2002

Month	Site	Nickel (mg/l)	Copper (mg/l)	Iron (mg/l)	Mercury (mg/l)
January	Sta–1	0.010	0.031	0.132	0.005
	Sta–2	0.019	0.052	0.201	0.008
February	Sta–1	0.009	0.027	0.127	0.006
	Sta–2	0.015	0.039	0.178	0.007
March	Sta–1	0.011	0.024	0.147	0.002
	Sta–2	0.024	0.042	0.219	0.005
April	Sta–1	0.014	0.019	0.251	0.006
	Sta–2	0.029	0.032	0.329	0.008
May	Sta–1	0.014	0.017	0.321	0.004
	Sta–2	0.026	0.029	0.402	0.006
June	Sta–1	0.012	0.018	0.380	0.009
	Sta–2	0.023	0.033	0.452	0.010
July	Sta–1	0.020	0.023	0.410	0.004
	Sta–2	0.040	0.042	0.532	0.008
August	Sta–1	0.021	0.031	0.443	0.007
	Sta–2	0.038	0.048	0.538	0.009
September	Sta–1	0.032	0.028	0.489	0.004
	Sta–2	0.045	0.051	0.556	0.005
October	Sta–1	0.036	0.036	0.538	0.003
	Sta–2	0.052	0.049	0.679	0.006
November	Sta–1	0.026	0.038	0.390	0.008
	Sta–2	0.057	0.053	0.680	0.009
December	Sta–1	0.018	0.035	0.398	0.002
	Sta–2	0.060	0.056	0.731	0.028
General standards for discharge of effluents (marine coastal areas)		5.0	3.0	3.0	0.01

Sta–1: Upstream of the river; Sta–2: Downstream of the river.

High concentration of alkalinity confirms the eutrophic nature of this water. The present work is in conformity with the result of Khan and Seenyga (1985).

The hardness of waters caused by multivalent metallic cations varies considerably from place to place. In general, surface waters are softer than groundwaters. During the investigation total hardness in the samples was found in the range of 81 mg/l. to 149 mg/l. This result is partially in agreement with the work of Prasad (2003).

Chlorides in reasonable concentration are not harmful to humans. At concentration above 250 mg/l they give a salty taste to water, which is objectionable to many people. In the present study the amount of chlorides was to be in the range of 11.7 mg/l to 14.9 mg/l, which were found within desirable limit. The present observation is in conformity with the work of Prasad (2003).

The biological oxygen demand (BOD) was 9 mg/l to 49 mg/l, which was within permissible limits as prescribed by World Health Organisation (WHO) and Ministry of Works and Housing for drinking water. The present result is in conformity with the work of Khedkar and Dixit (2003).

The chemical oxygen demand (COD) is the measure of oxygen consumed during the oxidation of the oxidisable organic matter by a strong oxidising agent. The high concentrations of suspended and dissolved solids are responsible for higher BOD and COD. The water is amenable for biological treatment (Deshmukh *et al.*, 1984). Though higher phosphate is suitable for fish production, the other parameters like dissolved solids, dissolved oxygen, biological oxygen demand, chemical oxygen demand etc. do not allow fish population in the water bodies (Jameel, 1998).

Distribution of trace metal in Andaman Sea has been studied by Sanzgiry and Braganca (1981). The general standards for discharge of effluents to marine coastal areas of Ni, Cu, Fe and Hg as prescribed by World Health Organisation (WHO) is 5.0 mg/l, 3.0 mg/l, 3.0 mg/l and 0.01 mg/l respectively. When nickel exceeds its safe level concentration it causes illness in organisms. When copper exceeds its safe level concentration, it causes hypertension, sporadic fever, uremias, coma, etc. There is a possibility of contamination of mercury in the aquatic environment through the atmosphere. The mercury concentration varied from 0.002 to 0.010 mg/l, which is lower than the safe level concentration. Although most of the heavy metals are essential for the biological process, any element above optimal level is known to cause serious damage to the ecosystem in general and aquatic organisms in particular (Nayak *et al.*, 1993).

References

APHA, AWWA and WPCF, 1998. *Standard Methods for Examination of Water and Wastewater,* 20th edn. American Public Health Association, Washington, DC.

Deshmukh, S.B., Gadgil, J.S. and Subramaniam, P.V.R., 1984. Treatment and disposal of wastewaters from synthetic drugs plant (I.D.P.L.), Hyderabad, *Indian J. Environ. Hlth.,* 26(1): 20–28.

Dixit, A.M., Lailman, S. and Srivastava, S.K., 1986. Effect of cardboard factory effluent on seed germination and early seedlings growth of rice seeds (*Oryza sativa*). *Seeds Res.,* 14: 66–71.

Gillespie, N.J., 1985. Pulp and paper effluent management. *J. Water Pollut. Cont. Fed.,* 57: 587–590.

Jameel, A.A., 1998. Physico-chemical studies in Uyyakondan channel water of river Cauvery. *Poll. Res.,* 24: 140–148.

Khan, Md. M. and Seenayga, G., 1985. Ecology of planktonic, *i.e.* blue green algae in profiles of industrially polluted Hussain Sagar lake, Hyderabad, India. *Phykos,* 24: 140–148.

Khadkar, D.D. and Dixit, A.J., 2003. Effects of wastewaters on growth pattern of Spinach (*Spinacea oleracea* L.). *Nature Environment and Pollution Technology,* 2(4): 441–445.

Nayak, L., Sahu, K.C. and Sahu, D.K., 1993. Heavy metals in some commercial fishes from Gopalpur coast, Orissa, India. *Bull. Env. Sci.,* 11: 33–35.

Nayak, L. and Behera, D.P., 2004. Seasonal variation of some physio-chemical parameters of the Chilika lagoon (east coast of India) after opening of new mouth, near Sipakuda. *Ind. J. Mar. Sci.,* 33(2): 206–208.

Neely, V.R. and Dhondizal, L.P., 1985. Observations on the possibility of using industrial effluent water for raising forest plantation. *J. Trop. Foresty,* 1: 132–139.

Prasad, G.B., 2003. Status of subsurface quality in relation to some physico-chemical parameters. *Nature, Environment and Pollution Technology*, 2(4): 423–428.

Rout, G.R., 1990. Studies on paper mill effluent on *Phaseolus aureus*. *Andhra Agric. J.*, 37: 24–27.

Sanzgiry, S. and Braganca, A., 1981. Trace metals in the Andaman Sea. *Indian Journal of Marine Sciences*, 10(2): 238–240.

Sexana, R.M, Kewal, P.F, Wadav, R.S. and Bhatnager, A.K., 1986. Impact of tannery effluents on some pulse crops. *Ind. J. Environ, Hlth.*, 28: 345–348

Strickland, J.D.H and Parsons, T.R., 1972. A practical handbook of seawater analysis. *Bull. Fish. Res. Bd., Canada, Ottawa*, 167: 1–310.

Surabhi, S., 1994. Limnological studies of Dighi Pond at Darbhanga (Bihar). *Ph.D. Thesis*, Phycological Laboratory, University Department of Botany, B.R.A., Bihar University, Muzaffarpur.

Upadhyay, S., 1988. Physico-chemical characteristics of the Mahanadi Estuarine Ecosystem, East Coast of India. *Ind. J. Mar. Sci.*, 17: 19–23.

Verma, S.R. and Shukla, G.R., 1970. The physico-chemical conditions of Kamla Nehuru Tank, Muzaffarnagar (U.P) in relation to the biological productivity. *Environ. Health*, 12: 110–128.

Zingde, D.M., Chander, S., Rokede, M.A. and Desai, B.N., 1981. Baseline water quality of the river Narmada (Gujarat). *Indian Journal of Marine Sciences*, 10(2): 161–164.

Author Index

Chapter 3

Chapter 4

Chapter 18

Heron, 1961, 405

Das and Dehadrai, 1986, 405

Bhaumelg, 2000, 405

Munnar, 1970, 405

Glodman, 1972, 405

Hiclking, 1962, 405

Chu, 1943, 405

Michael, 1964, 405

Smith, 1945, 405

Hutchisow and Bowen, 1950, 405

Riglar, 1956, 405

Onle, 1938, 406

Matida, 1956, 407, 1958, 406

Hepher, 1958, 406

Banerjee and Mandal, 1965, 406

Chakraborthy *et al.*, 1959, 406

Das and Srivastava, 1956, 406

Bhowning, 1968, 406

Moitra and Mandal, 1970, 406

Mandal, 1972, 406

Moyle, 1946, 406

Banerjee, 1967, 406

Hutchisow, 1957, 406

Mathew, 1969, 406

Chakraborti, 1980, 1960, 406

Jana, 1973, 406

Glodman, 1972, 406

Jhingran, 1977, 407

Bhowning, 1968, 407

Mathew, 1969, 407

Chakraborthy, 1980, 407

Das and Srivastava, 1956, 407

Jana, 1968, 407

Mandal, 1972, 407

Banarjee, 1967, 407

Schaperclaus, 1933, 407

Lakshnanan *et al.*, 1967, 407

Schaperclaus, 1933, 408

Lakshnanan *et al.*, 1967, 408

Benarjea, 1967, 408

Mandal, 1972, 408

Saha *et al.*, 1971, 408

Thomas, 1962, 408

Mc Intire and Bond, 1962, 408

Bhowning, 1968, 408

Chakravorti, 1980, 408

Banerjee, 1967, 408

Hucthinsow, 1944, 408

Jhingraw, 1977a, 409

Benarjee, 1967, 409

Sreenivasan, 1967a, 1967b, 409

Dendy, 1963, 409

Chang and Kaclspm, 1957, 409

Jhingram, 1977, 409

Skrgunan *et al.*, 2000, 409

Hickhing, 1962, 409

Prowse, 1968, 409

Lin and Chew, 1967, 410

Ohle, 1937, 410

Gulterman, 1967, 410

Benarjee and Mandal1, 965, 410

Ghosh, 1970, 410

Bhowning, 1968, 410

Mandal, 1972, 410

Subject Index